Smart Systems in Biotechnology

Editor

Munishwar Nath Gupta

Former Emeritus Professor
Department of Biochemical Engineering and Biotechnology
Indian Institute of Technology Delhi
New Delhi, India

CRC Press
Taylor & Francis Group
Boca Raton London New York

CRC Press is an imprint of the
Taylor & Francis Group, an **informa** business

A SCIENCE PUBLISHERS BOOK

Cover credit: Image used on the cover is drawn by the authors of Chapter 6. Reproduced by kind courtesy of the authors.

First edition published 2024
by CRC Press
2385 NW Executive Center Drive, Suite 320, Boca Raton FL 33431

and by CRC Press
4 Park Square, Milton Park, Abingdon, Oxon, OX14 4RN

CRC Press is an imprint of Taylor & Francis Group, LLC

Library of Congress Cataloging-in-Publication Data (applied for)

ISBN: 978-1-032-35808-6 (hbk)
ISBN: 978-1-032-35838-3 (pbk)
ISBN: 978-1-003-32891-9 (ebk)

DOI: 10.1201/9781003328919

Typeset in Times New Roman
by Prime Publishing Services

Preface

My interest in smart materials resulted from few serendipitous happenings. Many decades back, I was fascinated by what Prof Alan Hoffman of Washington University, USA was publishing on smart hydrogels. Around the same time, I happened to listen to Prof Bo Mattiasson who gave a lecture at IIT Delhi. I was familiar with his extensive work in many areas of biotechnology. We had a long meeting later on. It turned out that he was on his way to USA to meet Prof Hoffman! To cut a long story short, Bo and I had a collaboration which lasted for about a decade. The focus was on the use of reversibly soluble-insoluble polymers as he was interested in developing a technique called affinity separation. These polymers were stimuli-sensitive or smart polymers. I wanted to read a bit more about these materials and hence ended up writing a review on their applications in biological sciences. It is gratifying to note that this review published in 2003, continues to still be cited.

This book is the result of my trying to revisit this area.

To a layperson, the adjective smart is mostly associated with phones or wearables. While both kinds of devices are beginning to be used in a limited way in biotechnology, this book does not cover either of those developments.

Smart materials has become such a vast area that any selection of contents for a book devoted to them has to be a matter of subjective choice. I have aimed at covering some basics so that the book is useful as a standalone resource to anybody who has no prior knowledge of the smart materials. The first two chapters are a good introduction. Of all the diverse stimuli used, change in temperature continues to be the most common in case of these materials. The first chapter introduces smart materials but extensively focuses on design and synthesis of thermosensitive polymers. The second chapter provides a more broad brush treatment while highlighting the rich interface between nanomaterials and smart materials. The remaining chapters are an eclectic mix. The third and fourth chapters cover microfluidics and electrochemical biosensors in the context of smartness.

The fifth chapter describes how electrostatic stimuli can be used for aptamer-covered surfaces for some neat applications. The sixth chapter describes how smart dressings are proving useful in wound treatment. The seventh chapter discusses multistimuli-based micelles which act as prodrugs in anti-cancer therapy.

The last chapter is about the impact of smart hydrogels in tissue engineering. Tissue engineering had a slow take-off with few bumps on the way. It is very impressive how this technology has evolved into an approach to repair us! After an adequate introduction to tissue engineering, this eighth chapter describes its applications to tissues of oral cavity.

I thank all authors who agreed to contribute to this collective effort and patiently agreed to consider my editing suggestions. I take full responsibility for any shortcomings.

Finally, thanks to Mr Raju Primlani of Science Publishers imprint of CRC for his encouragement, support and guidance. I also thank all at CRC who are involved in bringing out this volume.

19th September, 2023

Munishwar Nath Gupta
Delhi, India

Contents

Chapter 1

Thermoresponsive "Smart" Polymer Systems for Drug Delivery, Gene Therapy and Tissue Engineering

Somdeb Jana, Daniel Stöbener and *Richard Hoogenboom**

Introduction

Natural living systems are often governed by adaptive and responsive behaviour to survive, which is mostly driven by biological functions of proteins often mediated by their conformational alterations. Inspired by such living systems, polymer researchers have designed and developed a wide range of synthetic responsive polymer materials that can respond with a property change to a large variety of chemical and physical stimuli such as pH, temperature, light, redox reactions, ionic strength, glucose, carbon dioxide (CO_2), shear stress, enzymes and so forth (Wei et al. 2017, Stuart et al. 2010, Gao et al. 2017, Cabane et al. 2012). Polymer materials that undergo non-continuous changes, such as phase transitions, conformational variations, chemical modifications, colour changes, and shape transformations as well as wetting properties of surfaces, topography, porosity or swelling, in response to an external signal/stimulus are often referred to as "smart" (Hoffman et al. 2000), or "intelligent" (Kikuchi and Okano 2002) materials and are very promising in the field of responsive bio-interfaces that mimic natural surfaces (Senaratne et al. 2005), controlled drug or nucleic acid delivery and their release to the desired site (Hoffman 2008, Bayer and Peppas 2008, Hoffman 2013, Indermun et al. 2018), tissue engineering (Municoy et al. 2020), and many other applications (Zhang et al. 2017, Mendes 2008, Tokarev

Supramolecular Chemistry Group, Centre of Macromolecular Chemistry (CMaC), Department of Organic and Macromolecular Chemistry, Ghent University, Krijgslaan 281-S4, 9000 Ghent, Belgium.
* Corresponding author: richard.hoogenboom@ugent.be

and Minko 2009, Zhang et al. 2021, Xu et al. 2022). However, considering biomedical applications, stimuli-responsive polymer systems only permit a very limited range of stimuli, such as ionic strength, pH and others, which restricts the biomedical applications of such systems due to the fact that proteins and cells may become damaged if the stimuli are beyond the physiological limits. On the other hand, many proteins and cells can withstand moderate temperature changes in between the freezing point of water and 42°C for limited time periods without any deterioration. Polymers which exhibit thermoresponsive properties in this temperature range are therefore highly interesting and have been, for example, exploited for many years to fabricate coatings on cell culture dishes in order to harvest single cells as well as confluent cell monolayers with an intact extracellular matrix (ECM) without digestive enzyme treatment by simply decreasing the temperature (Doberenz et al. 2020, Sponchioni et al. 2019). Furthermore, thermoresponsive polymers are highly interesting for the development of drug or nucleic acid/gene delivery vehicles if the transition temperature is close to body temperature, allowing to prepare formulations that are soluble at room temperature and undergo gelation upon injection (Pasparakis and Tsitsilianis 2020, Calejo et al. 2013, Hogan and Mikos 2020, Abulateefeh et al. 2011, Twaites et al. 2005). Therefore, thermoresponsive polymers that undergo a temperature-induced solubility phase transition (Aseyev et al. 2011, Weber et al. 2012, Seuring and Agarwal 2012, Vancoillie et al. 2014, Zhang and Hoogenboom 2015, Roy et al. 2013) in aqueous solutions and buffers have received significant interest because of their broad application potential in various interesting fields, such as temperature-triggered drug or nucleic acid/gene delivery, tissue engineering, protein chromatography, sensing devices, switchable surfaces, controlled/reversible protein adsorption and regenerative medicine (Pasparakis and Tsitsilianis 2020, Vanparijs et al. 2017, Bordat et al. 2019, Gandhi et al. 2015, Kim and Matsunaga 2017, Ward and Georgiou 2011, Trzebicka et al. 2017, Tan et al. 2012, Ng et al. 2018). In general, two types of thermoresponsive polymers that undergo a demixing phase transition in aqueous solution exist, namely those that undergo a de-mixing phase transition upon heating and those that undergo de-mixing upon cooling (Fig. 1).

In general, such thermoresponsive polymers exhibit a phase transition at a certain critical temperature, which causes a rapid change in their solvation state (Vancoillie et al. 2014). Polymers that are soluble at low temperatures and become insoluble upon heating display a so-called lower critical solution temperature (LCST)-type transition (Crespy and Rossi 2007), and polymers that are insoluble at low temperatures and become soluble upon heating display an upper critical solution temperature (UCST)-type transition (Seuring and Agarwal 2012). The LCST and UCST are defined as the respective minimum and maximum critical temperature points in the entire polymer-water phase diagrams, below and above which the polymer and solvent molecules are completely miscible (Fig. 1), resulting in a transparent homogeneous solution. The LCST represents the temperature at which the binodal coexistence curve exhibits a minimum in the phase diagram (Fig. 2). Below this minimum, only a single phase exists independent of the polymer concentration, as schematically shown in Fig. 2. On the contrary, the UCST represents the temperature

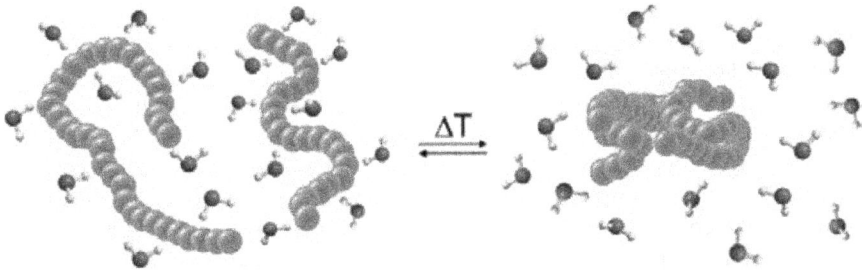

Fig. 1. Schematic illustration of a thermoresponsive polymer that undergoes phase transition from a completely dissolved homogeneous state (hydrated; left side) to a two-phase de-mixed system (dehydrated; right side) in aqueous medium. The scheme shows an LCST transition if $\Delta T > 0$ and an UCST if $\Delta T < 0$. Reprinted from Ref. (Vancoillie et al. 2014), with permission from ELSEVIER.

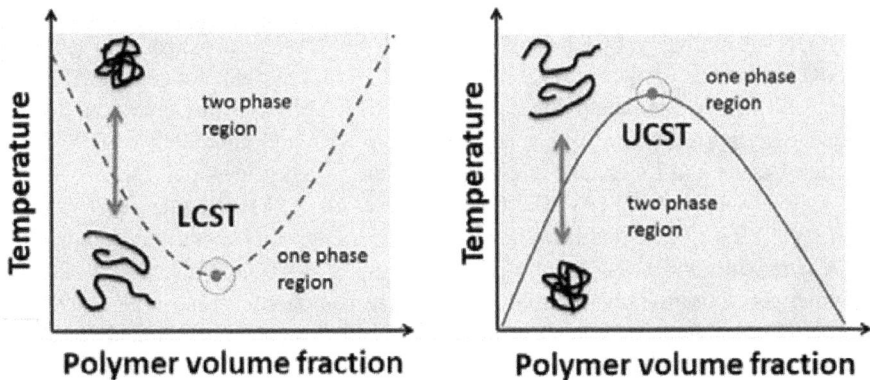

Fig. 2. Schematic representation of the phase transition behaviour of LCST- and UCST-type thermoresponsive polymers. The left panel shows the behaviour of an LCST polymer, such as poly(N-isopropylacrylamide) (PNiPAAm) in H_2O, and the right panel shows the behaviour of a UCST polymer, such as poly(N-acryloyl glycinamide) (PNAGA) in H_2O. Reprinted from Ref. (Kocak et al. 2020).

at which the binodal coexistence curve exhibits a maximum, above which one single phase exists (Fig. 2).

In both cases of LCST and UCST, there is a delicate balance between polymer-polymer interactions and polymer-water interactions that is altered upon changing the temperature. Polymers that exhibit LCST characteristics are soluble in solution due to strong hydrogen-bonding (H-bonding) interactions with surrounding solvent molecules (such as water) and restricted intra- as well as inter-molecular H-bonding between polymer chains. Upon heating, the entropy penalty for solvation of the polymer chains, due to fixation of the solvent molecules, increases and at the LCST, the gain in H-bonding with solvent molecule is no longer larger than the entropy penalty leading to a positive Gibbs free energy and desolvation of the polymer and, thus, phase-separation (Hocine and Li 2013). Following this principle, the transparent solution macroscopically starts to become turbid due to the phase separation of the high polymer concentration phase upon heating (Fujishige et al. 1989). This thermoresponsive phenomenon is reversible, meaning that when the temperature is decreased below the critical point, the solvation of the polymer is restored, and the polymer chains become soluble again. Nonetheless, the temperature at which the polymer solubilizes again may be slightly different from the heating curve resulting in a hysteresis of the system due to kinetically hindered rehydration of the polymer chains (Lu et al. 2010). The thermoresponsive phase transition is often called a coil-to-globule transition and the phase transition temperature at a specific polymer concentration is often referred to as cloud point temperature (T_{cp}). Of all known thermoresponsive polymers, poly(N-isopropylacrylamide) (PNiPAAm) is the most widely investigated LCST polymer in water and is generally considered to be the gold standard because of its sharp and pronounced phase transition and LCST at around 32°C, which is close to the human physiological temperature, making it a prime candidate for *in vivo* biomedical applications (Schild 1992, Rzaev et al. 2007). Other important classes of synthetic thermoresponsive polymers exhibiting a LCST in H_2O include poly(2-alkyl-2-oxazolines) (PAOx) (Hoogenboom and Schlaad 2017, Glassner et al. 2018, Jana and Hoogenboom 2022), poly(N,N-dimethyl aminoethyl methacrylate) (PDMAEMA) (Manouras et al. 2017), poly(N-vinylpyrrolidone) (Yan et al. 2010), poly(N-vinyl caprolactam) (PNVCL) (Cortez-Lemus and Licea-Claverie 2016), certain polypeptides, poly(acrylamides), poly(vinyl methyl ether) (PVME) (Moerkerke et al. 1998), and many others (Dai et al. 2009). The T_{cp} of thermoresponsive polymers depends on various parameters, such as polymer molecular weight (Hoogenboom et al. 2008, Plunkett et al. 2006), polymer concentration, incorporation of different comonomers (Liu et al. 2009, Feil et al. 1993), polymer tacticity (Katsumoto and Kubosaki 2008), chemical nature of the polymer end groups (Furyk et al. 2006, Roth et al. 2010), solvent quality, and salt concentration (Bloksma et al. 2010, Heyda and Dzubiella 2014, Jana et al. 2018).

Compared with polymers having LCST in aqueous solution, polymers exhibiting UCST behaviour in water are relatively less common. Copolymers of poly(acrylic acid)/poly(acrylamide) are a well-studied system that display a UCST-type phase transition in aqueous media (Echeverria et al. 2009). There are also some other synthetic polymers exhibiting UCST-type behaviour, such as poly(N-acryloylglutamineamide)(PNAGA)(Seuring and Agarwal 2010), zwitterionic poly(sulfobetaine)s (Li et al. 2020, Schulz et al. 1986, Huglin and Radwan 1991), poly(acrylic acid-*co*-acrylonitrile) (Zhao et al. 2019), poly(6-(acryloxyloxy-methyl) uracil)(Aoki et al. 1999) and ureido-derivatized polymers (Shimada et al. 2011, Mishra et al. 2014). UCST-type polymers typically have a pair of interactive sites leading to strong polymer-polymer interactions that cause the polymers to be aggregated and become insoluble at lower temperature due to intra- and inter-molecular interactions (such as H-bonding and electrostatic attraction) between polymer chains, which can be disrupted at higher temperature due to intensified dynamic molecular motion within the polymer chains, resulting in hydratation and solubilization of the polymer. Several methods have been proposed in the literature to characterize and determine the phase transition type of thermo-sensitive polymers and in particular their T_{cp} at a specific concentration, including turbidimetry measurements, nuclear magnetic resonance (NMR) analysis and calorimetric measurements, mainly, and differential scanning calorimetry (DSC). For a comprehensive description of the characterization methods to determine the T_{cp}, the reader is referred to the review reported by our group (Zhang et al. 2017).

An in-depth literature survey clearly tells that the number of scientific contributions to synthetic polymer materials with thermoresponsive characteristics increased drastically over the last few decades (Doberenz et al. 2020, Sponchioni et al. 2019, Pasparakis and Tsitsilianis 2020, Vanparijs et al. 2017, Bordat et al. 2019, Sarwan et al. 2020, Flemming et al. 2021, Avila-Salas and Duran-Lara 2020, Teotia et al. 2015). The development of new thermoresponsive synthetic polymer systems and new ways of functionalization have been facilitated by the introduction of advanced polymerization methods. Among them, reversible deactivation radical polymerization (RDRP) techniques, such as reversible addition-fragmentation chain-transfer (RAFT) polymerization (Moad et al. 2005, 2008, Perrier 2017), atom-transfer radical-polymerization (ATRP) (Matyjaszewski and Xia 2001, Braunecker and Matyjaszewski 2007) and nitroxide-mediated polymerization (NMP) (Nicolas et al. 2013) allow to synthesize complex and well-defined macromolecular architectures with narrow dispersity and high chain-end fidelity. Different kinds of ring-opening polymerization (ROP) techniques also enable the preparation of well-defined polymers (Nuyken and Pask 2013). Such above-mentioned advanced polymerization methods allow to finely tune the chain length/molecular weight of the resulting polymers and the impact of this parameter on thermoresponsive property has been extensively studied in the literature. In addition, post-polymerization modification of polymeric precursors via various "*click*" strategies are also found to be another appealing approach to introduce desired functional moieties at the polymer end groups or within the backbone (Gauthier et al. 2009). Thus, well-defined thermoresponsive polymeric materials with a variety of functionalities can be designed and prepared easily by using the above-mentioned polymerization/post-polymerization modification techniques. Overall, designing well-defined polymer systems is highly desirable when focusing on application opportunities towards the biomedical field.

This chapter is intended to report the emerging application possibilities of water-based thermoresponsive polymer materials in the biomedical field, especially in drug delivery, gene delivery and tissue engineering. The discussion will refer to selected reports on LCST-type thermoresponsive polymers that have been used to fabricate thermoresponsive nanocarriers as well as surfaces for several of the above-mentioned biomedical applications.

LCST-type thermoresponsive polymers

Among the water-soluble thermoresponsive polymers, those exhibiting an LCST are the most common and widely investigated for different applications including the biomedical field. In case of LCST polymers, the phase separation occurs due to the loss of solvation of the polymer chains by water molecules and the polymer-polymer interactions becoming thermodynamically favoured. Considering the free energy of the system using the Gibbs equation, i.e., $\Delta G = \Delta H - T\Delta S$

(H: enthalpy, S: entropy and G: Gibbs free energy), the reason phase separation is more favourable when increasing the temperature is mainly due to the entropy of the total system. Especially, the major driving force for heating-induced phase separation is the increasing entropy loss for the solvating water molecules that have less degrees of freedom than water molecules in solution. It is therefore noteworthy to mention that the LCST is an entropy-driven phenomenon resulting in a sharp phase transition (Halperin et al. 2015). Consequently, the LCST behaviour originates from a critical balance between the hydrophilic and the hydrophobic segments of the polymer and, as one could assume, is strongly dependent on the ability of a polymer to form H-bonds with solvating water molecules (Klouda 2015). It is therefore worth mentioning that the cloud point temperature (or phase transition temperature) can be enhanced by incorporating a hydrophilic monomer into a statistical (co)polymer. The effect is to increase the extent of H-bonding and, therefore, to shift the phase transition towards higher temperatures and, thus, to increase the LCST. Conversely, the incorporation of hydrophobic monomers into a statistical (co)polymer weakens the extent of H-bonding with solvating water molecules and, thus, reduces the LCST. The (co)polymerization of monomers that results in LCST-type thermoresponsive polymers upon polymerization with hydrophilic or hydrophobic functional monomers is, therefore, a common and straight-forward approach to tune the LCST to the desired value as well as to introduce distinct functionalities in the polymer chains (Liu et al. 2009, Eeckman et al. 2004).

Examples of some selected LCST-type thermoresponsive polymer classes

Poly(N-alkyl-substituted acrylamides)

Poly(N-alkyl-substituted acrylamide)s (Fig. 3) are the most studied thermoresponsive polymer class revealing a LCST-type phase transition behaviour in aqueous media. Among them, poly(N-isopropylacrylamide) (PNiPAAm) is the most prominent candidate although a second polymer in this category has a nearly identical transition temperature, i.e., poly(N,N-diethylacrylamide) (PDEAAm) (Idziak et al. 1999). However, in contrast to PNiPAAm, the phase transition temperature of PDEAAm depends on the polymer tacticity (Matsumoto et al. 2018). Poly(N-isopropylacrylamide) (PNiPAAm) is particularly widely exploited for biomedical applications due to its LCST at around 32°C, which is close to the human physiological temperature, making PNiPAAm a very interesting and promising material, e.g., for controlled drug delivery application. The LCST of PNiPAAm is almost independent on the molecular weight, the concentration and pH of the medium (Fujishige et al. 1989, Halperin et al. 2015), but it can be changed or altered substantially upon altering the hydrophilic/hydrophobic balance. However, a major limitation in the use of PNiPAAm is the hysteresis between the heating/cooling cycles which can be explained by the formation of intra- and inter-molecular H-bonds in the polymer chain after the polymer collapse above the T_{cp}, which in turn retards the polymer rehydration and dispersion when the temperature is reduced (Deshmukh et al. 2014, Sambe et al. 2014). Furthermore, acrylamide

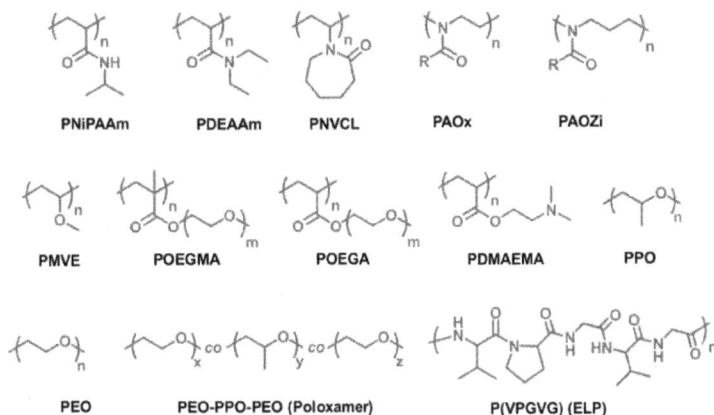

Fig. 3. Chemical structures of typical thermoresponsive polymers that exhibit LCST-type phase transition in aqueous medium.

monomers show acute cytotoxicity, which requires a vigilant purification of the synthesized poly(arylamide)s, specifically when aimed at biomedical applications.

Poly(N-vinyl caprolactam) (PNVCL)

PNVCL (Fig. 3) is an LCST polymer that has been applied in biomedical and environmental applications, in cosmetics and as an anti-clogging agent (kinetic hydrate inhibitor) in pipelines (Cortez-Lemus and Licea-Claverie 2016, Mohammed et al. 2018). PNVCL is usually described as a more biocompatible alternative to PNiPAAm as both polymers (PNVCL and PNiPAAm) exhibit LCST behaviour close to the physiological temperature (Cortez-Lemus and Licea-Claverie 2016). As a misconception, PNVCL has been often described as a thermoresponsive polymer with a well-defined LCST at 32°C, even though the LCST has been observed in broad temperature ranges between 25 to 50°C, depending on polymer molar mass and concentration. In contrast, PNiPAAm exhibits a LCST at about 32°C, which is almost independent of the polymer molecular weight. On the other hand, PNVCL miscibility can be described by the Flory–Huggins theory (Zhao et al. 2010) and depends on polymer molecular weight, salts as well as protein concentration. Due to this dependency, the control of both molecular mass and dispersity (Đ) during PNVCL synthesis is of crucial importance for any application. The first "*in vitro*" evaluation of PNVCL cytotoxicity on intestinal Caco-2 and pulmonary Calu-3 cell lines was reported in 2005 and demonstrated great cell tolerance, although PNVCL exhibited toxic effects above its LCST (Vihola et al. 2005). Afterwards, the cytocompatibility of PVCL has been confirmed in several different cell lines, including different types of human carcinomas (Cortez-Lemus and Licea-Claverie 2016). Functional PNVCL has been reported for the preparation of PNVCL-based drug delivery vehicles in combination with biocompatible polysaccharides, such as chitosan (Prabaharan et al. 2008) or dextran (Feng et al. 2008). PNVCL was also established as a suitable environment for cell proliferation and manipulation (Shakya et al. 2014).

Poly(2-alkyl-2-oxazoline)s (PAOx)

PAOx (Fig. 3) have recently emerged as a promising alternative to PNiPAAm as well as poly(ethylene glycol) (PEG). Thermoresponsive PAOx do not suffer hysteresis issues during the LCST phase transition (i.e., from heating to cooling) as they only have hydrogen bond accepting groups and, interestingly, they exhibit protein repulsion (anti-fouling) characteristics similar to PEG which renders them attractive for the biomedical field. In the past few decades, the use of PAOx in biomedical applications has evolved because of their biocompatibility as well as their stealth characteristics (Adams and Schubert 2007, Hoogenboom 2009, Viegas et al. 2011). PAOx is a synthetic class of pseudo-peptidic polyamides that are prepared by cationic ring-opening polymerization (CROP) of 2-substituted-2-oxazoline monomers, yielding the corresponding ring-opened polymers that have a tertiary amide structure of which just the nitrogen atom is incorporated in the polymer backbone (Fig. 3). The 2-substituent of the 2-oxazoline monomer can be varied allowing precise control over the hydrophilic-hydrophobic balance of the resulting PAOx (Hoogenboom et al. 2005, Rossegger et al. 2013). Tuning of the solution properties of PAOx, from fully hydrophilic, e.g., poly(2-methyl-2-oxazoline) (PMeOx), to completely hydrophobic, e.g., poly(2-^{n}butyl-2-oxazoline) (P^{n}BuOx), is possible simply by changing their side chain structure. Interestingly, PAOx that contain intermediate hydrophobic side chains, such as ethyl, n-propyl and iso-propyl, are thermoresponsive with tunable LCST behaviour (Weber et al. 2012, Hoogenboom and Schlaad 2017, Glassner et al. 2018). The LCST of PEtOx in water varies from 61°C to above 100°C, depending on the polymer concentration and polymer molar mass (Hoogenboom et al. 2008), and can be further lowered to as low as 20°C in the presence of sodium sulfonate as strong salting out salt (Bloksma et al. 2010). Amongst the thermoresponsive PAOx, those with an LCST close to physiological temperature are of special interest for applications in drug delivery or bioengineering. Interestingly, PAOx with propyl side-chains, namely, P^{i}PrOx (T_{cp}: 35°C), P^{c}PrOx (T_{cp}: 30°C) and P^{n}PrOx (T_{cp}: 25°C) belong to this class of polymers. However, the phase transition in case of P^{i}PrOx is found to be irreversible after keeping the phase separated dispersion for a longer time above the T_{cp} due to isothermal crystallization of the (partially) dehydrated polymer chains (Katsumoto et al. 2012, Demirel et al. 2007). In contrast to P^{i}PrOx, the phase transitions of amorphous P^{c}PrOx and P^{n}PrOx are fully reversible, even when kept above T_{cp} for an extended time as they are fully amorphous polymers. Thermoresponsive PAOx are extensively reviewed in the literature. Very recently, synthetic amines as well as ether side-chain functional PAOx with amines and ether side chains were found to also exhibit interesting pH and thermoresponsive behaviours (Jana and Hoogenboom 2022).

Poly(methyl vinyl ether) (PMVE)

PMVE (Fig. 3) has an LCST around 37°C, which makes it very promising for biomedical applications. PMVE can be prepared by cationic polymerisation. Additionally, when an aqueous solution of PMVE is irradiated with electrons or γ-rays, the solution is transformed into a hydrogel (Arndt et al. 2001). This cross-linking is happening through the formation of radicals from water molecules, which attack PMVE molecules. The hydrogel possesses thermoresponsive properties like those of the PMVE in aqueous solution. It swells at temperatures below the LCST and shrinks above this critical temperature. Such kind of stimuli-responsive hydrogels that undergo abrupt changes in volume in response to temperature near 37°C have emerging application potential in the creation of 'smart' materials for drug delivery and tissue reconstruction.

Poly(oligoethylene glycol (meth)acrylate)s (POEG(M)A)

Polymers bearing a short oligo ethylene glycol (OEG) side chain have been shown to combine the biocompatibility of polyethylene glycol (PEG) with a versatile and controllable LCST behaviour (LCST ranges from ca. 10–100°C) (Vancoillie et al. 2014, Lutz 2008, Lutz et al. 2006). In this context, thermoresponsive poly[oligo(ethyleneglycol)methylethermethacrylate]s (POEG$_n$MA) (n is the number of OEG units) is nowadays attracting significant attention (Fig. 3). Comb-like POEG$_n$MA structures can be easily obtained from the radical or anionic polymerization of methacrylate monomers comprising an oligo(ethylene glycol) (OEG) side-chains. OEG chains are basically linked to the polymer backbone through an ester linkage and the number of ethylene glycol repeating units in the pendent heavily influences the LCST of the resulting polymer. For example, the LCST values of POEG$_2$MA, POEG$_3$MA, POEG$_4$MA and POEG$_8$MA are found to be 26, 54, 62 and 90°C, respectively (Lutz and Hoth 2006, Mertoglu et al. 2005). In addition, the LCST decreases with increasing molar mass and, by selecting the number of repeating units in the PEG comb and the molar mass, the desired LCST can easily be obtained. Furthermore, by statistically copolymerizing such PEG monomers (OEG$_n$MA) with another hydrophilic or hydrophobic monomers, it is also possible to easily tune the LCST to any desired value without any significant hysteresis between heating and cooling cycles (Lutz 2008, Lutz et al. 2006, 2009). However, despite the above-mentioned advantages, recent investigation revealed that PEG has some medical limitations including non-specific interactions with blood, hypersensitivity, allergic reactions, accelerated blood clearance among others (Knop et al. 2010, Kong and Dreaden 2022, Bigini et al. 2021). Also, enzymatic oxidation under physiological environmental conditions generates various toxic compounds, such as aldehyde, ether and peroxides (Konradi et al. 2012). Another drawback is that OEG$_n$MAs can coordinate with heavy metal ions which could impose certain limitations for their use. Consequently, PEG-containing polymers are often unsuitable for several biomedical applications and, therefore, new biocompatible as well as thermoresponsive polymer alternatives (e.g., poly(2-oxazoline)s) must be developed to overcome these limitations.

Elastin-like polypeptides (ELPs)

Thermoresponsive ELPs are basically a sort of arranged repetitious proteins with the general amino acid sequence (aPGbG)$_n$ as repeating unit (where a can be 'I', i.e., isoleucine or 'V', i.e., valine, 'G' is glycine, 'P' is proline, 'b' is any amino acid except 'P' and n is the number of repeating units) (Aluri et al. 2009, Andrew Mackay and Chilkoti 2008, Meyer et al. 2001). ELPs are mostly originated from a motif discovered in human tropoelastin. Tropoelastins are a set of water-soluble proteins that are cross-linked together to form the insoluble elastin found in the extracellular matrix. Elastin, as its name indicates, is an elastic material that retrieves its initial shape after being subjected to mechanical stress (Bellingham et al. 2003). Because of their important functional role as component of the native extracellular matrix, ELPs have attracted increasing interest in cancer therapy (Dreher et al. 2003, Furgeson et al. 2006) and tissue engineering (Li et al. 2005, Annabi et al. 2009). Unlike other LCST-type thermoresponsive polymers, ELPs undergo reversible inverse temperature responsive transition (McDaniel et al. 2013) (T_t) where they are present in a completely soluble state below T_t and then collapsed to the form of coacervates/aggregates state above the T_t (Urry 1992). This type of transition above the T_t can be ascribed to the stabilization of a secondary supramolecular structure (also known as β-spiral or β-sheet) of polypeptides via hydrophobic bonding and subsequent assembly (Urry 1997). For example, an elastic protein-based polymer, namely, poly(VPGVG) (Fig. 3) fully dissolves in water below 25°C. However, upon increasing the temperature to 37°C (i.e., above the T_t), it undergoes inverse temperature phase transition (ITT) and transforms into a well-arranged state with the development of a reversible insoluble coacervate.

Such reversible thermosensitive ELPs possess potential applications in the biomedical field including the avoidance of post-surgical and post-trauma gluing, drug delivery (Dreher et al. 2003, Furgeson et al. 2006), tissue engineering (Li et al. 2005, Annabi et al. 2009), wound closures and bandages, various usages for the eye, and vascular grafts, due to several reasons (Chilkoti et al. 2002, Cui et al. 2017). First, they are biocompatible, biodegradable and non-immunogenic (Urry et al. 1991). Second, it is possible to adjust the T_t of poly(aPGbG)$_n$-type ELPs precisely by altering n, a, b (McDaniel et al. 2013). Thirdly, ELPs can be genetically encoded, which means it is possible to obtain monodisperse ELPs with specific amino acid sequence as well as controlled molar mass, even in a heterologous host (such as in a eukaryotic cell or bacteria). ELPs can be prepared by direct synthesis or through recombinant expression. Also, one can design and synthesize the polypeptide for a particular application of interest with tailored characteristics. However, elastin from natural resources remains costly owing to tedious purification and the inexorable use of harsh chemicals.

Other LCST-type polymers

Apart from the above-mentioned polymers, several other polymers also have H-bonding sites and, therefore, exhibit LCST-type thermoresponsiveness in water. These polymers include poly((2-dimethylamino)ethyl methacrylate) (PDMAEMA; exhibiting pH-dependent LCST-type thermoresponsiveness in the range from 32 to 53°C) (Manouras et al. 2017, Agut et al. 2010), poly(2-ethyl-2-oxazine) (PEtOzi; $T_{cp} \sim 56$°C) (Bloksma et al. 2012), poly(2-n-propyl-2-oxazine) (P^{n}PrOZi; $T_{cp} \sim 11$°C) (Bloksma et al. 2012), hydroxypropyl cellulose ($T_{cp} \sim 45$°C) (Xia et al. 2003), few poly(ethylene oxide) (PEO; showed LCST-type thermoresponsiveness in the range from 100 to 150°C) (Rackaitis et al. 2002), as well as poly(propylene oxide) (PPO; showed LCST-type thermoresponsiveness in the range from 10 to 45°C) within a certain molecular weight range (Schild and Tirrell 1990). The LCST of PDMAEMA is pH-dependent, which can be attributed to the presence of ionizable tertiary amine groups in the side chains. However, the cytotoxicity of PDMAEMA is a major concern for its clinical application (You et al. 2007, Agarwal et al. 2012). PEtOzi and P^{n}PrOZi were found to be thermoresponsive, exhibiting LCST-type thermoresponsiveness in water and their T_{cp} values are lower than for PAOx with similar side chains. However, in comparison to PAOx, the iso-structural poly(2-oxazine)s are more water soluble, indicating that the location of the hydrophilic as well as hydrophobic groups plays an important role for the LCST behaviour of polymers, in addition to the hydrophobic/hydrophilic balance.

Poloxamers or Pluronics are commercially available co-polymers composed of PEO and PPO with the block structure PEO-PPO-PEO (Dumortier et al. 2006). These materials also showed LCST-type phase transition behaviour and their T_{cp} can be adjusted in between 10 to 100°C simply by the changing the compositions of PEO and PPO (Alexandridis and Hatton 1995). PEO is water soluble up to temperatures of 85°C, while PPO is hydrophobic (De las Heras Alarcón et al. 2005). By preparing co-polymers with different ratios of PEO to PPO, the transition temperature and solubility can be adjusted. Due to their strong solvation (less susceptible to be broken), these materials can also maintain their thermoresponsiveness over a broad range of pHs and ionic strengths. Interestingly, at the LCST, solutions containing a critical amount of certain Poloxamers undergo a remarkable increase in viscosity. This phenomenon is termed reverse thermal gelation (RTG) (Cohn et al. 2003). Below the LCST, the viscosity of the transparent solution containing PEO-PPO-PEO copolymer is lower, which is indeed favourable for injection. However, above the LCST, the viscosity increases dramatically, and the solution transformed to a semi-solid gel (Cohn et al. 2003). PEO-PPO-PEO polymers with such RTG characteristics are interesting and have been largely exploited in the biomedical field such as tissue engineering, controlled drug release (through preparing nano-formulations as well as pharmaceutical excipients), wound dressing among others (El-Aassar et al. 2016, Sosnik et al. 2003, Bhattacharjee et al. 2016). More comprehensive details on the non-ionic aqueous thermoresponsive LCST polymers can be found in the excellent review written by Winnick and colleagues (Aseyev et al. 2011).

In the following, the major applications of LCST-type thermoresponsive polymers are described to reflect the current state-of-the-art in the biomedical field, including drug delivery, gene transfection and tissue engineering.

Applications

LCST-type thermoresponsive nanocarriers for drug delivery

One of the widely investigated applications of thermoresponsive polymer systems in the field of biomedical applications is the design of "smart" micellar nanocarriers to encapsulate active pharmaceutical compounds (drug/payload) and control its release both in space and in time (Pasparakis and Tsitsilianis 2020, Zhang et al. 2017, Akimoto et al. 2014). Typically, two kinds of thermoresponsive micellar systems have been reported in the literature, which are schematically depicted in Fig. 4. In the first system, a hydrophilic polymer segment has been coupled to the thermoresponsive polymer segment, thus leading to the formation of a (co)polymer architecture that is fully water-soluble below the phase transition temperature ($< T_{cp}$) and forms micelles/polymersomes upon heating above the phase transition of the thermoresponsive block/segments ($> T_{cp}$) (Yan et al. 2008). This strategy can be employed to facilitate the loading of a hydrophobic therapeutic agent into polymer micelles, which is encapsulated while micelles are generated upon heating and, therefore, could be interesting in the preparation of nano-formulations for intravenous administration. In the second system, a thermoresponsive polymer segment can be attached to a hydrophobic polymer segment to generate amphiphilic (co)polymer architectures that are able to form micelles/polymersomes stabilized by the thermoresponsive polymer segments (corona) below the critical temperature ($< T_{cp}$) (Sponchioni et al. 2016). Heating above its T_{cp} leads to collapse of the micellar thermoresponsive corona over the core, leading to the formation of shrunk micelles or micellar aggregates/agglomerates. The micellar disruption in response to the thermoresponsive corona phase transition escalates the release rate of the therapeutic agent/payload. Therefore, this approach can be conveniently utilized to localize the drug release at the target site/cells. This way, drug bioavailability can be enhanced, and the relatively cytotoxic side-effects related to the drug dispersion in healthy cells or tissues cells can be reduced. In this approach, the heat required to control the drug release can be employed from the outside, such as via local thermal heating, photo-illumination or microwave irradiation. In terms of drug delivery application, the LCST of the thermoresponsive micelles should preferably be in the range from 31°C, which is higher than room temperature but lower than the physiological body temperature, to 41°C, that is slightly higher than the body temperature (corresponding to mild hyperthermia).

The influence of thermally-induced hydrophilic to alter the hydrophobic property of thermoresponsive polymer segments on micellar interactions with cells and tissues has been investigated by Okano and colleagues (Akimoto et al. 2009). The authors have prepared diblock copolymers comprising fluorescently tagged poly (*N*-isopropylacrylamide-*co*-*N*,*N*-dimethylacrylamide) (PNiPAAm-*co*-PDMAAm) as thermoresponsive segments and poly(d,l-lactide) (PLA) as hydrophobic segments by the combination of RAFT and ROP strategy (Fig. 5). The intracellular uptake of linear PNiPAAm-*co*-PDMAAm is extremely low at temperatures both below and above the LCST. Unlike thermoresponsive linear PNiPAAm-*co*-PDMAAm chains, thermoresponsive (PNiPAAm-*co*-PDMAAm)-*b*-PLA block copolymer micelles revealed time-dependent and increased intracellular uptake (Fig. 5) at 42°C, i.e., above the micellar LCST ($T_{cp} \sim 39°C$) due to the enhanced interactions between cells and micelles mediated through the thermoresponsive phase transition of the micellar coronas. (PNiPAAm-*co*-PDMAAm)-*b*-PLA micelles also showed negligible cytotoxicity to the cultured cells (bovine carotid endothelial cells) and, therefore, can be useful as intracellular delivery tools for anticancer drugs, genes,

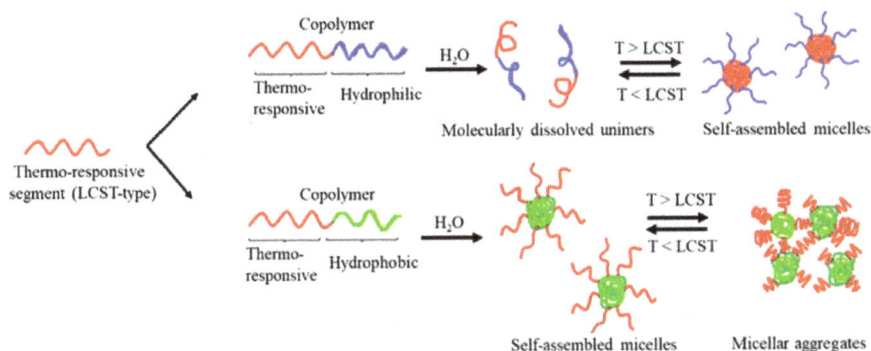

Fig. 4. Schematic representation of different strategies towards the fabrication of thermoresponsive (co)polymer micelles. Reprinted from Ref. (Sponchioni et al. 2019), with permission from Elsevier.

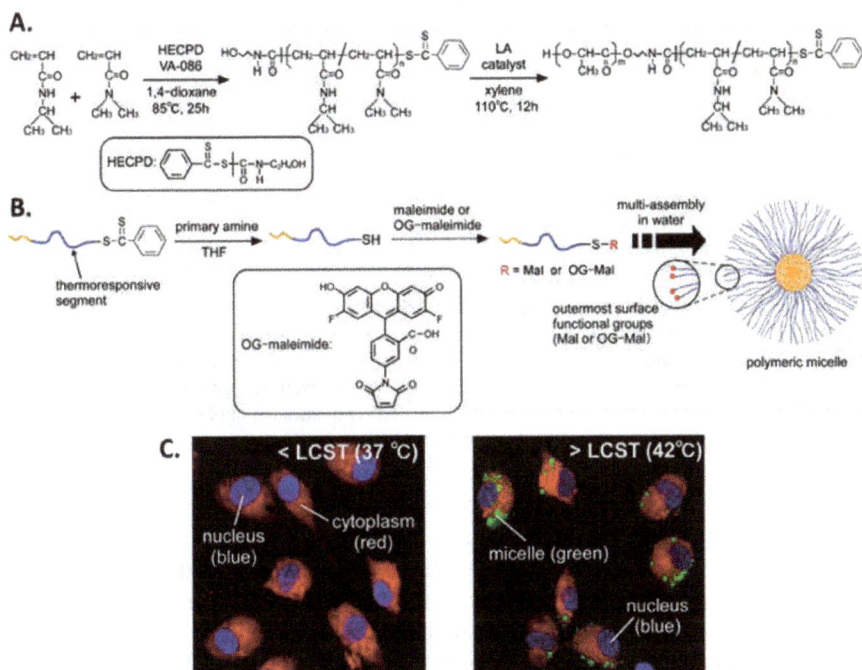

Fig. 5. Synthesis of fluorescently tagged (PNiPAAm-co-PDMAAm)-b-PLA diblock copolymers and schematic illustration of (PNiPAAm-*co*-PDMAAm)-*b*-PLA micelles formation (A and B). Confocal laser scanning microscopy (CLSM) showed the enhanced intracellular uptake of polymer micelles above LCST (C). CLSM images of polymeric micelles localized within cultured cells after incubation for 9 h (a) below the LCST (37°C) and (b) above the LCST (42°C) in 10% serum culture media. The nuclei and cytoplasm were stained with Hoechst 33258 (blue) and Cell Tracker Red (red), respectively. Green fluorescence was derived from OG-maleimide-labelled micelles. Reprinted from Ref. (Akimoto et al. 2009), with permission from ACS.

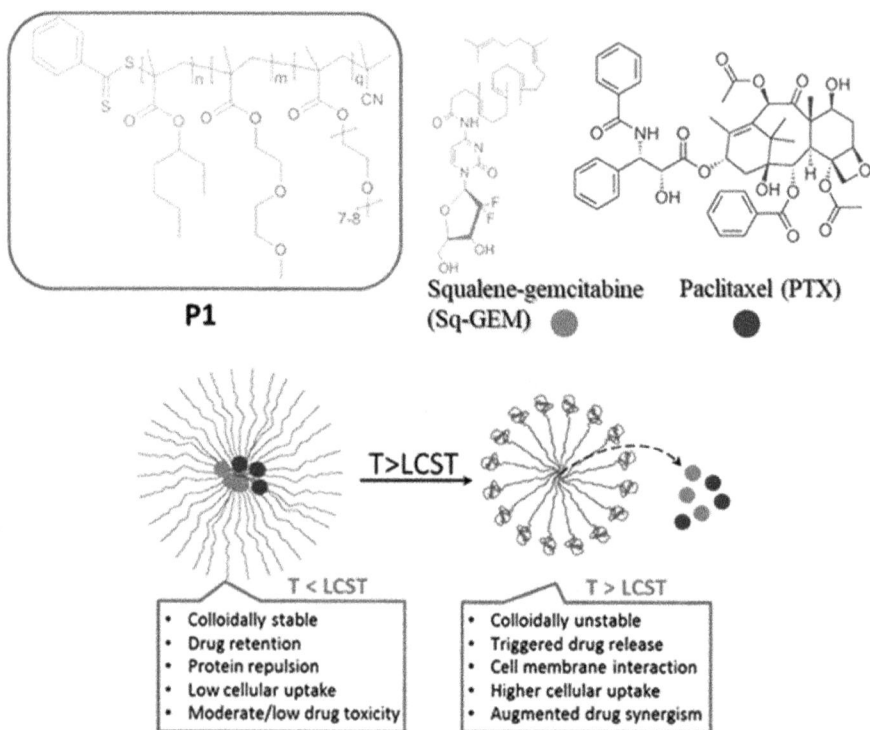

P1

Squalene-gemcitabine (Sq-GEM)

Paclitaxel (PTX)

T>LCST

T < LCST
- Colloidally stable
- Drug retention
- Protein repulsion
- Low cellular uptake
- Moderate/low drug toxicity

T > LCST
- Colloidally unstable
- Triggered drug release
- Cell membrane interaction
- Higher cellular uptake
- Augmented drug synergism

Fig. 6. P(PEGMA)-based thermoresponsive block copolymers, namely, poly(2-ethylhexyl methacrylate)-*b*-poly(di(ethylene glycol)-oligo(ethylene glycol)methyl ether methacrylate) (P1), synthesized by RAFT polymerization and their applications as micellar dispersions for hydrophobic drug(s) loading. LCST-driven disruption leads to increased drug release rates as well as higher cellular uptake due to enhanced cell membrane interactions. Reprinted from Ref. (Emamzadeh et al. 2018), with permission from RSC.

and peptides by combination with clinically applied local heating systems. Such a temperature-dependent interaction of the micellar outer shell with the cell membranes was further exploited by Pasparakis and colleagues with the employment of P(OEGMA) as the thermoresponsive segment, which further enabled the co-delivery of multiple drug(s) combinations both in the form of micelles (Emamzadeh et al. 2018) (Fig. 6) or polymer-coated liposomes (Emamzadeh et al. 2019). In another report, Li and colleagues synthesized a triblock copolymer consisting of a poly(l-lactide) (PLLA) as central block and 2-(2-methoxyethoxy) ethyl methacrylate (OEG$_2$MA) and oligo(ethylene glycol) methacrylate (OEG$_n$MA) as alternative thermoresponsive terminal blocks (Hu et al. 2015). These amphiphilic P(OEG$_2$MA-*co*-OEG$_n$MA)-*b*-PLLA-*b*-P(OEG$_2$MA-*co*-OEG$_n$MA) tri-block copolymers self-assembled into micelles in aqueous media with an LCST at ~ 45°C (blank micelles) and were used to encapsulate the anticancer drug curcumin. Upon drug loading, the LCST of the drug-loaded micelles is reduced to 38–41°C and revealed higher drug release above the LCST values than that at 37°C. However, these micelles did not present a burst-like release and the authors attributed this to strong hydrophobic interactions of curcumin with the hydrophobic chains of the PLLA core.

In recent decades, attempts have been made to fabricate active targeting thermoresponsive micellar nanocarriers (Fig. 7). In an attempt to tune the LCST of PNiPAAm to target the temperature-induced drug release in the tumour, a folate-decorated amphiphilic block copolymer, namely, [poly(ε-caprolactone)-*b*-(poly(*N*-isopropylacrylamide-*co*-acrylamide)-*b'*-methoxy poly(ethylene glycol)/poly(ethylene glycol)-folate)] ([PCL-*b*-(P(NIPAAm-*co*-AAm)-*b'*-MPEG/PEG-FA)] (PCIAE-FA)), was synthesized and subsequently employed to formulate paclitaxel (PTX)-loaded micelles (Rezaei et al. 2012). For the localization of the delivery systems at the targeted tissue/site, some targeting moieties/agents such as folate, biotin, saccharide, peptide, among others, are covalently attached to the nanoparticle surface, which can recognize and bind to the specific receptors *in vivo* that are only distinctive to the tumour cells (Liu et al. 2010, Tai et al. 2010). The micelles are formed by self-assembly in H_2O and exhibited an LCST at 39°C. This thermoresponsive characteristic imparted a capacity for a temperature-induced release (after 100 h) of PTX drug *in vitro* with ~ 90% release of the PTX from micelles at 40°C (i.e., above the LCST), in contrast to ~ 50% release at 37°C (i.e., below the LCST). The PCIAE-FA polymer (without drug) does not show any apparent cytotoxicity (measured polymer concentration range: 0.1 to 1 mg/mL, against AE cell lines). Incubation of HeLa cells with these thermoresponsive PTX-PCIAE-FA micelles at 40°C significantly enhanced the cellular uptake and cytotoxicity compared to the treatment at 37°C. Interestingly, PTX-PCIAE-FA showed higher cytotoxicity than the Tarvexol (commercial PTX formulation; at equivalent PTX concentration in each formulation) at both the investigated temperatures (at 37 and 40°C). These thermoresponsive micellar nano-formulations could offer a promising "smart" carrier to improve the delivery efficiency and tumour specificity of hydrophobic chemotherapeutic drugs.

In another work, poly(d,l-Lactic acid) (PLA) was covalently bound to a hydrophilic yet responsive poly(ethylene oxide)-poly(propylene oxide)-poly(ethylene oxide) (PEO-PPO-PEO; commercially available as Pluronic F127) polymer to form an amphiphilic block copolymer (F127-PLA), which was subsequently modified with folate (FA) to target folate-over expressing cancer cells (Fig. 7). (Guo et al. 2014). The amphiphilic F127-PLA copolymer formed micelles with an LCST of 39°C. Therefore, micelles are stable at normal physiological temperature (37°C), while rapid release of encapsulated anticancer drug (DOX) was noticed under hyperthermia (40°C). These polymeric micelles (DOX free) possessed excellent cytocompatibility, and the FA-decorated F127-PLA could actively target folate receptor (FR)-over expressed tumour cells. The FA-decorated F127-PLA micelles (DOX loaded) were more cytotoxic (against HeLa cell lines) than non-target (FA free) micelles at both 37°C and 40°C which can be attributed to the enhanced internalization by FR-mediated endocytosis. Furthermore, the IC_{50} (half maximal cell inhibitory concentration) of FA-F127-PLA was decreased by fivefold at 40°C compared to 37°C due to the faster drug release above their LCST. Therefore, these

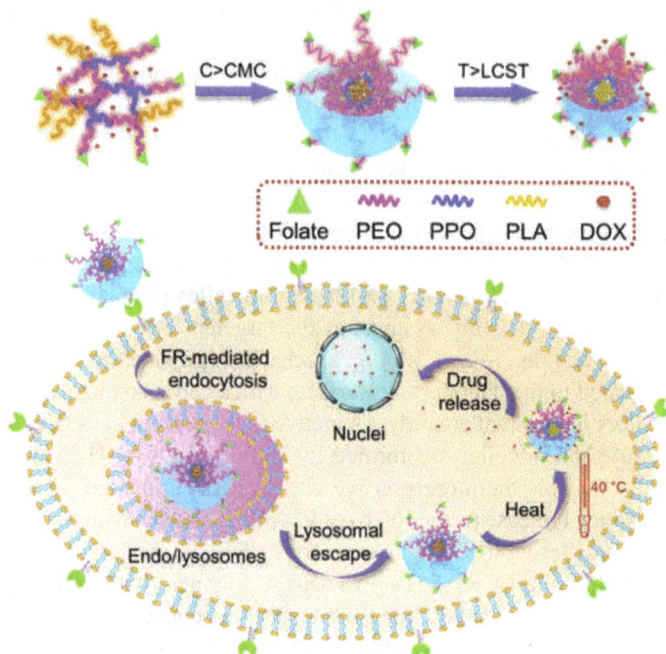

Fig. 7. Schematic illustration of thermoresponsive nanocarrier working as a targeted drug delivery system. Reprinted from Ref. (Guo et al. 2014), with permission from ACS.

temperature-responsive micelles show promising potential as a drug nanocarrier for targeted anticancer therapy.

Chattopadhyay and colleagues employed PNiPAAm and another thermoresponsive polymer, poly(*N*-vinylcaprolactam) (PNVCL), to prepare star block (co)polymers with pentaerythritol as core having polycaprolactone-*b*-poly (*N*-isopropylacrylamide) or polycaprolactone-*b*-poly(*N*-vinylcaprolactam) arms to fabricate DOX-loaded thermoresponsive micelles functionalized with folic acid targeting units (PE-PCL-*b*-PNiPAAm-FA and PE-PCL-*b*-PNVCL-FA) (Panja et al. 2016). The PE-PCL-*b*-PNiPAAm-FA and PE-PCL-*b*-PNVCL-FA polymers exhibited molecular weight-dependent LCST values at around 30–32°C and 38–39°C, respectively. DOX-loaded PE-PCL-*b*-PNiPAAm-FA/PE-PCL-*b*-PNVCL-FA micelles were able to maintain a slow release ($< 20\%$ after 41 h) of the drug below their LCST values and it shifted to a faster release rate ($> 60\%$ after 24 h) above their LCST's due to the collapse of the thermoresponsive coronas (PNiPAAm/PNVCL) and subsequent deformation/shrinkage of the micelles (Fig. 8). The cellular uptake study revealed selective internalization of these folic acid-modified micelles into C6 glioma cancerous cells. DOX-loaded PE-PCL-*b*-PNVCL-FA (also abbreviated as DOX + FA + PM) was used for the *in vivo* assays and showed increased accumulation of DOX in a C6 glioma rat tumour model which drastically inhibited the tumour growth by $\sim 84\%$ (with respect to the control) without any systemic toxicity (Fig. 8). Overall, these targeted micelles enhance the pharmacokinetics of the drug in rats and, thus, indicate good efficacy on tumour growth inhibition.

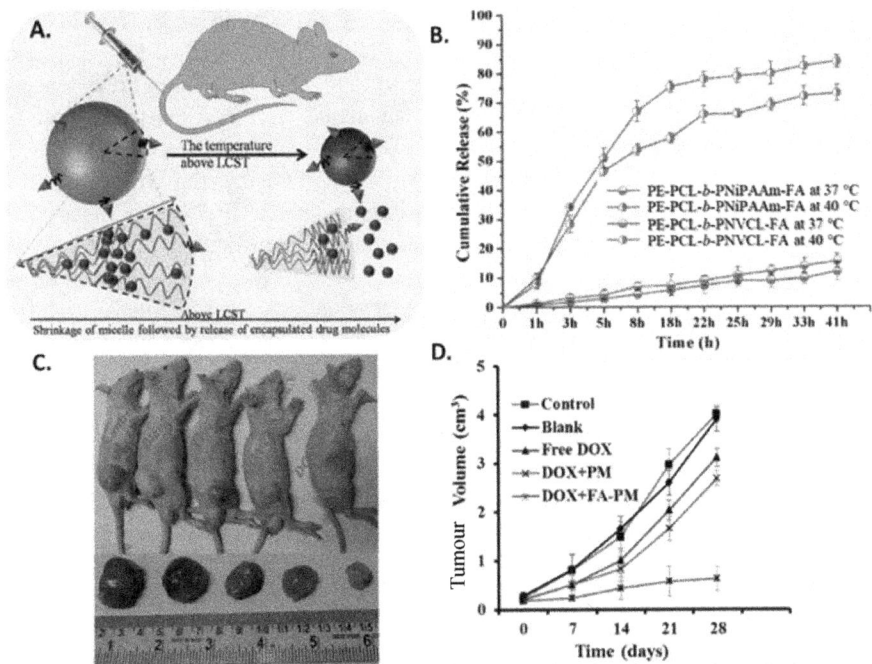

Fig. 8. Schematic illustration of temperature-induced shrinkage of PE-PCL-*b*-PNiPAAm-FA/PE-PCL-*b*-PNVCL-FA micelles followed by targeted drug release (A). Release profile of the DOX-loaded PE-PCL-*b*-PNiPAAm-FA/PE-PCL-*b*-PNVCL-FA polymeric micelle, indicating enhanced release of DOX from both polymeric micelles (PM) above the LCST (B). Tumour volume of the rat (C) and variation of tumour volume with time (D) after the treatment with different samples mentioned in the figure. The results showed that, except for blank FA-PM, the free DOX, DOX-loaded PM, and DOX-loaded FA-PM exhibits the inhibition of tumour volume compared to that of the control. Interestingly, enhanced inhibition of tumour growth (83.9%) is only observed with the use of DOX-loaded folic acid decorated polymeric micelles (FA-PMs). The FA guides the FA-PM to reach selectively into the tumour site and shows enhanced therapeutic efficiency. Reprinted from Ref. (Panja et al. 2016), with permission from ACS.

The impact of the nature of the hydrophobic blocks on the physicochemical properties of drug-loaded thermoresponsive polymeric micelles has been systematically investigated by Pietrangelo and colleagues (Sun et al. 2015). They studied a series of thermoresponsive block-copolymer micelles with identical thermoresponsive PNiPAAm coronas and distinguishable hydrophobic PNP (poly(*N*-acryloyl-2-pyrrolidone)), PMNP (poly(*N*-acryloyl-5-methoxy-2-pyrrolidone)), or PBNP (poly(*N*-acryloyl-5-butoxy-2-pyrrolidone)) cores to identify the influence of slight alteration of the core-segment structures on the micellar physicochemical properties and drug delivery efficacies. Three different block-copolymers (i.e., $PNiPAAm_{72}$-*b*-PNP_{29}, $PNiPAAm_{72}$-*b*-$PMNP_{29}$ and $PNiPAAm_{72}$-*b*-$PBNP_{26}$) were synthesized by sequential RAFT polymerization and their LCST's were found to increase with hydrophobicity, which warrants further investigation. More interestingly, they observed that the more hydrophobic the block, the better the doxorubicin (DOX) encapsulation (e.g., 20, 24 and 26 wt.% DOX loading for PNiPAAm$_{72}$-b-PNP$_{29}$, PNiPAAm$_{72}$-b-PMNP$_{29}$ and PNiPAAm$_{72}$-b-PBNP$_{26}$ micelles, respectively) and the lower the drug release (e.g., 40, 30 and 25% release were calculated for PNiPAAm$_{72}$-b-PNP$_{29}$, PNiPAAm$_{72}$-b-PMNP$_{29}$ and PNiPAAm$_{72}$-b-PBNP$_{26}$ micelles, respectively) even at temperatures above the LCST. While this work is promising, the DOX-unloaded micelles themselves exhibited toxicity (measured against MCF-7 breast cancer cell lines) even at low concentrations (0.13 mg/mL), making it difficult to translate to the clinical trial in their current form.

Thermoresponsive poly(*N*-isopropylacrylamide-*co*-acrylamide)-*b*-poly(n-butyl methacrylate) (P(NIPAAm-*co*-AAm)-*b*-PBMA) copolymer micelles (T_{cp} ~ 40°C) for targeting anticancer drug (methotrexate) towards solid tumours have been discussed by Ma and colleagues (Sun et al. 2014). A significant temperature-triggered release of loaded methotrexate from P(NIPAAm-*co*-AAm)-*b*-PBMA micelles (drug loading: 15 wt%) with improved *in vitro* anti-tumour activity (measured using LLC cell lines) has been reported. Such a finding could be interesting as it might open new possibilities for the utilization of hyperthermia to enhance site-specific drug accumulation and to achieve both spatial and temporally controlled release of anti-carcinogens. Synthesis of another thermoresponsive amphiphilic block copolymer, namely, poly(NiPAAm-*co*-N-hydroxymethylacrylamide)-*b*-polycaprolactone (P(NIPAAm-*co*-NHMAAm)-*b*-PCL) is reported by Wang and colleagues (Wang et al. 2014). P(*N*-hydroxymethylacrylamide) was introduced to improve/enhance the LCST of PNiPAAm as hydrophilic outer shell. The P(NIPAAm-*co*-NHMAAm)-*b*-PCL micelles (T_{cp} ~ 40°C) showed a sustained and temperature-triggered DOX release. DOX-unloaded micelles exhibited much less toxicity for QBC939 cells and, in contrast, DOX-loaded micelles (DOX loading: 6.3 wt%) effectively inhibited the growth and induced apoptosis of QBC939 cells (Fig. 9). Furthermore, *in vivo* experiments indicated that the DOX-loaded micelle significantly repressed tumour growth in nude mice through the induction of apoptosis by 21.49% (p < 0.05) (Fig. 9). These results indicate that these thermoresponsive amphiphilic polymer micelles have great potential for improving chemotherapy towards cholangiocarcinoma.

In another report, Cheng et al. introduced supramolecular interactions into thermoresponsive polymers to fabricate physically crosslinked and fairly stable thermoresponsive micelles in aqueous solution (Cheng et al. 2016). For example, PNiPAAm containing pendant N-(6-(3-(2,4-dioxo-3,4-dihydropyrimidin-1(2H)-yl)propanamido)pyridin-2-yl)undec-10-enamide (U-DPy) groups (PNI-U-DPy) in aqueous solution showed extremely high micellar stability (thermodynamic) with low critical micellization concentration (CMC) and a very high DOX loading capacity (up to 16.1 wt%). U-DPy units can form strong sextuple H-bonded dimers with high association constant (K_a > 10^7 M^{-1}) (Cheng et al. 2011). Such supramolecular interactions promote the PNI-U-DPy polymer self-assembly (at low CMC) into highly stable physically crosslinked network-like micellar structures at room temperature, thus making the loading process extremely stable and efficient.

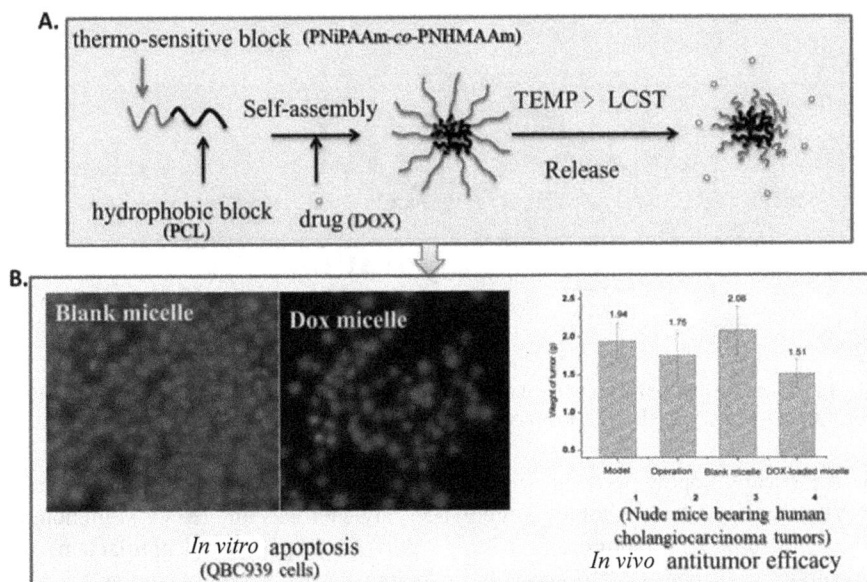

Fig. 9. Schematic illustration of temperature-induced shrinkage of DOX-loaded (P(NIPAAm-*co*-NHMAAm)-*b*-PCL micelles followed by drug release (A). Fluorescence microscopy images of the apoptosis induced by blank micelles group and DOX-loaded micelles. After treatment with the DOX-loaded micelles, QBC939 cells exhibited typical features of apoptosis. On the other hand, treatment with blank micelles did not result in cell apoptosis. Effective inhibition of QBC939 cell growth with DOX-loaded micelles could be attributed to the released drug from the micelles above LCST (B; left panel). Comparison of the weight of tumours treated with different groups after 14 days. To evaluate the anti-tumour efficiency of DOX-loaded (P(NIPAAm-*co*-NHMAAm)-*b*-PCL micelles *in vivo*, nude mice bearing human cholangiocarcinoma tumours were treated with a single injection of PBS (operation group), no treatment (model group), blank micelles, or DOX-loaded micelles. The results indicated that DOX-loaded micelle group exhibited significant inhibition of tumour growth *in vivo* (B; right panel). Reprinted from Ref. (Wang et al. 2014), with permission from ELSEVIER.

PNI-U-DPy micelles are thermoresponsive ($T_{cp} \sim 34°C$) and DOX-unloaded PNI-U-DPy micelles demonstrated excellent biocompatibility against HEK 293 and HepG2 cells. DOX-loaded PNI-U-DPy micelles are efficiently endocytosed by the cancer cells and reported to exert dose-dependent cytotoxicity behaviour against HepG2 cancer cell lines in vitro at the physiological temperature of 37°C. Such supramolecular polymer network (SPN) micelles may serve as multifunctional nanocarriers/vehicles for the effective delivery of anticancer drugs to primary tumours and metastases.

Despite considerable attention towards temperature-responsive micellar block-copolymer nanocarriers, copolymer micelles with double LCST have been rarely investigated and are usually limited to the systems outside the physiological temperature range (Kermagoret et al. 2013, Liu et al. 2014, Beija et al. 2011). Liang et al. first developed a block-copolymer with a precisely controlled double LCST within the physiological temperature range (Fig. 10) (Liang et al. 2015). The authors have prepared the first thermoresponsive segments (with LCST1 at ~ 19–26°C)

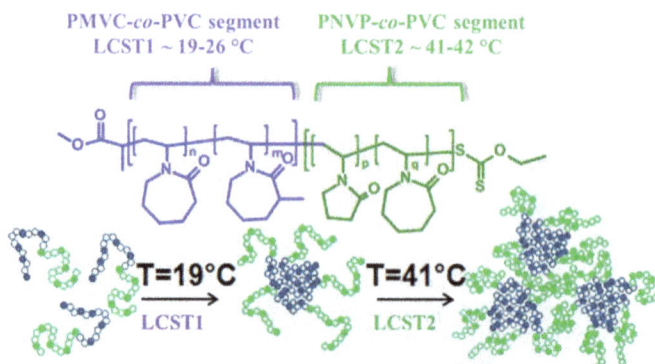

Fig. 10. Schematic representation of the double LCST behaviour of (PMVC-*co*-PVC)-*b*-(PVC-*co*-PVP) copolymers with LCST2 > LCST1. Reprinted from Ref. (Liang et al. 2015), with permission from ACS.

by statistically copolymerizing (via RAFT) thermoresponsive *N*-vinylcaprolactam (NVC) with a hydrophobic 3-methyl-N-vinylcaprolactam (MVC) monomer. The resultant thermoresponsive poly(3-methyl-*N*-vinylcaprolactam)-*co*-(*N*-vinylcaprolactam) (PMVC-*co*-PVC) copolymer was then employed as a RAFT macroinitiator to introduce another thermoresponsive statistical copolymer segment (with LCST2 at ~ 41–42°C) comprising of thermoresponsive PVC and hydrophilic poly(*N*-vinylpyrrolidone) (PVP). At LCST1, (PMVC-*co*-PVC)-*b*-(PVC-*co*-PVP) di-block copolymers self-assemble into micelles loaded with DOX, whereas at LCST2 (at ~ 40°C, i.e., at a slightly elevated physiological temperature), these micelles collapse, resulting in the burst release of the drug. For example, 85% DOX released was reported from these micelles at 42°C, as compared to only 40% release at 30°C (after 24 h). The second elevated temperature (LCST2) is typical for tumours and can trigger the drug-loaded micelle accumulation within the tumour tissues, highlighting the potential as novel types of drug career for tumour therapy.

Literature reports showed that commonly used anti-cancer drugs, like DOX and epirubicin, can have several side-effects. Such anti-cancer drugs generate metal complexes (e.g., with iron) that have the potential to react with oxygen and form reactive oxygen species (ROS), which can oxidize lipids, proteins, and DNA in cells ultimately leading to cardiac damage (Gutteridge 1984). Antioxidants such as polyphenols can react with the free radicals (ROS) and suppress the damage to lipids, proteins, and other cell parts. Co-administration of DOX and antioxidants could therefore reduce the toxic side effects of DOX (Al Fatease et al. 2019). Also, combination loading is an effective strategy to increase drug loading (Washington et al. 2018). Combination loading of DOX and a polyphenol-type antioxidant, namely, quercetin (Que) in thermoresponsive as well as biodegradable poly(γ-oligo(ethylene glycol)-ε-caprolactone-*b*-poly(γ-benzyloxy-ε-caprolactone) (PME$_x$CL-*b*-PBnCL) polymeric micelles for increased loading efficiency and efficacy has been recently demonstrated by Stefan and colleagues (Soltantabar et al. 2020). The γ-substituted oligo(ethylene) glycol (OEG) poly(ε-caprolactone)s (PME$_x$CL) segments in the PME$_x$CL-*b*-PBnCL copolymer are thermoresponsive and it is possible to tune the LCST-type T_{cp} of the copolymer micelles by varying the OEG chain length. PME$_x$CL-*b*-PBnCL

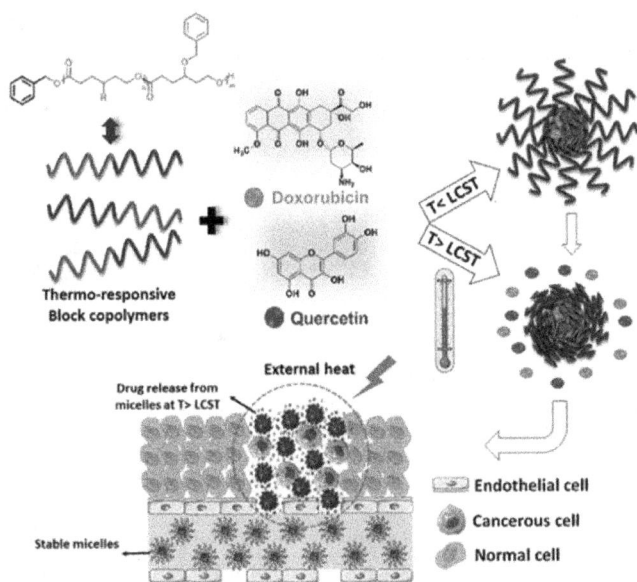

Fig. 11. Schematic illustration of the preparation and loading of DOX- and Que-loaded PMExCL-*b*-PBnCL micelles and temperature-triggered drug release. The images showed that hyperthermia could facilitate drug release as well as uptake and therefore showed more cytotoxic effect towards cells. Such observation can be attributed to the enhanced interaction of micelles with the cell membrane at higher temperatures or higher drug release in the extracellular microenvironment of cells at elevated temperature and its uptake by the cells. Reprinted from Ref. (Soltantabar et al. 2020), with permission from ACS.

block-copolymer micelles showed thermodynamic stability having CMC values in the order of 10^{-5} mg/mL. Combination loading demonstrated improved loading of DOX and Que. The increased loading can be attributed to the favourable H-bonding interactions between DOX and Que as well as π stacking interactions with benzyl substituents of PME$_x$CL-*b*-PBnCL diblock-copolymer micelles. The PME$_3$CL-*b*-PBnCL micelles showed LCST value at around 41°C and hyperthermia could induce both drug release and cellular uptake as more DOX and Que was found within the HepG2 cells at a higher temperature (42°C) (Fig. 11). Cell viability studies showed that the empty/blank micelles did not show any cytotoxicity to HepG2 or H9c2 cell lines (measured polymer concentration range: 0.02 to 0.25 mg/mL). A total of 20% higher cytotoxicity on HepG2 cells were observed from the coloaded (DOX and Que) micelles at a higher temperature (42°C; in comparison to 37°C), which is attributed to a higher release of anticancer drugs at elevated temperatures above LCST (Fig. 11). The employment of such biodegradable, cytocompatible and thermoresponsive (co)polymer systems along with the combination loading approach is indeed a good strategy for developing "smart" drug delivery systems.

Amphiphilic ABA triblock-copolymers with two hydrophilic poly(2-methyl-2-oxazoline) (PMeOx) A blocks and one central thermoresponsive poly(2-isopropyl-2-oxazoline)-*co*-poly(2-butyl-2-oxazoline)   (P^{i}PrOx-*co*-P^{n}BuOx) copolymer as B block were investigated regarding their thermoresponsive micelles as radionuclide delivery systems (Hruby et al. 2010). A phenolic group was incorporated to allow radionuclide labelling with [125]iodine. ^{125}I-loaded micelles showed good labelling yield and *in vitro* stability. Copolymers were molecularly dissolved below their T_{cp} and, upon increasing the temperature (above the cloud point), radionuclide-loaded micelles were formed due to the collapse of the middle block that could benefit from the enhanced permeation and retention (EPR) effect for passive targeting of nanocarriers to tumours. The toxicity of all the copolymers was considered essentially non-existing as confirmed by haemolytic activity studies (less than 3.5% haemolysis even at $1\,mg\,mL^{-1}$ copolymer concentration). Hence, the particular system shows potential for utilization in radio diagnostic protocols. In another example, P^{i}PrOx-*co*-P^{n}BuOx has also been combined with pluronic F127 to produce thermoresponsive micelle-type nanoparticles (with an appropriate size that are suitable for the EPR effect) for potential use in solid tumour diagnostics (Panek et al. 2012, Bogomolova et al. 2013). It was also demonstrated that P^{i}PrOx-*co*-P^{n}BuOx (co)polymers are more stable against small doses of β-radiation than other potential thermoresponsive polymers for therapeutics, including PNiPAAm and PNVCL (Sedlacek et al. 2016).

Some naturally occurring and biocompatible polymers such as hydroxypropyl cellulose (HPC) also possess LCST-type thermoresponsive properties ($T_{cp} \sim 45°C$) and can be utilized in the fabrication of thermoresponsive micelles for drug delivery applications (Jing and Wu 2013). For example, Bagheri et al. designed a hydroxypropyl cellulose-based amphiphilic polymer (HPC-PEG-Chol) that contained poly(ethylene glycol) (PEG) and cholesterol-containing moieties (Chol) with specific degrees of substitution (Bagheri et al. 2014). The resulting synthesized polymer was subsequently linked to a biotin conjugate (HPC-PEG-Chol-biotin) to develop a drug delivery carrier, specifically against cancer cells. The authors used the as-synthesized conjugate (HPC-PEG-Chol-biotin) to form paclitaxel (PTX)-loaded micelles (PTX-HPC-PEG-Chol-biotin) with an LCST of 39.8°C. These micelles presented a drug release which was about two times faster at 41°C (i.e., above LCST) than at 37°C (i.e., below LCST), indicating that they would be able to partially avoid PTX leakage/release in the normal bloodstream or in normal tissues during circulation, and favour accumulation as well as site-specific release in the tumour microenvironment at a temperature above 37°C. However, substantial drug release at 37°C severely hampers further clinical translation as most drug will be released during circulation, before reaching the tumour. The cytotoxicity measurement indicated that the polymer and PTX-loaded micelles had no apparent cytotoxicity against normal cells, and PTX-loaded micelles showed higher toxicity against tumour cells than PTX alone. Also, they showed very strong adsorption to both HeLa and MDA-MB-231 cancer cell lines compared to the free PTX drug owing to their nanostructure. Therefore, such "intelligent" thermoresponsive micelles will not only improve the intracellular uptake by increasing the attachment/interaction between the cells and micelles, but also may have great potential to be safely used in tumour targeting chemotherapy.

Another approach to further enhance the localized effect of a payload/drug is to apply multiple stimuli to trigger the site-specific release. Temperature- and pH-responsive (co)polymers are among the most interesting and, therefore, extensively studied as dual responsive nanocarriers for drug delivery. Many pH- and temperature-responsive (co)polymers are designed and synthesized by introducing pH-sensitive blocks/segments (such as weak acids) into the thermoresponsive polymer (e.g., PNiPAAm), which results in copolymers with pH-dependent LCST. The tumour microenvironment is slightly acidic (pH 5.6 to 6.8) due to glycolysis in cancer cells, hypoxia, and deficient blood perfusion (Liu et al. 2015). Considering pH and temperature are different from healthy tissues in the tumour microenvironment, the development of dual pH- and thermoresponsive (co)polymer micelles can promote a better treatment outcome. In this context, fine-tuning of the LCST by subtle pH-changes is highly desirable towards the development of pH- and thermoresponsive drug delivery nanocarriers for tumour therapy (Soppimath et al. 2005, Lo et al. 2005). Such systems can be very efficient due to their ability to differentiate between pathological and healthy tissues. For example, Qiao and colleagues developed thermo- and pH-responsive PHis-PLGA-PEG-PLGA-PHis copolymer micelles composed of poly(ethylene glycol) (PEG), poly(D,L-lactide-*co*-glycolide) (PLGA) and poly(L-histidine) (PHis) for DOX encapsulation and triggered release (Hong et al. 2014). The copolymer self-assembled into core-shell micelles with inner core-forming hydrophobic PLGA-PHis blocks surrounded by a hydrophilic PEG shell in aqueous medium. PLGA-PEG-PLGA confers thermoresponsiveness, while PH is segment in the micelles is pH-responsive. *In vitro* drug release experiments revealed an enhanced DOX release from the micelles at pH 6.5 and 41°C as compared to pH 7.4 at 37 or 41°C, indicating that the dual (pH and temperature) stimuli may promote a better response than the single stimulus. Also, *in vitro* cytotoxicity (evaluated in MCF-7 cell lines) at different pH values revealed lower cell viability under acidic pH, whereas no difference was noticed in the effect of free DOX within the studied pH range. These results suggest that the synergistic two-step delivery system with the increment of intracellular uptake with external temperature stimulation and fast release of DOX with internal acidic endosomal pH environment was effective in tumour therapy.

Design and synthesis of thermoresponsive materials based on poly(amino acid)s is an interesting approach as they provide good biocompatibility and low toxicity. Furthermore, this type of polymer materials can be degraded into amino acids or short peptide chains via the cleavage of peptide (secondary amide) bonds through enzymatic reactions and/or microbial cleavage and, therefore, has attracted increasing attention for the fabrication of biodegradable polymers for biomedical applications (Qiao et al. 2010). In this context, thermo- and pH-responsive poly(amino acid)-based biodegradable amphiphilic copolymer micelles with a pH-sheddable hydrophilic PEG corona have recently been developed by Liu and colleagues (Liu et al. 2015). A thermoresponsive poly(amino acid) was first prepared through the conjugation of *N,N*-diisopropylaminoethyl amine groups to poly(aspartic acid) (PASP) via polysuccinimide (PSI) ring opening reaction (aminolysis). Afterwards, DOX and PEG chains were chemically grafted to the peptide chain via acid-cleavable hydrazone

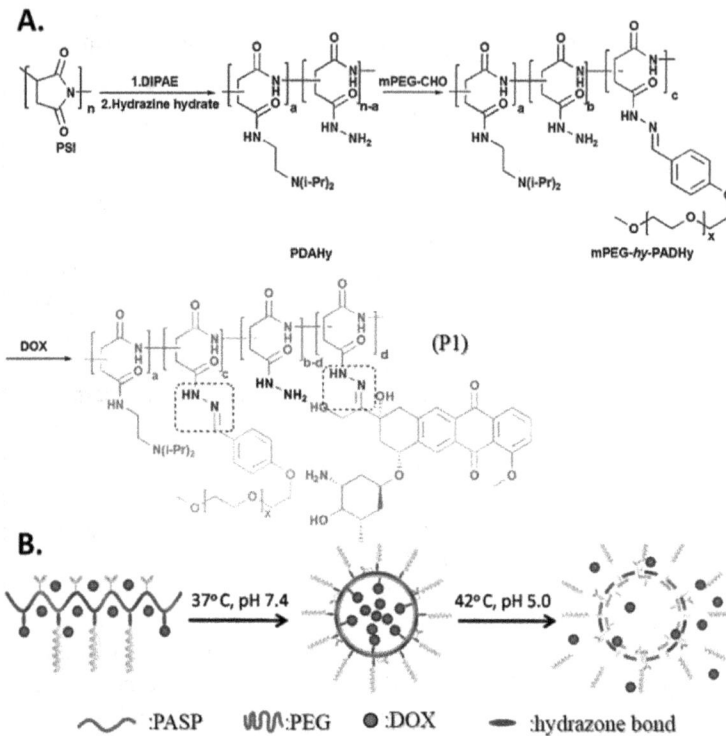

Fig. 12. Synthetic procedure for the preparation of pH- and thermoresponsive biodegradable poly(amino acid)-based drug delivery system (P1) (A). Suggested pH- and temperature-dependent assembly/disassembly processes of P1 (B). Reprinted from Ref. (Liu et al. 2015), with permission from ELSEVIER.

linkages. Conjugated DOX basically acts as hydrophobic moiety to facilitate the poly(amino acid) self-assembly into nanoparticles. On the other hand, PEG provides a removable hydrophilic shield to the nanoparticles which not only protects the drug-loaded micelle-type nanoparticles from unwanted interaction with blood protein, but also reduces the premature clearance of nanoparticles in the bloodstream. Free DOX molecules could also be encapsulated (via hydrophobic interactions as well as π–π stacking) into the self-assembled nanoparticles. Drug release can be triggered by the temperature with a significantly enhanced release under acidic conditions (Fig. 12). Furthermore, the DOX-loaded nano-assemblies exhibit cytotoxic characteristics (against NIH-3T3 cells) with high efficacy and therefore might serve as a potential drug delivery system for DOX.

The imidazole containing amino acid histidine (His) is also an attractive molecule for pH-responsive delivery systems and this was employed by Hsiue and colleagues to fabricate temperature- and pH-responsive mixed micelles for triggered drug release and enhanced passive targeting (Chen et al. 2012a). In this report, DOX loaded temperature and pH-responsive mixed micelles were developed from mPEG-*b*-P(HPMA-Lac-co-His) and mPEG-PLA. Micelles were also conjugated with Cy5.5 fluorescent probe to visualize tumour accumulation. Cytotoxicity of DOX encapsulated micelles, free DOX as well as blank micelles were monitored against human breast carcinoma (ZR-75-1, MCF-7), human cervical carcinoma (HeLa) and human lung large cell carcinoma (H661) cells using the methylthiazoletetrazolium (MTT) assay with different time intervals. Blank micelles did not exhibit any cytotoxicity towards all the investigated cell lines even after 72 h of incubation. However, DOX and DOX-micelles were more cytotoxic and in particular, DOX-loaded micelles revealed enhanced cytotoxicity compared with free DOX. These results indeed confirmed endocytosis of DOX encapsulated micelles and burst release with higher drug concentration at pH 5.4 and 37°C similar to the acidic endosomal component. Confocal laser scanning microscope (CLSM) analysis also confirmed efficient release of DOX from DOX encapsulated micelles into the cell nuclei effectively. Cy5.5-tagged DOX encapsulated micelles injected into Balb/c nude mice bearing human cervical tumours showed stronger fluorescence intensity at the tumour site than at normal tissues (Fig. 13), thus suggesting the ability of the mixed micelles to preferably target the tumours by passive tumour targeting (EPR effect) (Fig. 4). To examine *in vivo* antitumor activity of the dual responsive micelles, Balb-c/nude mice bearing human cervical tumours were treated with saline, blank micelles, free DOX and DOX encapsulated micelles via intravenous injections and the growth of subcutaneously implanted tumours was assessed over a period of 43 days. A 17- and 12-fold enhancement in the tumour volumes was observed for the control (untreated) and free DOX treated mice, respectively. Empty micelles displayed no anti-tumour activity and interestingly, mice treated with DOX loaded micelles showed slow enhancement (5-fold) in the tumour volumes after 43 days. These results indicated that such amino acid-based DOX-loaded dual pH- and thermoresponsive micelles possess specific targeting efficiency and high anti-tumour activity.

Zhang and colleagues also reported temperature- and pH-responsive mixed-shell polymeric micelles prepared via complimentary as well as multiple hydrogen bond-induced self-assembly of diaminotriazine-terminated PCL (DAT-PCL), uracil-terminated PEG (MPEG-U), and uracil-terminated PNVCL (PNVCL-U) at room temperature (Wu et al. 2016). In the formed micelles, PCL acted as the core and MPEG/PNVCL as the mixed shell at room temperature. However, upon increasing the temperature above the LCST, thermoresponsive PNVCL collapsed and enclosed the PCL core. Hydrophilic mPEG formed the corona and the chains of this corona embedded into the PNVCL shell. These mPEG chains acted as a protective barrier and a channel against aggregation of micelles (Fig. 14). *In vitro* drug release studies of DOX-loaded micelles showed an accelerated release at pH 5.0 compared to physiological pH. This accelerated release could be ascribed to the hydrogen bond dissociation process. In this study, authors have also proved strong responsiveness of the complex micelles to mildly acidic pH and capability of micelles to release DOX molecules inside the cells (Fig. 14) to achieve significantly enhanced drug efficacy.

In another report, two amphiphilic as well as partially biodegradable copolymers, namely, poly(*N*-isopropylacrylamide-*co*-dimethylacrylamide)-

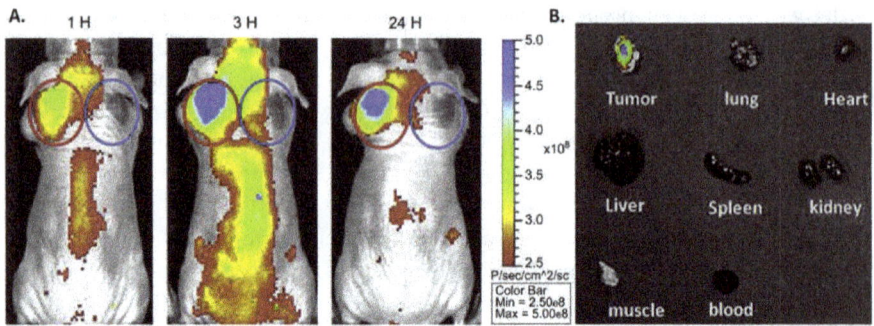

Fig. 13. Optical fluorescence imaging *in vivo* of HeLa tumours xenografted into nude mice and treated with Cy5.5-DOX-micelles. The red (ROI1) and blue (ROI2) cycles indicate the locations of tumour and normal tissues, respectively (A). Photoimaging of organs and tumours harvested 24 h posttreatment with 2 mg/kg of DOX-micelles. Reprinted from Ref. (Chen et al. 2012a), with permission from ELSEVIER.

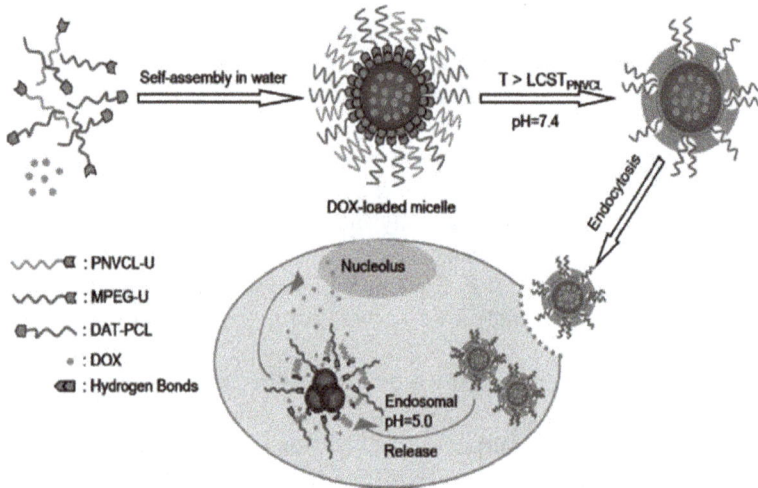

Fig. 14. Schematic illustration of the structure and functioning mechanism of DOX-loaded mixed shell micelle. Reprinted from Ref. (Wu et al. 2016), with permission from John Wiley and Sons.

block-PLA (P(NiPAAm-*co*-DMAAm)-*b*-PLA) and poly(*N*-isopropylacrylamide-*co*-dimethylacrylamide)-*block*-PCL (P(NiPAAm-*co*-DMAAm)-*b*-PCL) were synthesized by Li and colleagues to study the dual pH- and thermo-triggered release of encapsulated adriamycin (ADR) (Li et al. 2011). P(NiPAAm-*co*-DMAAm)-*b*-PLA/P(NiPAAm-*co*-DMAAm)-*b*-PCL showed an LCST-type phase transition at around 40°C and their self-assembled micelles were stable under normal physiological conditions (pH = 7.3, T = 37°C). However, micelles were found to be unstable when the pH and temperature were altered to that of tumour cell conditions (pH = 5.3 and 40°C) which is attributed to the degradation of the hydrophobic PLA or PCL. Compared to the normal physiological conditions (pH = 7.3, T = 37°C), the ADR release and intracellular uptake were greatly enhanced at pH = 5.3 and

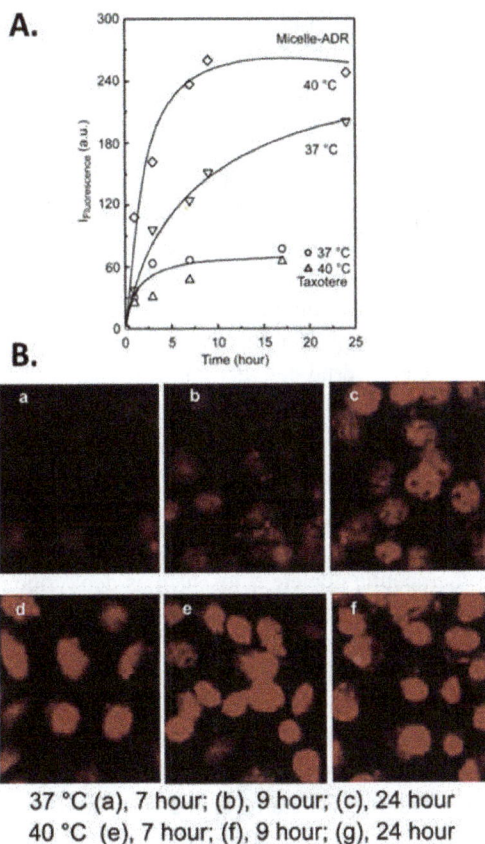

A.

B.

37 °C (a), 7 hour; (b), 9 hour; (c), 24 hour
40 °C (e), 7 hour; (f), 9 hour; (g), 24 hour

Fig. 15. The time dependence of N-87 stomach cancer cell uptaking ADR formulation Taxotere and micelle-ADR at 37 and 40°C (A). The bottom picture shows the fluorescence images for micelle-ADR formulation at 37 and 40°C (B). For the ADR-micelle system, the cellular uptake at 40°C is much larger than that at 37°C. Reprinted from Ref. (Li et al. 2011), with permission from ELSEVIER.

40°C (simulated tumour tissue condition). Flow cytometry as well as fluorescent microscopy revealed much higher ADR transportation by these micelles in N-87 stomach cancer cell as compared to the Taxotere® (commercially available ADR formulation). It was also shown that the intracellular uptake was enhanced above the phase transition temperature (Fig. 15). The results suggested that such dual pH- and thermoresponsive biodegradable micelles could be promising for anticancer therapy.

Hiruta et al. also reported dual pH- and thermoresponsive micelles based on a P(NiPAAm-*co*-DMAAm)-*b*-poly[2-(diisopropylamino)ethyl methacrylate] (P(NiPAAm-coDMAAm)-*b*-PDPA) block-copolymer (Hiruta et al. 2017). The P(NiPAAm-coDMAAm)-*b*-PDPA micelles are composed of a thermoresponsive P(NIPAAm-coDMAAm) as corona segments and a pH-responsive PDPA as core. The authors also prepared thermoresponsive P(NIPAAm-co-DMAAm)-*block*-poly(butyl methacrylate) (P(NiPAAm-coDMAAm)-*b*-PBMA) micelles with a non-pH-responsive PBMA core for comparison. The resultant micelles exhibited a

LCST-type phase transition at a temperature slightly higher than body temperature ($T_{cp} \sim 39°C$). Such a feature enabled the suppression of intracellular uptake with normal body temperature and the increment of intracellular uptake upon local heating. Significantly higher cytotoxicity was noticed for HeLa cell lines after the treatment with DOX-loaded P(NiPAAm-coDMAAm)-*b*-PDPA micelles at 42°C compared to 37°C. However, no difference in the cytotoxicity was detected at both investigated temperatures (42 and 37°C) when the cells were treated with DOX-loaded P(NiPAAm-coDMAAm)-*b*-PBMA, indicating that the absence of a pH-stimulus delayed the DOX release from the endosomes.

As these examples highlight, the potential for the dual responsiveness gives an advantage over the use of a single stimulus for the polymeric micelles and therefore bears high potential for cancer drug delivery applications.

LCST-type thermoresponsive polymers for gene therapy

Gene therapy has appeared as a new and efficient modality to combat acquired as well as genetic diseases such as cancer, genetic disorders, neurological disorders, cardiovascular diseases, diabetes mellitus among others (Kay 2011). In gene therapy, a vector containing therapeutic nucleic acids such as deoxyribonucleic acid (DNA) and ribonucleic acid (RNA) is introduced into target cells to modify the gene expression via altering defective genes, modifying missing genes, and silencing mutated genes. In this context, gene-delivery vectors play a key role because they can deliver therapeutic nucleic acids/genes into targeted cells, unpacking genes intracellularly, and assuring that the genes conduct eventual transfection.

To date, gene-delivery vectors can be approximately categorized into two types: non-viral and viral vectors. Regardless of their high gene-transfection efficacy, the development of viral vectors towards gene delivery is significantly impeded because of several drawbacks in terms of safety aspects such as high immune responses, inflammatory responses, toxicity issues and insertional mutagenesis. In contrast, non-viral vectors have garnered significant interest for gene delivery in recent decades. Non-viral vectors, e.g., cationic polymers, lipids, and inorganic nanoparticles, among others, offer several advantages including non-immunogenicity, low(er) cost for high-scale manufacture, improved DNA/RNA loading capacity and high chemical versatility. However, the transfection efficacy is usually lower compared to viral vectors. Of these non-viral vectors, cationic polymer-based non-viral gene delivery systems can be distinguished from other non-viral vectors by their high chemical versatility and have therefore gained the greatest interest for use as an alternative to viral vectors (Guo and Huang 2012, Samal et al. 2012, Pack et al. 2005, Chen et al. 2012b, Lee and Kataoka 2012).

In general, positively charged polymers can form polyplexes with negatively charged therapeutic nucleic acids, such as siRNA and plasmid DNA (pDNA), via electrostatic interactions (Fig. 16). These polyplexes (with a size around ranging from 50 nm to a few hundred nanometers) are then taken up by cells through endocytosis. Sequentially, polyplexes dissociate, presumably via the proton-sponge mechanism and, consequently, their gene payloads, such as siRNA and pDNA, are released into the cytoplasm. The unpacked siRNA is then degraded into sequence-specific messenger

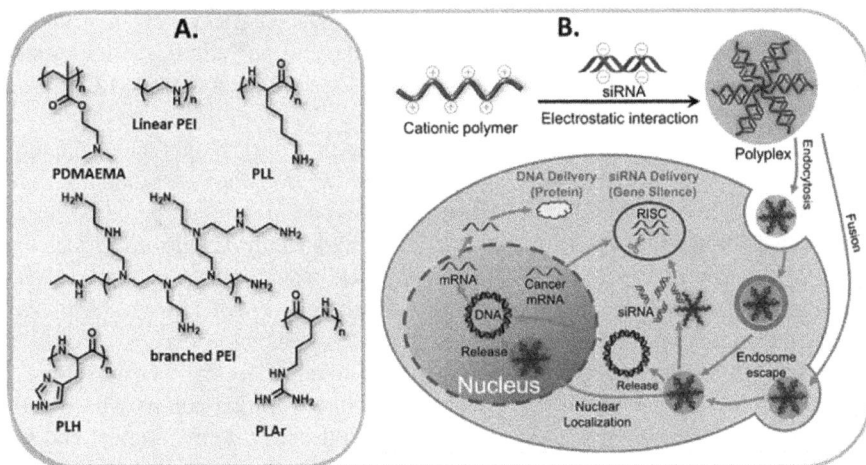

Fig. 16. Chemical structures of conventional cationic polymers used for gene therapy (A). Gene delivery mechanism of polyplexes in the presence of cationic polymers (B). Reprinted from Ref. (Chen et al. 2020b; doi.org/10.2147/IJN.S222419), with permission from Taylor and Francis.

RNA (mRNA) through the formation of RNA-induced silencing complexes (RISC) in cytosol, followed by inhibiting the activity of pathogen proteins. However, in case of the released pDNA, it must import to the nucleus for the subsequent transcription and protein expression process. The use of stimuli-responsive polymers in gene therapy is considered particularly beneficial since, by applying a physical stimulus, such as light and temperature, complex/polyplex association or dissociation can be controlled. Additionally, one can control the site, timing as well as the duration of the gene expression (Yokoyama 2002). Thermoresponsive polymers are versatile in this context as the association/dissociation of the polymer-polynucleotide complex can be controlled by simply changing the temperature. Thermoresponsive polymers with LCST slightly below or close to body temperature can also be used to enhance the gene transfection efficiency not only by condensing the polyplex structure to an appropriate size for cell internalization, but also by protecting the polyplex structures during intracellular delivery from the nuclease and other interfering anionic substances such as chondroitin sulfate (CS) and heparan sulfate (HS) located in the cellular surface and extracellular matrix (ECM) (Belting and Petersson 1999, Kurisawa et al. 2000, Yokoyama et al. 2001, Wong et al. 2007).

The proximity of the LCST of PNiPAAm (T_{cp} ~ 32°C) to physiological temperature makes it a particularly interesting polymer for drug and gene delivery applications. Also, it is well-known in the literature that the T_{cp} of PNiPAAm can be altered simply through co-polymerization of NiPAAm with other hydrophilic or hydrophobic (co)monomers. PNiPAAm copolymers have been extensively studied in literature in terms of novel non-viral vectors for gene transfection, including poly[2-(dimethylamino)ethyl methacrylate-*co*-[*cis*-butenedioic anhydride-poly [(*N*-isopropylacrylamide)-*co*-(butyl methacrylate)]]] (Ma et al. 2010), *N*-*N,N*-trimethyl chitosan chloride-*g*-(*N*-isopropylacrylamide)) (Mao et al.

2007), poly(l-lysine)-g-poly(N-isopropylacrylamide) (Oupicky et al. 2003), PNIPAAM-b-polyethyleneimine (Turk et al. 2004), and polyethyleneimine-g-poly [2-(2-methoxyethoxy)ethyl methacrylate]-b-poly(2-hydroxyl methacrylate) (Yang et al. 2010).

The synthesis of a PNiPAAm-based cationic block copolymer, namely, Poly(N-isopropylacrylamide)-b-poly((3-acrylamidopropyl)trimethylammonium chloride) (PNiPAAm-b-PAMPTMA) and its evaluation in terms of potential for *in vitro* transfection of HeLa cells has been investigated by Calejo et al. (Calejo et al. 2013). The presence of the PNiPAAm block provided the copolymers with thermoresponsive features, while the positively charged PAMPTMA block allows for the interaction with the negatively charged DNA, leading to the formation of a polymer-DNA complex (polyplex). Copolymers with relatively short, charged blocks were highly associative at physiological temperature, forming compact core-shell structures that were able to protect the DNA from the external environment and led to impressive transfection efficiencies. The authors have further studied transfection efficiency and cytotoxicity of the polyplexes as a function of polymer block lengths and N/P ratio among others. The N/P ratio is the ratio of nitrogen atoms in the cationic polymer over the phosphorus atoms in the gene. The transfection efficiency was found to increase with N/P ratio as the overall presence of positive charges on the complexes was believed to facilitate cellular uptake, leading to higher transfection efficiencies. However, increasing the molar mass of the charged PAMPTMA led to an increased degree of charge repulsion between the polymer molecules, obstructing to some extent the thermoresponsive association of PNiPAAm which led to the formation of polyplexes with an 'open' structure. The low compactness of these polyplexes was ascertained as the key factor for the lower transfection efficiency. For long, charged blocks, a very long PNiPAAm block was therefore found essential to ensure the formation of compact nanoparticles that lead to successful *in vitro* gene delivery. The PNiPAAm$_{65}$-b-PAMPTMA$_{20}$ formulation showed maximum transfection efficiency (80%) at an N/P ratio of 30 while, at the same time, the formulation showed acute cytotoxicity (cell viability $< 40\%$). Low cytotoxicity levels and negligible transfection efficiency were observed for the formulations at low N/P ratios of 1/5 and 5/1. This study indicates that the modification and fine-tuning of this thermoresponsive gene delivery systems is crucial to achieve high transfection efficiency and good biosafety (minimal toxicological implications) at the same time.

Programmable temperature-triggered DNA condensation and enhanced gene transfection by polyethylenimine-g-(poly(2-(2-methoxyethoxy) ethyl methacrylate)-b-poly(2-hydroxyethyl methacrylate)) (PEI-g-(PMEO$_2$MA-b-PHEMA)) copolymer-based nonviral vectors have been investigated by Liang and colleagues (Yang et al. 2010). Thermoresponsive diblock poly[2-(2-methoxyethoxy) ethyl methacrylate]-b-poly(2-hydroxyethyl methacrylate) (PMEO$_2$MA-b-PHEMA) copolymers were synthesized by ATRP and PEI (M_w = 1200) was grafted onto 1,1'-carbonyldiimidazole (CDI)-activated as-synthesized PMEO$_2$MA-b-PHEMA to fabricate PEI-g-(PMEO$_2$MA-b-PHEMA) (PEIMH) copolymer vectors (Fig. 17). At 37°C, PEI-g-(PMEO$_2$MA-b-PHEMA) condensed DNA more effectively due to the shielding effect of collapsed PMEO$_2$MA chains

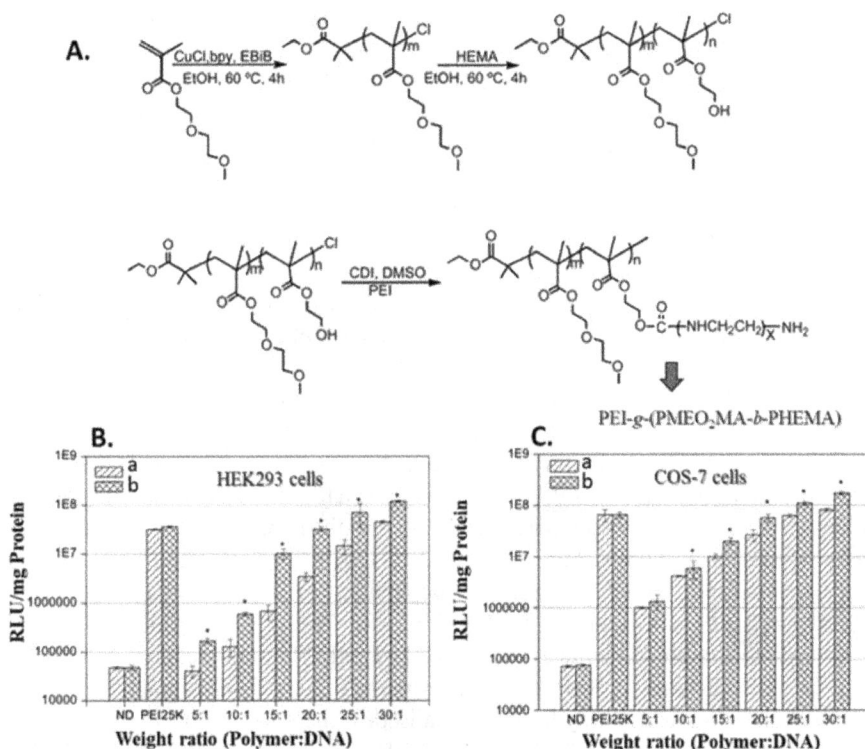

Fig. 17. Synthesis procedure of PEI-*g*-(PMEO2MA-*b*-PHEMA) copolymer (A). Thermoresponsive transfection efficiency of PEI-*g*-(PMEO2MA-*b*-PHEMA)/pDNA complexes in comparison with that of PEI25K or naked DNA (ND) in HEK293 (B) and COS-7 (C) cells in the absence of serum. Route a: 37°C (48 h); Route b: 37°C (21 h) → 20°C (3 h) → 37°C (24 h). Compared with constant temperature route (a), variable temperature (b) contributes to a 2.6–15 fold increase in transfection efficiency of HEK293 cells. There is a statistically significant difference between the transfection levels achieved by the two protocols. The improvement in transfection efficiency by variable temperature mode presumably results from temporary cooling-induced unpacking gene from carrier, which probably promotes/enhances the gene transcription. Reprinted from Ref. (Yang et al. 2010), with permission from ELSEVIER.

($T_{cp} \sim 32.5°C$) and the contracted PMEO$_2$MA chains led to more exposure of surface positive charges of PEI-*g*-(PMEO$_2$MA-*b*-PHEMA)/pDNA complexes, which was assumed to be favourable for gene transport. The low transfection level with the PEI1200 formulation was ascribed to the poor DNA condensation ability of this low molecular weight polycation. The authors have employed two approaches to investigate the thermoresponsive PEI-*g*-(PMEO$_2$MA-*b*-PHEMA)-mediated gene transfection. One is the constant temperature protocol where the incubation temperature was fixed at 37°C for 48 h and the other one is the variable temperature approach where the transfected cells were incubated for 21 h at 37°C, 3 h at 20°C followed by additional 24 h at 37°C. As compared to constant temperature route, the variable temperature approach was found to be prominent in improving gene expression level in HEK293 cells (Fig. 17). The improvement in transfection efficiency through the variable temperature protocol was believed to be a result of temporary cooling-induced unpacking of the gene from the carrier, which consequently promotes the gene transcription process. The results also demonstrate that the transfection efficiency of the PEI-*g*-(PMEO$_2$MA-*b*-PHEMA)/pDNA formulation is increased with vector/DNA (N/P) charge ratio and at an optimal N/P ratio (25:1 and 30:1), the efficiency of PEI-*g*-(PMEO$_2$MA-*b*-PHEMA)/pDNA (T_{cp} ~ 32.5°C) formulation is almost equivalent or even superior to that of PEI25K (Fig. 17). This non-cytotoxic and thermoresponsive PEI-*g*-(PMEO$_2$MA-*b*-PHEMA) copolymer could be a promising candidate as a potential nonviral vector with a function of temperature-tuned gene transfection.

In another report, Osawa, Kataoka and colleagues prepared thermoresponsive, rod-shaped polyplex micelles by polyplexation of plasmid DNA (pDNA) with ABC triblock copolymers (PEtOx-*b*-PnPrOx-*b*-PLys) consisting of hydrophilic poly(2-ethyl-2-oxazoline) (PEtOx, A), thermoresponsive poly(2-*n*-propyl-2-oxazoline) (P^{n}PrOx, B) and cationic poly(L-lysine) (PLys, C) towards improving the stability of polyplex micelles for efficient gene transduction (Osawa et al. 2016). The thermoresponsive property of P^{n}PrOx was employed to fabricate a thermo-switchable hydrophilic middle layer after packaging pDNA in the core of the polyplex micelles below the LCST of P^{n}PrOx. Above the LCST (~ 37°C), P^{n}PrOx became hydrophobic, forming a palisade between the pDNA core compartment and the hydrophilic outer PEtOx shell of the polyplex micelles (Fig. 18). These hydrophilic-hydrophobic doubly protected polyplex micelles provided remarkable improvements in stability against both nuclease (DNase I) and polyanion (CS) attacks compared to those polyplex micelles without hydrophobic palisades, thereby significantly protecting the genetic material from degradation, and also leading to improved transfection with increased cellular uptake.

In a follow up study, Vermonden and colleagues synthesized poly(*N*-isopropylacrylamide)-*b*-poly(ethylene glycol)-*b*-poly(2-(dimethylamino) ethyl methacrylate) (PNiPAAm-*b*-PEG-*b*-PDMAEMA) triblock copolymers and demonstrated that the introduction of a thermoresponsive PNiPAAm segment in the cationic terpolymer enables the formation of thermoresponsive polyplex micelles with pDNA that shield the charge of the polyplexes to a much larger extent than the simple PEG-*b*-PDMAEMA diblock copolymers at physiological temperature (i.e., higher than the LCST of PNiPAAm segments) (Fliervoet et al. 2019). The presence of thermoresponsive blocks in PNiPAAm-*b*-PEG-*b*-PDMAEMA-based polyplexes resulted in improved cytocompatibility compared to PEG-*b*-PDMAEMA-based polyplexes with similar transfection efficiencies into HeLa cells, even in the presence of serum proteins. As a control, a linear polyethylenimine (L-PEI) formulation was also investigated and revealed about 3–10 times higher transfection compared to PNiPAAm-*b*-PEG-*b*-PDMAEMA or PEG-*b*-PDMAEMA formulations. However, a decrease in cell viability was observed for L-PEI formulation as well. Hydrophobic and electrostatic interactions have been investigated very clearly in this report by altering the molecular weight of cationic and thermoresponsive segments in the terpolymer and the results demonstrated that, proper adjustment of the cationic segment and thermoresponsive segment in the terpolymer is required to achieve a favourable cytotoxicity-transfection efficiency balance in thermo-sensitive

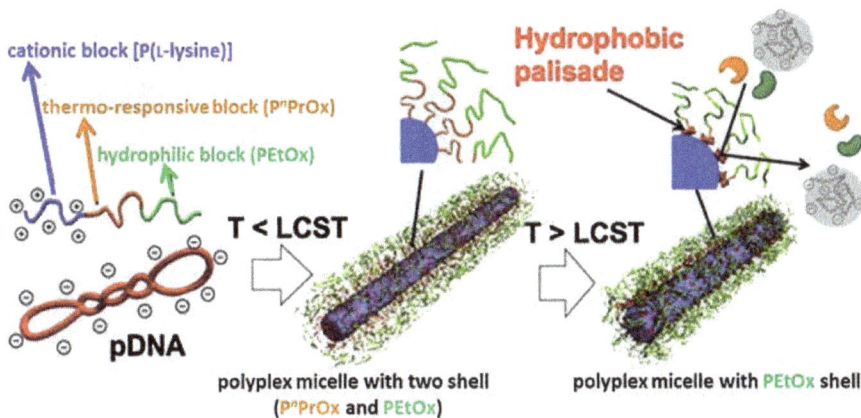

Fig. 18. Schematic representation of thermoresponsive polyplex micelles (obtained from the electrostatic interaction of P(L-lysine)-*b*-PnPrOx-*b*-PEtOx and pDNA) with hydrophilic-hydrophobic double protective compartments constructed from hydrophilic PEtOx shell and thermoresponsive PnPrOx palisade. Above the LCST (~ 37°C), PnPrOx became hydrophobic, forming a palisade between the pDNA core compartment and the hydrophilic outer PEtOx shell of the polyplex micelles, thereby providing extra stability against nuclease. Reprinted from Ref. (Osawa et al. 2016), with permission from ACS.

polyplexes. This result suggests the high potential of thermoresponsive "smart" polyplex micelles as nonviral gene vectors used in physiological environment for gene therapy.

With the aim of further improving the stability of polyplex micelles in order to promote their gene transduction efficiency, a mixture of two cationic block copolymers composed of an identical cationic block, poly(*N*-[*N*-(2-aminoethyl)-2-aminoethyl]aspartamide) (PAsp(DET)), but varying shell-forming blocks, poly[2-(2-methoxyethoxy) ethyl methacrylate] (PMEO$_2$MA), and poly[oligo(ethylene glycol)methyl ether methacrylate] (POEGMA), was synthesized to complex with plasmid DNA (pDNA) to prepare spherical polyplex micelles with mixed (PMEO$_2$MA and POEGMA) shells (MPMs) at 20°C (Li et al. 2014). The thermoresponsive characteristics of the PMEO$_2$MA segment allows its distinct phase transition from hydrophilic to hydrophobic by increasing the temperature from 20 to 37°C, which results in a distinct heterogeneous corona containing hydrophilic (POEGMA) and hydrophobic (PMEO$_2$MA) regions at the surface of the polyplex micelles (MPMs) (Fig. 19). As compared to polyplex micelles formed from a single POEGMA-*b*-PAsp(DET) block copolymer (referred to as SPMs), mixed shell polyplex micelles (MPMs) revealed high stability in salty and protein solution and superior tolerance to nuclease degradation. Moreover, the proposed mixed shell polyplex micellar system (MPMs) exhibited prolonged *in vivo* blood circulation, remarkably high cell transfection efficiency, especially at low N/P ratios, and negligible cytotoxicity (Fig. 19), thus representing a promising candidate for systemic gene therapy applications.

Similar improvements were observed by Mori and colleagues (Kanto et al. 2021). A mixture of the two block copolymers, poly(vinyl amine)-*b*-

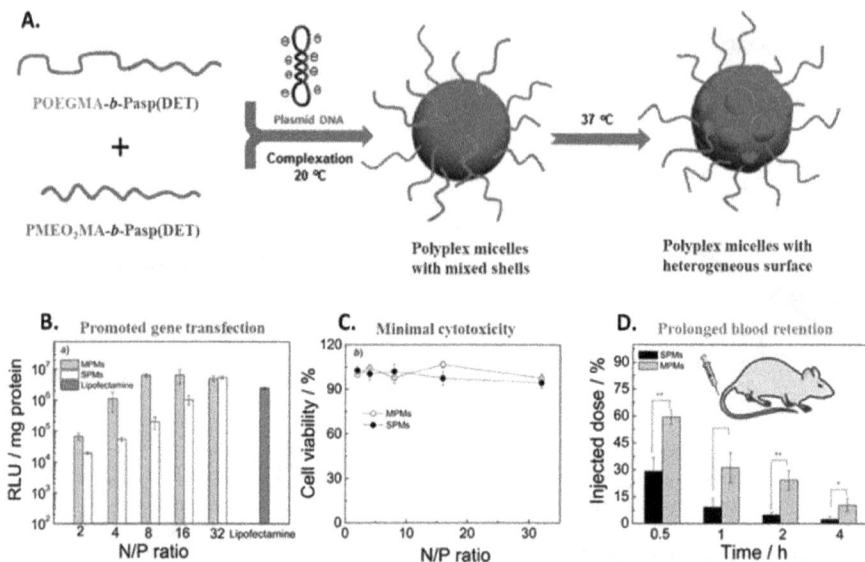

Fig. 19. Schematic illustration of preparation of mixed polyplex micelles with distinct heterogeneous coronas for enhanced stability as gene delivery vectors (A). *In vitro* luciferase gene transfection efficiency and cytotoxicity mediated by MPMs, and SPMs prepared at various N/P ratios against HeLa cells (B & C). Blood retention of polyplex micelles loading FITC-labelled pDNA in the bloodstream after intravenous injection of MPMs and SPMs at an N/P ratio of 8 (D). These results showed that the as-fabricated MPMs exhibited prolonged *in vivo* blood circulation and remarkably high cell transfection efficiency, especially at low N/P ratios (e.g., N/P = 8) and negligible cytotoxicity. Reprinted from Ref. (Li et al. 2014), with permission from ACS.

poly(*N*-acryloyl-l-lysine) (PVAm-*b*-PALysOH) and poly(vinyl amine)-*b*-poly (*N*-isopropylacrylamide) (PVAm-*b*-PNiPAAm), which have the same cationic PVAm chain but different shell-forming segments (PNiPAAm/PALysOH), were used to construct mixed shell polyplex micelles with DNA. The cationic PVAm segment in both block copolymers showed site-specific interactions with DNA as confirmed by an agarose gel electrophoresis assay. The mixed shell polyplex micelles revealed temperature-induced stability originating from the hydrophobic PNiPAAm chains upon heating and high stability under salty conditions owing to the presence of the zwitterionic PALysOH chain on the polyplex surface. However, further investigations on *in vivo* transfection as well as cytotoxicity experiments of these polyplexes is necessary to translate these findings to the clinical trial process.

In another report, Ge and colleagues (Li et al. 2015) constructed novel rod-shaped mixed polyplex micelles (MPMs) via complexation between the mixed block copolymers of poly(ethylene glycol)-*b*-poly(*N*'-(*N*-(2-aminoethyl)-2-aminoethyl)aspartamide) (PEG-*b*-PAsp(DET)) and poly(*N*-isopropylacrylamide)-*b*-PAsp(DET) (PNiPAAm-*b*-PAsp(DET)) and plasmid DNA (pDNA) at room temperature (25°C), exhibiting distinct temperature-responsive formation of a hydrophobic PNiPAAm intermediate layer between PEG shells and

PAsp(DET)/pDNA cores through facile temperature increase from room temperature to body temperature (37°C). As compared to polyplex micelles formed from a single PEG-*b*-PAsp(DET) block copolymer (SPM), rod-shaped MPMs showed enhanced stability/tolerability against nuclease and polyion exchange reaction as well as strong resistance against protein adsorption. Enhanced *in vitro* gene transfection efficiency was observed from these MPMs due to the synergistic effect of improved colloidal stability and low cytotoxicity, which was attributed to efficient cellular uptake and endosomal escape. Furthermore, *in vivo* performance investigation after intravenous administration confirmed that MPMs accomplished prolonged blood circulation, high tumour accumulation, and enhanced gene expression in tumour tissue. Also, MPMs loading with EGFP pDNA (therapeutic pDNA encoding an anti-angiogenic protein) remarkably suppressed tumour growth of H22 tumour-bearing mice, thus suggesting such MPMs may have great potentials as systemic non-viral gene vectors for cancer gene therapy. Feng et al. further utilized the above-mentioned efficient non-viral MPM-based gene delivery system to deliver therapeutic pDNA for nucleus pulposus regeneration toward disc degeneration disease. The authors have found that high expression of heme oxygenase-1 (HO-1) in nucleus pulposus (NP) cells (also known as disc cells) transfected by these MPMs loading with HO-1 pDNA significantly decreased the inflammatory response caused by interleukin-1β (IL-1β) and simultaneously increased the extracellular matrix production in NP (Feng et al. 2015). Therefore, genetic modification of disc cells through controlled and site-specific delivery of genetic materials (DNA or RNA) could be a promising approach to treat disc degeneration-associated diseases (Nishida et al. 1999). The progression of disc degeneration in this particular report (Feng et al. 2015) was confirmed to be significantly slowed down in rat tail discs and provides better therapeutic efficacy than SPMs, as investigated in this report, indicating that HO-1 enzyme expressed by MPMs showed great potency for NP regeneration.

Thermoresponsive poly(2-oxazoline)-based cationic copolymers, namely poly(2-oxazoline)-*co*-poly(ethylenimine) (PAOx-*co*-PEI) resulting from partial hydrolysis of the PAOx, are also attractive materials for gene delivery, albeit somewhat under-researched and under-utilized (Fernandes et al. 2013, Jeong et al. 2001, Hsiue et al. 2006, Haladjova et al. 2020a). The combination of non-ionic poly(2-oxazoline) (PAOx) moieties, exhibiting water soluble or thermoresponsive properties, excellent biocompatibility and stealth characteristics with cationic polyethyleneimine (PEI) moieties into a single copolymer is a promising approach to balance the toxicity and transfection efficacy. PAOx-*co*-PEI copolymers can be obtained via partial and controlled acidic/basic hydrolysis of PAOx and the combination of PAOx/PEI properties could be promising and attractive approach for the design and development of non-viral vectors for gene transfection (Lambermont-Thijs et al. 2010, de la Rosa et al. 2014, Mees and Hoogenboom 2018, Brissault et al. 2003). The present state of research in the area of PAOx-*co*-PEI copolymer-based non-viral gene delivery systems focusing particularly on thermoresponsive PAOx has been recently reviewed by Haladjova et al. (Haladjova et al. 2020b).

LCST-type thermoresponsive polymers for tissue engineering

Tissue engineering strategies, e.g., cell-sheet engineering, evolved from the field of biomaterials development and refers to the implementation of combining scaffolds, cells, and biologically active molecules into functional tissues (Langer and Vacanti 2016). This is an interdisciplinary field that integrates both the principles of engineering and biology towards the construction of biological substitutes (e.g., tissues or cellular products) that restore/repair or improve tissue function (Langer and Vacanti 1993, Place et al. 2009). For example, the main goal of cell-sheet engineering is to assemble functional constructs that restore, maintain, or improve damaged tissues or generate replacement organs for a wide range of medical issues such as heart diseases, cirrhosis, osteoarthritis, spinal cord injury and disfiguration (Langer and Vacanti 1993, Khademhosseini and Langer 2016, Place et al. 2009, Vert 2007, Khademhosseini et al. 2009). The advantages as well as the benefits of cell sheet transplantation compared to single cell injection towards the treatment of many diseases (e.g., cardiac repair, limbal stem-cell deficiency, bone fractures or as lung air leak sealants, among others) has already been demonstrated in numerous reports by many research groups, first and foremost, Okano and colleagues (Pham et al. 2006, Sekine et al. 2006, Shimizu et al. 2006, Yang et al. 2006). Therefore, there is an increasing research interest towards the possibility of obtaining immunologically compatible, confluent cell sheets for medical treatments which led to the development of different cell harvesting technologies (Patel and Zhang 2013).

The traditional tissue engineering process involves the use of a material scaffold within which cells are seeded and consequently tissue will develop. This requires the use of a biocompatible scaffolds, for example natural materials like proteins or synthetic polymers, with the appropriate 3D structure that will provide sufficient mechanical support and has the ability to transport both nutrients and growth factors to the encapsulated cells (Place et al. 2009). In contrast, cell-sheet engineering relies on the bottom-up construction of tissues from 2-dimensional cell monolayers/sheets. The fabrication and harvesting of cell sheets can be achieved by growing and detaching confluent cell monolayers from responsive tissue culture substrates, e.g., thermoresponsive coatings. The traditional route of cell harvesting involves enzymatic digestion by trypsin to cleave the bonds between the membrane receptors and the surface on which the cells are cultured. However, this type of treatment has some disadvantages as it causes the degradation of the other important cell membrane or surface proteins which are essential for both the cell-cell adhesion and cell-environment interactions (Brun-Graeppi et al. 2010, Tang et al. 2012). In addition, the degradation of important surface proteins leads to the death of a significant number of cells during the harvesting process.

The use of LCST-type thermoresponsive biocompatible polymer coatings as a cell culture substrate is considered an important alternative to enzymatic treatments in the cell seeding and recovery processes (Fig. 20) (Tang et al. 2012, Nithya et al. 2011). The thermoresponsive property of the polymer substrate, e.g., a polymer coating immobilized on tissue culture material, is exploited in this case to regulate the cell attachment and detachment process from a surface (Nagase et al. 2018). The cells attach to the thermoresponsive surface as long as they are kept above the

A.   37 °C   **B.**   **Room Temperature**
Cell attachment   Spontaneous cell detachment
and proliferation

T < LCST

Tissue culture   Dehydrated thermo-   Hydrated thermo-responsive
Polystyrene dish   responsive layer   layer
(above LCST)   (below LCST)

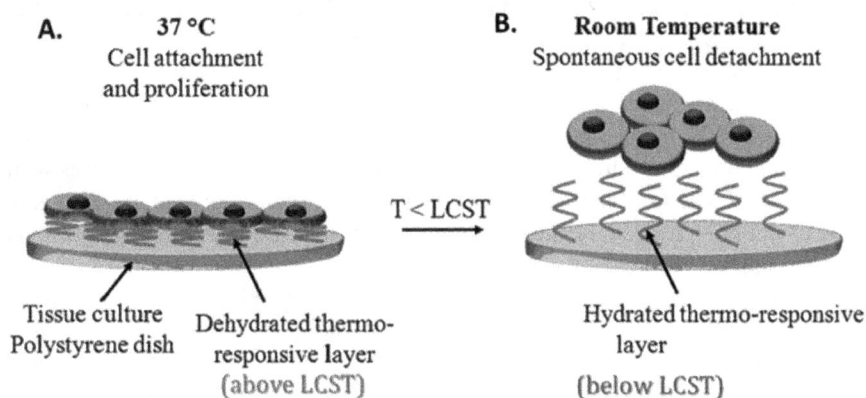

Fig. 20. Schematic illustration of cell sheet on thermoresponsive polymer modified surface. (A) At 37°C, cells adhere on dehydrated thermoresponsive polymer layer; (B) At room temperature, cells detach from hydrated thermoresponsive polymer layer, and form an intact cell sheet. Reprinted from Ref. (Sponchioni et al. 2019), with permission from ELSEVIER.

phase transition temperature, since at this temperature the thermoresponsive layer is in a dehydrated, hydrophobic collapsed state to which the cells like to adhere. In contrast, when the temperature is reduced below the phase transition temperature, the polymer chains become re-hydrated, generating extended non-fouling structures in water which leads to the detachment of the cells (Fig. 20) (Sponchioni et al. 2019, Cunliffe et al. 2003, Kumashiro et al. 2010, Varghese et al. 2010, Reed et al. 2010, Nitschke et al. 2007).

Thermoresponsive polymers can be grafted onto different substrates such as silicon, glass, quartz, and polyethyleneterephthalate (PET) sheets, but are most grafted on tissue culture grade polystyrene (TCPS) dishes. The choice of substrate, method of application of thermoresponsive polymer on the substrate and thickness of the grafted polymer have a major impact on cell attachment and detachment (Elloumi-Hannachi et al. 2010). Different types of polymerization strategies can be employed for grafting the thermoresponsive surfaces, such as electron beam irradiation (Yamato et al. 2001), gamma radiation (Kumar et al. 2007), plasma polymerization (Canavan et al. 2005), UV irradiation (Nagase et al. 2009), controlled radical polymerization (e.g., atom transfer radical polymerization, reversible addition fragmentation chain transfer polymerization) (Mizutani et al. 2008), solution casting method (Varghese et al. 2010), and oxygen plasma treatment (Shimizu et al. 2010).

It has been reported for PNiPAAm that a grafting thickness of 15 to 20 nm is a prerequisite for efficient cell attachment and detachment (Akiyama et al. 2004). A detailed investigation of the PNiPAAm layer on polystyrene surfaces was conducted by measuring the thickness of the thin PNiPAAm layer, and the relationship between thickness and cell adhesion/detachment was studied (Akiyama et al. 2004). PNiPAAm hydrogel-modified TCPS dishes were fabricated with different amounts of PNiPAAm by an electron beam irradiation-induced polymerization process. The thickness of PNiPAAm on TCPS surfaces was analysed by atomic force microscopy (AFM) for dry-state PNiPAAm and were found to be 15.5 and 29.3 nm. The thicker PNiPAAm layer is more hydrophilic than the thinner one. A cell adhesion experiment in the presence of bovine endothelial cells showed that cells did not adhere to the thicker PNiPAAm surface, even at 37°C, while cells adhered and detached at 37°C and 20°C, respectively, on the thinner PNiPAAm grafted surface. This kind of difference is attributed to the molecular motion of the PNiPAAm chains, which increases with increasing distance from the TCPS interfaces (Fig. 21). Thus, the outermost region of the thick PNiPAAm layer tends to be hydrated, even at 37°C, and thereby prevents cell adhesion (Fig. 21).

The results revealed that a thick PNiPAAm hydrogel layer (approximately 29.3 nm) is not suitable for switchable cell adhesion. In contrast, relatively thin PNiPAAm (approximately 15.5 nm) surface coatings allow for thermally modulated cell adhesion (at 37°C) and detachment (at 20°C).

Thermoresponsive substrates designed for cell-sheet engineering have mainly used PNiPAAm and its (co)polymers for cell adhesion and detachment (Yang et al. 2020). The use of thermoresponsive substrates to harvest the confluent cell sheets without the use of conventional enzymatic treatments was first reported by Yoshizato, Takezawa and colleagues in 1990 (Takezawa et al. 1990). In this study, the authors used PNiPAAm as a substrate by conjugating it with collagen proteins for the culture of human dermal fibroblasts. These fibroblast monolayers were harvested at reduced temperature by the dissolution of the dish coating. Almost at the same time, Yamada, Okano and colleagues reported the successful culture of bovine hepatocytes cells on PNiPAAm grafted tissue culture polystyrene (TCPS) (Yamada et al. 1990). The authors demonstrated the functionalization of tissue culture polystyrene (TCPS) dishes with thermoresponsive PNiPAAm surfaces synthesized by *in-situ* electron beam-initiated radical polymerization (Fig. 22). The electron beam method facilitates the fabrication of thin, grafted surface coatings and the large-scale production of thermoresponsive surfaces for cell-sheet engineering is possible via this method, albeit being expensive. The contact angle of the PNiPAAm-grafted TCPS surface was found to be 48° at temperatures above the cloud point (37°C). However, upon cooling to 10°C (i.e., below to the cloud point), the contact angle is reduced to 30°, thus confirming the effective hydration of the surface. Bovine hepatocyte cells were cultured on both thermoresponsive PNiPAAm-functionalized TCPS dishes and unmodified dishes. The cell growth was found to be similar for both modified and unmodified dishes. However, as expected, almost 100% of the cultured cells could be detached and recovered from the PNiPAAm-grafted dishes just by lowering the temperature below the T_{cp}, without any enzymatic treatment, while only about 8% of the cells were able to be detached from the unmodified TCPS dishes. Mechanistic insights in the cell sheet detachment from thermoresponsive surfaces is also provided by Okano and colleagues, where they have investigated the dynamics of cell detachment from PNiPAAm-grafted surfaces upon lowering the temperature to different values (Okano et al. 1995). Interestingly, the authors demonstrated that at lower temperatures, despite a higher degree of hydration of the surface, the percentage of detached cells is found to be less as compared to the case of incubation at temperature a bit closer to the LCST of the polymer. For example, the maximum of

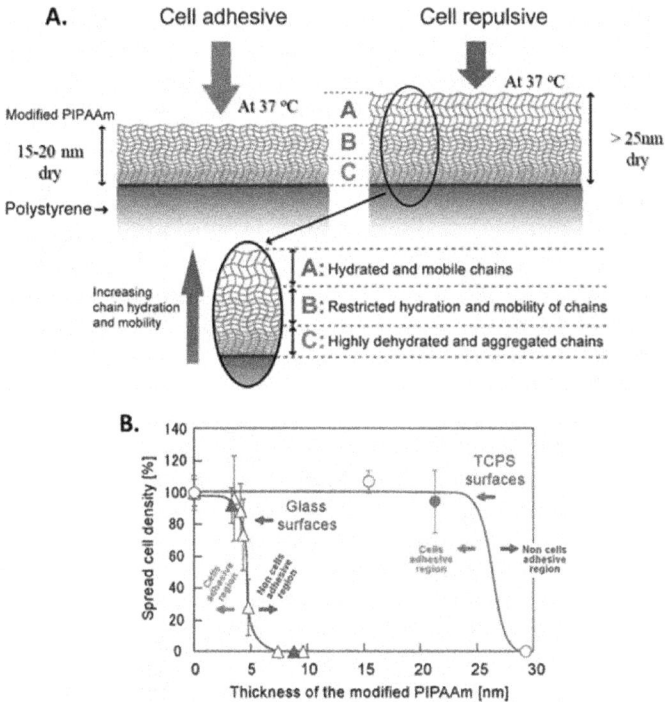

Fig. 21. Schematic drawings of the influence of molecular motion of grafted PNiPAAm chains on hydration of the polymer chains, when the grafted PNiPAAm gels are thin (left panel) and thick (right panel) at above the LCST (37°C) (A). Relationship between the spread cell density and thickness of the PNiPAAm hydrogel layer on substrate (B). Hydrophobic TCPS interfaces promoting aggregation and dehydration were represented as a black region in TCPS (A). Molecular motion of the grafted polymer chains becomes larger according to the distance away from TCPS interfaces. Therefore, outermost region of the thick PNiPAAm layer tended to hydrate, even at above the LCST (37°C) and prevents cell adhesion. Reprinted from Ref. (Fukumori et al. 2010; doi.org/10.1002/mabi.201000043), with permission from John Wiley and Sons.

detached hepatocyte cells could be obtained by reducing the temperature to 10°C for 30 min followed by incubating the cells at 25°C for 5 min. These results indicate that the surface hydration represents an important initial stimulus for cell detachment, which however further requires active metabolic processes and consumption of energy for the transformation from an initially spread to the typical round shape. Overall, a good adjustment in between the surface hydration and active metabolism is crucial to enhance cell detachment.

After these preliminary findings by Okano and colleagues, the use as well as fabrication of thermoresponsive surfaces in tissue engineering/cell sheet engineering has been extensively studied and reviewed by many research groups (Sponchioni et al. 2019, Cunliffe et al. 2003, Kumashiro et al. 2010, Varghese et al. 2010, Reed et al. 2010, Nitschke et al. 2007, Abraham et al. 2010). Copolymerization of NiPAAm with hydrophobic or hydrophilic monomer can modulate the LCST for

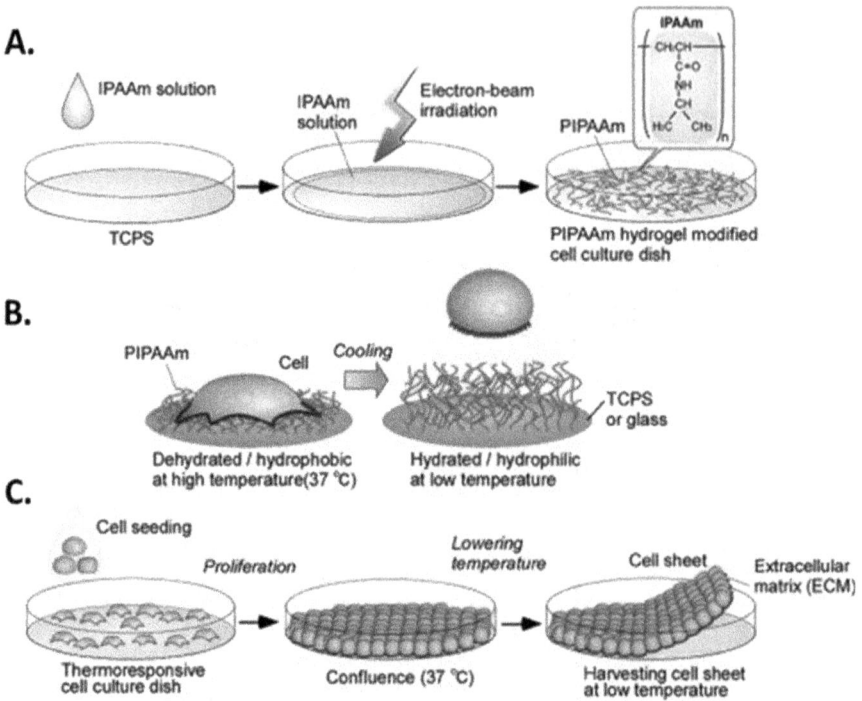

Fig. 22. Thermoresponsive cell culture dish for thermally induced cell adhesion and detachment as well as cell-sheet fabrication. Preparation of thermoresponsive cell culture dish by electron beam-induced polymerization (A). Temperature-dependent cell adhesion and detachment (B). Cell-sheet fabrication using thermoresponsive cell culture dish (C). Reprinted from Ref. (Nagase et al. 2018, doi.org/10.1016/j. biomaterials.2017.10.026), with permission from ELSEVIER.

the systematic regulation of cell attachment and detachment. Thermoresponsive copolymers, such as PNiPAAm-*co*-poly(glycidylmethacrylate) (Praveen et al. 2017, Madathil et al. 2014, Joseph et al. 2010), PNiPAAm-*co*-poly(methylmethacrylate) (Abraham et al. 2010), PNiPAAm-*co*-poly(butylmethacrylate) (Tsuda et al. 2004), PNiPAAm-*co*-poly(*N-tert*-butylacrylamide) (Moran et al. 2007), PNiPAAm-*co*-poly(*N*-vinylcaprolactum) (Lim et al. 2007), among others, were successfully used as thermoresponsive surfaces for cell-sheet engineering.

Copolymerization of PNiPAAm with glycidyl methacrylate could be interesting as it is identified as potential substrate for cell culture harvesting systems for generating 3D synthetic tissue. For example, Madathil, Kumar and colleagues have investigated the potential of thermoresponsive PNiPAAm-*co*-poly(glycidyl methacrylate) copolymer in corneal tissue engineering by evaluating its potential in culturing corneal endothelial cells (CEC) and retrieval of functional corneal endothelial cell sheets (Madathil et al. 2014). The results revealed the successful culture of corneal endothelial cells on the PNiPAAm-*co*-poly(glycidyl methacrylate) copolymer-modified TCPS substrate without the need for any additional ECM coatings. The cultures could be maintained for prolonged periods and intact cell sheets could be harvested by simple incubation below the LCST. The obtained cells sheets were found to have intact morphology and cell-cell contacts. Gene expression analysis and immunocytochemistry further confirmed the presence of functionally active endothelial cells in the cell sheet. These results demonstrate the use of the in-house developed thermoresponsive PNiPAAm-*co*-poly(glycidyl methacrylate) culture dishes as a potentially useful substrate for the generation of carrier-free intact corneal endothelial cell sheet towards transplantation for endothelial keratoplasty. The same authors have more recently reported the design and synthesis of a flexible thermoresponsive PNiPAAm-*co*-poly(glycidyl methacrylate)-based cell culture substrate for direct transfer of keratinocyte cell sheets (Praveen et al. 2017). In this report, a polyethylene terephthalate (PET) based overhead projection sheet (OHPS) was selected as the base material for the fabrication of an efficient thermoresponsive cell culture surface. The OHPS is first alkali modified to make it more flexible as well as hydrophilic and subsequently, PNiPAAm-*co*-poly(glycidyl methacrylate) is spin-coated using an indigenously designed spin coater. Human keratinocyte cells were cultured on this spin-coated surface and intact, viable as well as scaffold-free cell sheets were successfully harvested by simple variation of temperature, thus confirming the efficacy of the substrate to act as both a thermoresponsive surface and pliable transfer tool. Furthermore, the presence of unreacted epoxy rings within these copolymer structure/surface allows for the facile incorporation of diverse type of biomolecules which could modulate cellular response on the thermoresponsive substrates (Joseph et al. 2010).

The spontaneous/rapid detachment of cultured cell sheets is crucial for preserving their viability. As for example, PNiPAAm grafted onto porous membranes promotes the acceleration of cell sheet detachment (within 30 min) by providing rapid water diffusion between the interface of cell sheets and membrane surface (Fig. 23) (Kwon et al. 2000). In contrast, on the PNiPAAm-modified TCPS, water for hydration of grafted PNiPAAm penetrated only the cell sheet periphery, leading to slow hydration and cell sheet detachment (within 75 min). Co-grafting of a hydrophilic polymer (e.g., PEG) with the thermoresponsive PNiPAAm onto the porous membrane could be another elegant strategy to further accelerate the cell sheet detachment (Kwon et al. 2003). Another fast-responding thermoresponsive PNiPAAm-*co*-poly(diethyleneglycol methacrylate) (PNiPAAm-*co*-PDEGMA) substrates for the harvest of human corneal endothelial cell (HCEC) sheets has been reported by Nitschke et al. (Nitschke et al. 2007). Thin films of such fast-responding PNiPAAm-*co*-PDEGMA copolymer with a phase transition temperature of 32°C were prepared on fluorocarbon substrates by low pressure plasma immobilization. Compared to PNiPAAm homopolymer films attached via the same technique, PNiPAAm-*co*-PDEGMA layers showed a sharper and higher phase transition temperature closer to the physiological range. Beyond that, the cell detachment from the diethyleneglycol containing copolymer occurred faster and at higher temperatures (30°C), as compared to the PNiPAAm homopolymer (Fig. 24). Human corneal endothelial cells (HCEC) as well as L929 mouse fibroblasts were found to adhere, spread and proliferate on these substrates and showed a characteristic morphology.

A.

B.

C.

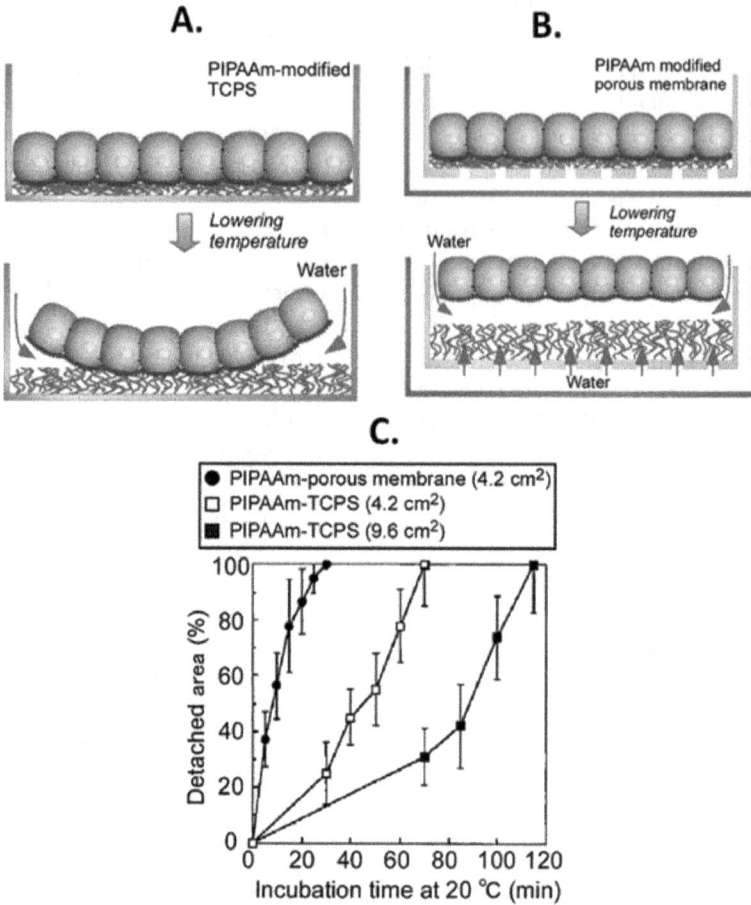

Fig. 23. Thermoresponsive polymer-modified porous membrane for rapid cell detachment. Schematic representation of cell sheet detachment from PNiPAAm-modified TCPS (A) and PNiPAAm-modified porous membrane (B). Detachment behaviour of BAEC sheets from PNiPAAm-modified porous membrane and PIPAAm-modified TCPS (C). Figure in parentheses represents culture surface area. The results showed that PNiPAAm grafted onto porous membranes promotes the acceleration of cell sheet detachment by providing rapid water movement between the interface of cell sheets and membrane surfaces. Adapted from Ref. (Kwon et al. 2000), with permission from John Wiley and Sons.

Functional HCEC and L929 mouse fibroblast sheets were harvested gently from PNiPAAm-*co*-PDEGMA surfaces simply by lowering the temperature to 30°C. Most of the extracellular matrix (ECM) can also be detached from the surface together with the cells because of the presence of ethylene glycol units and thus confirms the suitability of the substrates for repeated cell harvesting. These kinds of functional thermoresponsive PNiPAAm-*co*-PDEGMA coatings could be advantageous for the efficient generation as well as subsequent transplantation of HCEC sheets.

Various types of thermoresponsive ionic copolymers can also be employed in thermoresponsive culture dishes. By introducing ionic characteristics to

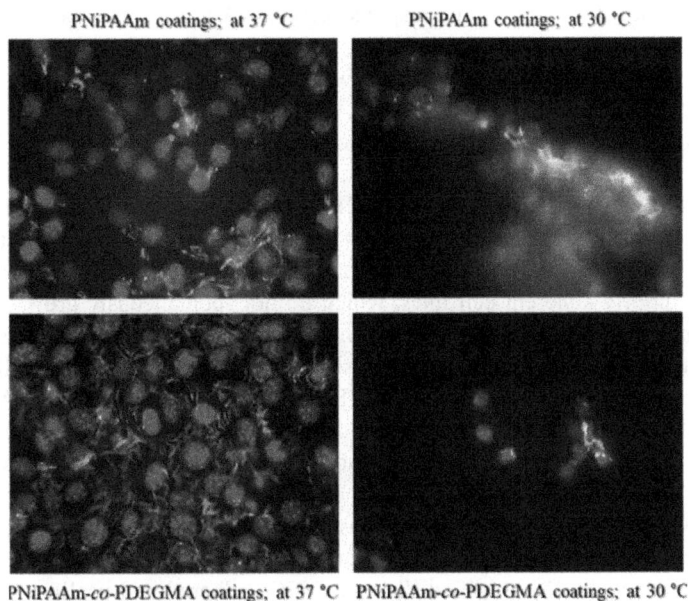

PNiPAAm coatings; at 37 °C   PNiPAAm coatings; at 30 °C

PNiPAAm-*co*-PDEGMA coatings; at 37 °C   PNiPAAm-*co*-PDEGMA coatings; at 30 °C

Fig. 24. Fluorescence microscopy images of L929 mouse fibroblasts on PNiPAAm (top) and PNiPAAm-co-PDEGMA (bottom) after standard cultivation at 37°C (left) and after cooling to 30°C (right). The results showed that the cell detachment on the hydrophilic diethylene glycol containing PNiPAAm-co-PDEGMA copolymer occurs faster and at higher temperatures (30°C), as compared to the PNiPAAm homopolymer. Reprinted from Ref. (Nitschke et al. 2007), with permission from John Wiley and Sons.

PNiPAAm-modified dishes, additional properties and functionalities, such as rapid cell detachment/recovery, can be achieved. As example of an ionic monomer, 2-carboxy-*iso*-propylacrylamide (CiPAAm) was introduced into the thermoresponsive PNiPAAm on the TCPS dishes through EB-induced polymerization (Ebara et al. 2003). PNiPAAm and PNiPAAm-*co*-poly(acrylic acid) modified TCPS dishes were also synthesized and cell adhesion/detachment profiles were investigated. At 37°C, bovine aortic endothelial cells (BAECs) adhered to the PNiPAAm-*co*-PCiPAAm dishes while BAECs did not adhere to the PNiPAAm-*co*-poly(acrylic acid) dishes because of its relatively strong hydrophilic characteristics. Additionally, cell detachment as well as cell sheet recovery were much faster (within 35 min) from the PNiPAAm-*co*-PCiPAAm dishes as compared to the PNiPAAm-modified dishes (within 60 min). Accelerated cell detachment in case of the PNiPAAm-*co*-PCiPAAm copolymer modified dishes is attributed to the presence of hydrophilic carboxyl groups that induce hydration of the modified copolymer. A similar study was also executed using 3-carboxy-*n*-propylacrylamide (CNPAAm) as a co-monomer of PNiPAAm. CNPAAm has a slight structural difference from CiPAAm, resulting in a higher charge density (lower pKa) and relatively strong hydrophilicity of the PNiPAAm-*co*-PCNPAAm copolymer as compared to PNiPAAm-*co*-PCiPAAm. On the PNiPAAm-*co*-PCNPAAm dish, BAEC detachment was found to be ~ 5% faster than on the PNiPAAm-*co*-PCiPAAm dish. Introduction of ionic PCiPAAm or

PCNPAAm polymers into thermoresponsive PNiPAAm polymer accelerate the rate of cell detachment from the corresponding dishes. Thus, these thermo-sensitive ionic polymer-modified dishes could also be beneficial for rapidly recovering cultured cells simply by lowering the temperature.

Another advancement in this field has been achieved by the incorporation of cell-adhesive peptide sequences, e.g., arginine-glycine-aspartic acid (RGD) or arginine-glycine-aspartic acid-serine (RGDS) sequence into the thermoresponsive polymer, to improve the cell adhesion on synthetic polymer surfaces (Ebara et al. 2008a). This is indeed very interesting as it empowers the temperature-mediated culture of poorly adherent cell lines, which do not spontaneously grow on simple PNiPAAm substrates. The first effort to introduce a cell-adhesive peptide sequence onto the thermoresponsive cell culture dish was accomplished by immobilizing the RGDS sequence onto PNiPAAm dishes (Ebara et al. 2004a). Firstly, CiPAAm and NiPAAm monomers were dissolved in isopropanol. Afterwards, the solution was added to the TCPS dish and electron beam irradiation was conducted, resulting in a PNiPAAm-*co*-PCiPAAm-grafted TCPS dish. Finally, RGDS was immobilized on the copolymer surface through a condensation reaction. Cell adhesion as well as detachment of human umbilical vein endothelial cells (HUVECs) were analysed on the as-fabricated RGDS-modified dish. HUVECs adhered and proliferated on the RGDS-modified dish at 37°C (above the LCST, of the copolymer) in serum-free media which is possible due to the immobilized RGDS on the thermoresponsive polymer layer. Moreover, cells spread on RGDS-immobilized surfaces at 37°C and detach spontaneously by lowering the culture temperature below the LCST as hydrated grafted copolymer chains dissociate immobilized RGDS from cell integrins. However, upon excess immobilization of RGDS on the cell culture dish, HUVECs did not detach at reduced temperatures below the LCST. These results suggested that the concentration of immobilized RGDS must be optimized for the efficient cell adhesion/detachment process. Using an RGDS-immobilized dish, a bovine aortic endothelial cell (BAEC) sheet can also be fabricated in serum-free culture, thus reducing the cost, and avoiding safety issues (Ebara et al. 2004b).

In another report, Smith and colleagues functionalized the TCPS surface with a PNiPAAm-*co*-poly(*N*-acryloxysuccinimide) copolymer and exploited the reactive succinimide group to covalently attach the RGD sequence via an amide bond using the amine-reactive *N*-acryloxysuccinimide (NASI) groups (Smith et al. 2005). The adhesion of bone morphogenetic protein-2 (BMP-2)-responsive C2C12 cells to RGD-grafted PNiPAAm-*co*-poly(*N*-acryloxysuccinimide) surfaces is significantly higher than adhesion on non-grafted NiPAM/NASI surfaces. The cell harvesting was still obtained by lowering the temperature below the copolymer LCST, although the presence of the peptide was expected to decrease the detachment efficiency due to a strong binding with the cell membrane integrin. Unlike cells on TCPS, cells harvested from PNiPAAm-*co*-poly(*N*-acryloxysuccinimide) copolymer surfaces exhibited an alkaline phosphatase (ALP) activity even without BMP-2 exposure, indicating that such copolymer surfaces were more conductive for the expression of ALP activity by the chosen cell model and no specific effect of RGD grafting was observed towards inducing ALP activity. These findings showed that

RGD-containing PNiPAAm-based thermoresponsive surfaces allow attachment and osteoblastic differentiation of BMP-responsive cells *in vitro*. However, investigation of intracellular mediators of BMP-2 signalling will be required to further elucidate the capability of NiPAAm surfaces to induce osteoblastic differentiation. The effect of extensible PEG tethers on shielding between grafted thermoresponsive polymer chains and integrin-RGD binding has also been investigated in another report by Ebara, Okano and colleagues (Ebara et al. 2008b). The report elegantly explains the shielding effect of the grafted polymer chains on the dissociation of integrin-RGD binding below the LCST. To assess the ability of the polymer-shielding, extensible poly(ethylene glycol) (PEG) tethers were introduced between peptides and the grafted poly(N-isopropylacrylamide-co-2-carboxyisopropylacrylamide) copolymer surface. The significant increment of cell adhesion on RGD-containing substrates was also observed at the "on" state above the LCST, further confirming that the cell-substrate interaction is highly RGD specific. However, when the temperature was decreased below the LCST (inactive "off" state), the time required to release cells from the surface was found to be longer when peptides were coupled to an extensible PEG tether end. PEG chains allow peptides to be tethered to surfaces via functional PEG end-groups. Therefore, masking efficiency of the hydrated coating is significantly reduced, leading to active "on" state even below the LCST. These findings suggest that the architectural changes, even on the nanometer scale length, are crucial for controlling integrin-RGD binding and one of the main factors causing cell detachment is the shielding effect of the grafted polymer chains.

Incorporation of cell growth factor proteins such as insulin, basic fibroblast growth factor (bFGF) and epidermal growth factor (EGF) on thermoresponsive cell culture dishes could also be interesting as it showed efficient proliferation of cultured cells, resulting in the effective fabrication of cell sheets (Hatakeyama et al. 2007). For example, insulin (well-known protein to act as a growth factor, induces cell proliferation) was immobilized on the PNiPAAm-*co*-PCiPAAm-grafted dish, and cell proliferation on the dishes was explored in the presence of bovine carotid artery endothelial cells (Hatakeyama et al. 2005). These insulin-modified dishes induced cell proliferation in a serum-supplemented culture medium. The effects of both the amount of CiPAAm carboxyl groups incorporated as well as the immobilized amount of insulin on endothelial cell proliferation and detachment was investigated in this report. Cell proliferation was found to increase by increasing the concentration of immobilized insulin. Cells grown on the insulin-immobilized surfaces could be recovered as contiguous cell monolayers simply by lowering the culture temperature (below the cloud point) without need for exogenous enzyme or calcium chelator additions. The findings reveal that cell adhesion, proliferation as well as detachment can be stimulated by an appropriate design of thermoresponsive surfaces with immobilized insulin. In another report, both insulin as well as RGDS were co-immobilized onto PNiPAAm-*co*-PCiPAAm-modified surface, and the proliferation of bovine endothelial cells was investigated (Hatakeyama et al. 2006). Co-immobilization of RGDS and insulin on thin PNiPAAm-*co*-PCiPAAm-modified dishes promoted initial cell adhesion and proliferation of bovine endothelial cells in the absence and presence of serum in cell culture medium. Therefore, co-immobilization of insulin and RGDS on the thermoresponsive copolymer-modified dish could be an effective strategy towards enhancing the cell sheet fabrication in the absence of serum.

In a more recent report, heparin was immobilized on PNiPAAm-*co*-PCiPAAm through the carboxyl group of CiPAAm (Arisaka et al. 2013). Afterwards, basic fibroblast growth factor (bFGF) was immobilized onto the heparin-modified PNiPAAm-*co*-PCiPAAm dishes based on the affinity between bFGF and immobilized heparin through the temperature-dependent conformational change of grafted PNiPAAm-*co*-PCiPAAm chains. At 37°C, mouse fibroblast (NIH/3T3) cells showed two- to three-fold faster proliferation in the presence of bFGF-bound heparin-thermoresponsive surfaces compared to the same cells cultured in the presence of bFGF-physiosorbed surface or PNiPAAm surface with soluble bFGF. Bound bFGF via heparin on the collapsed grafted PNiPAAm-*co*-PCiPAAm chains at 37°C was able to reinforce the formation as well as the stabilization of the bFGF-FGF receptor complex (Fig. 25), thus enhancing cell proliferation. In contrast, the proliferation efficiency of physiosorbed bFGF on PNiPAAm-grafted surfaces was decreased by non-specific and randomly oriented adsorption. At 20°C, the cultured NIH/3T3 cell sheet with bFGF detached from heparin-functionalized thermoresponsive surface. The release of bFGF from the surfaces was induced by lowering the affinity binding between bFGF and immobilized-heparin due to the increased mobility of the swollen grafted PNiPAAm-co-PCiPAAm chains below the T_{cp}. Such heparin-functionalized thermoresponsive cell culture surfaces could be useful for rapid cell sheet fabrication and manipulation.

Fabrication of thermoresponsive cell culture surface for the co-culturing of different cells could also be interesting and has recently gained great attention as it allows the formation/development of hetero-cell layers that can be utilized to replace defective/damaged tissues. In addition, it has been mentioned in the literature that heterotypic cell-cell interactions are pivotal to realize specific regulatory functions for organs and tissues (Song et al. 2019). For example, hepatocyte layers can restore most of the liver functions and hepatocyte-tissue is therefore an attractive option for the treatment of many liver diseases (Iansante et al. 2018, Bhatia et al. 2014). However, unfortunately, conventionally/normally cultured hepatocytes are prone to lose their morphology and biological function, possibly because of the poor interactions with the ECM and neighbouring cells (Oda et al. 2008). Therefore, many research groups have attempted to develop a cell sheet-based hepatocyte co-culture system that enables cultured hepatocytes to preserve their morphology and functions for a longer time. One such approach to maintain the structure and function of the cultured hepatocyte cells is its co-culture with endothelial cells (ECs) and the deposition of ECM in the narrow space in between the two cell layers. Kim, Okano and colleagues have developed a simple method (Kim et al. 2012) in which a uniform monolayer of endothelial cells was bedded on a layer of hepatocytes, thus allowing for a more stable and persistent viability of the hepatocytes. The authors first fabricated an endothelial cell (EC) monolayer on a PNiPAAm-grafted surface (Fig. 26). Upon lowering the temperature to 20°C, a contiguous endothelial cell sheet could be detached and retrieved by means of a gelatine-coated manipulator.

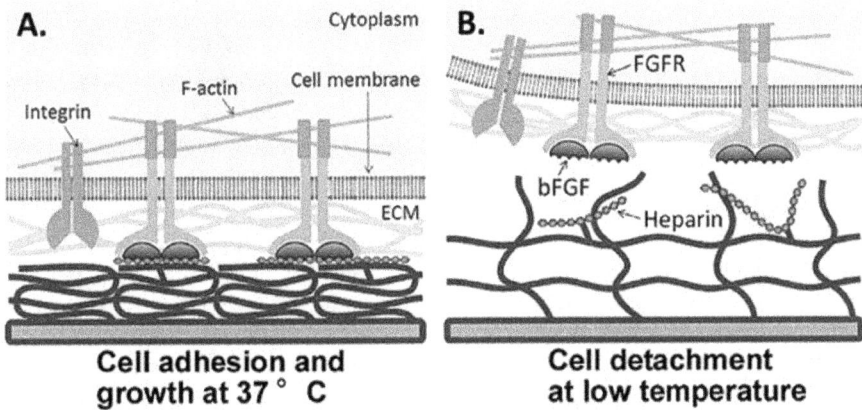

Fig. 25. Schematic illustration of cell growth/detachment behaviour by altering the temperature. During cultivation at 37°C, cultured cells adhered and grew on basic fibroblast growth factor (bFGF)-bound heparin-functionalized thermoresponsive cell culture surfaces (A). Upon lowering temperature to 20°C, cultured cells were detached by the reduction of the affinity interaction between bFGF and surface-immobilized heparin due to the increased steric hindrance of swollen grafted thermoresponsive polymer (B). Adapted from Ref. (Arisaka et al. 2013), with permission from ELSEVIER.

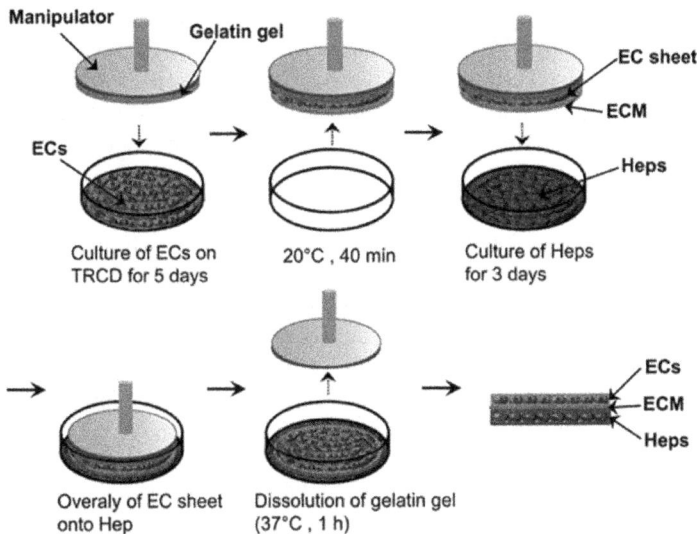

Fig. 26. Schematic representation of the process to stratify bovine endothelial cells (EC) to the hepatocyte sheet (Heps). ECs obtained from bovine carotid artery are cultured on a PNIPAAm-based temperature-responsive culture dish (TRCD) at 37°C. When ECs reach confluency at day 5, a gelatine-coated cell-sheet layering manipulator is placed on the cells. At this point, the culture temperature is reduced from 37°C to 20°C for 40 min. After incubating at the lower temperature, the EC sheet recovered with the manipulator is placed onto the 3-day cultured primary hepatocytes. The temperature is raised back to 37°C for 60 min, and this allows the gelatine gel to dissolve into the culture medium. Subsequent removal of manipulator resulted in the creation of a stratified hybrid cell sheet. Reprinted from Ref. (Kim et al. 2012), with permission from ELSEVIER.

The recovered endothelial cell monolayer was then placed over a monolayer of hepatocytes (Hep) by dissolution of the gelatine at 37°C (Fig. 26). While hepatocytes obtained from a conventional cell culture process rapidly lose their biological functions, the above-mentioned co-cultured hepatocytes (with endothelial cells) retained their cuboidal morphology and sustained albumin secretion and bile acid secretion for up to 28 days, suggesting the preservation of liver-like functions. In a similar fashion, Shimizu and colleagues also fabricated multilayered cardiomyocyte-based tissue, which was further successfully transplanted in the inferior vena cava of rats. The cardiac functions and pressure profile were retained up to eight weeks after transplantation, confirming the possibility of realizing a bioengineered heart assist pump.

Patterned thermoresponsive cell culture dishes have also been developed and investigated for the fabrication of co-cultured cell sheets consisting of different cell types (Tsuda et al. 2006). Tsuda, Okano and colleagues have reported the fabrication of patterned dual thermoresponsive cell culture dishes by the initial electron beam irradiation of a NiPAAm solution on a TCPS dish and subsequent partial co-grafting of poly(n-butyl methacrylate) (PBMA) to the PNiPAAm main chain through a second electron beam irradiation in the presence of a photomask with a 1-mm diameter hole (Tsuda et al. 2005). Using the patterned dual thermoresponsive culture dish, endothelial cells and hepatocytes were co-cultured (Fig. 27). At 27°C, hepatocytes were seeded on the micro-patterned dish and exclusively adhered to the hydrophobic, dehydrated PNiPAAm-PBMA co-grafted domains (1-mmØ area), but not onto neighbouring hydrated PNiPAAm domains. Sequentially, endothelial cells were seeded on the cell culture dish and adhered selectively to hydrophobic PNiPAAm domains upon increasing the temperature to 37°C, achieving patterned co-cultures. Reducing the culture temperature to 20°C promoted hydration of both domains, resulting in the recovery of the co-cultured, patterned cell monolayers as continuous cell sheets with heterotypic cell interactions. Patterning of surfaces with a variety of domain sizes, area fractions and shapes could also be achieved by using arbitrarily designed masks. Additionally, patterned bio-functionalized thermoresponsive surfaces provide access to achieve spatiotemporal control over cell adhesion, growth as well as temperature-induced detachment (Hatakeyama et al. 2007).

In addition to cell attachment/detachment, thermoresponsive PNiPAAm (co)polymer surfaces have also been investigated in recent days for cell separation (Sung et al. 2020, Nagase et al. 2015, 2020a, 2020b). For many cell-based tissue engineering and therapy protocols, cell separation is also needed to generate cellular tissues or cell suspensions. For example, in a sample of cells collected from a human, various contaminant cells are mixed with target cells. In stem cell culture, different cells co-exist with stem cells and, therefore, targeted stem cells must be separated and purified from other contaminant cells before they can be employed for stem cell therapy or other applications. Compared with the PNIPAAm homopolymer brush, ionic PNIPAAm copolymer-grafted substrates showed a relatively higher adhesion selectivity for mesenchymal stem cells (MSCs). However, the amount of ionic monomer incorporation is always limited because the thermoresponsive property of PNIPAAm is lost at high ionic content. Very recently, Nagase and colleagues

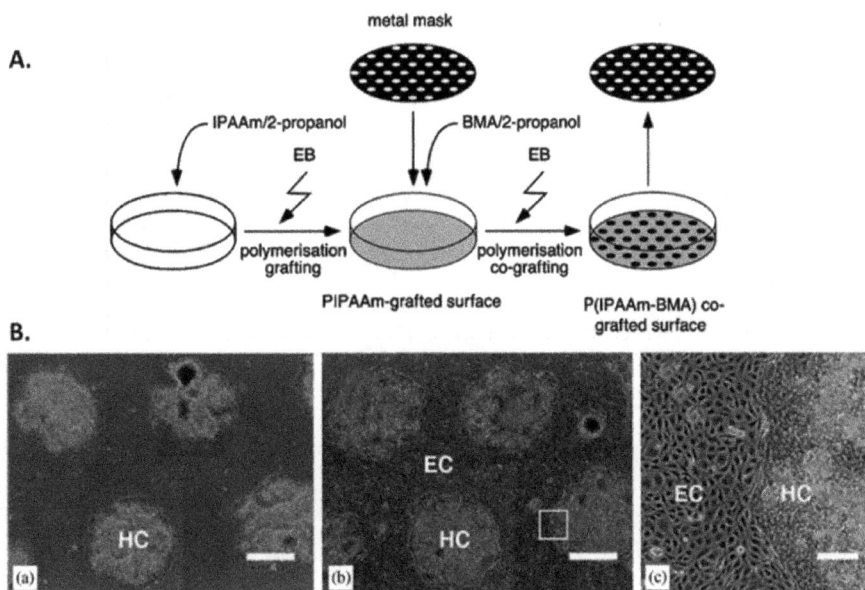

Fig. 27. Schematic diagram for preparation of patterned dual thermoresponsive PIPAAm/P(IPAAm-BMA) polymer-grafted cultural dishes for the fabrication of co-cultured cell sheets of hepatocytes and endothelial cells (A). Patterned co-culture of HCs and ECs (B). (a): Selective adhesion of rat primary HCs onto P(IPAAm–BMA) co-grafted domains cultured at 27°C for 2 days and then at 37°C for additional 2 days. (b): Sequentially seeded ECs adhere to hydrophobised PIPAAm regions and co-culture with pre-seeded HCs at 37°C into organised patterns. (c): Magnified view of the periphery of patterned co-cultures (square region in (b)). Reprinted from Ref. (Tsuda et al. 2005), with permission from Elsevier.

have developed mixed polymer brushes composed of thermoresponsive PNIPAAm and cationic poly(N,N-dimethylaminopropylacrylamide) (PDMAPAAm) via a combination of surface-initiated RAFT polymerization and ATRP (Fig. 28) (Nagase et al. 2023). The as prepared mixed polymer brushes revealed increased cationic nature with grafted PDMAPAAm length. Interestingly, the surface cationic property can be adjusted by tuning the temperature because of the temperature-induced shrinking and extension of exposed PNIPAAm and concealed PDMAPAAm, respectively. At 37°C, the mixed polymer brushes showed enhanced cell adhesion through the electrostatic interactions between cells and PDMAPAAm, which consequently shortened the required cell adhesion time. Interestingly, authors have separated mesenchymal stem cells (MSCs) from adipocytes and HeLa cells by simply changing the temperature of prepared thermoresponsive mixed polymer brush. Only HeLa cells are found to adhere and detached from simple PNiPAAm coated surface at 37°C and 20°C, respectively, whereas bone marrow mesenchymal stem cells (BMMSCs) and adipocyte-BM did not. On the other hand, bone marrow mesenchymal stem cells (BMMSCs) adhered on the mixed polymer brushes at 37°C and detached at 20°C. HeLa cells also adhered on the mixed polymer brushes at 37°C, but did not detach

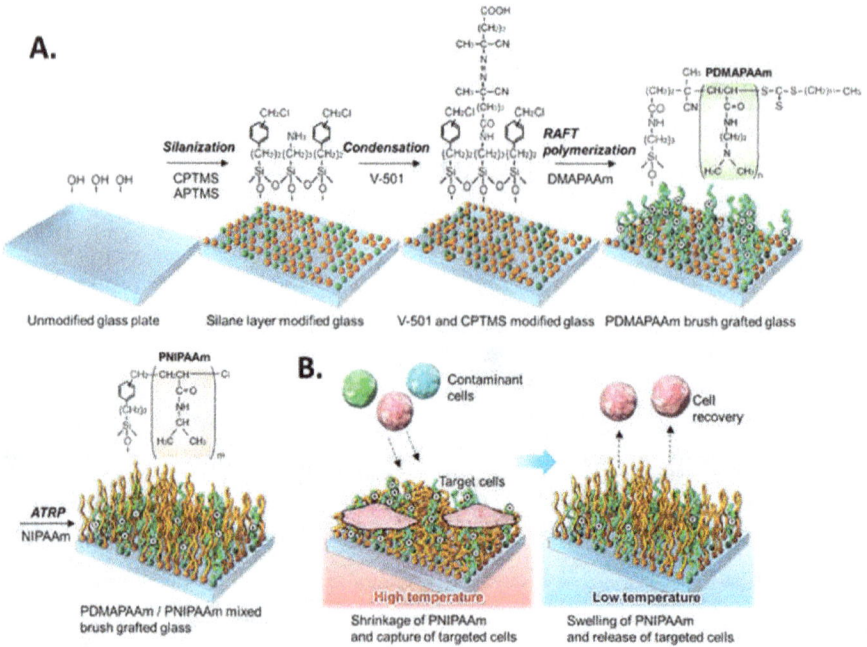

Fig. 28. Schematic illustration of the synthesis of PNiPAAm-based thermoresponsive cationic PNiPAAm-co-PDMAPAAm mixed polymer brush for the selective capture of cells (A). Schematic illustration of temperature-controlled cell capture and separation from mixture of cells using the as-prepared PNiPAAm-co-PDMAPAAm mixed polymer brush (B). Reprinted from Ref. (Nagase et al. 2023), with permission from ELSEVIER.

at 20°C, whereas adipocytes did not adhere on the mixed polymer brushes at 37°C and therefore could be separated at 37°C. Thus, by using the mixed polymer brushes, a high amount of purified BMMSCs can be separated from the mixture of cells (BMMSCs, HeLa and adipocytes) at 20°C. These results suggested that cationic PDMAPAAm can influence the attachment/detachment behaviour of the cells and that such mixed brushes can be used to separate undifferentiated and differentiated stem cells or remove cancer cells from stem cells by changing the temperature. In another report, the same authors also investigated PNiPAAm-based thermoresponsive anionic P(NIPAAm-co-tBAAm-co-acrylic acid) copolymer brush-grafted glass surfaces for the separation of a mixture of cells (human umbilical vein endothelial cells and normal human aortic smooth muscle cells) used in cardiovascular tissue engineering (Nagase et al. 2020b).

Not only PNiPAAm, but also other thermoresponsive polymers, such as, poly(oligo(ethylene glycol)methacrylate)s (POEGMA) and pseudopeptidic poly(2-oxazoline)s (PAOx) have recently been investigated towards the fabrication of thermoresponsive coatings for tissue engineering, which is due to their already proven advantages over PNiPAAm and the opportunity to finely tune/alter their LCST-behaviour (Hoogenboom and Schlaad 2017, Lutz 2011, Jana and Uchman 2020). Several investigations employing POEGMA and PAOx to regulate the cell attachment/detachment behaviour are reported in the literature (Park et al. 2010, Uhlig et al. 2012, Anderson et al. 2017, Desseaux and Klok 2014, Dworak et al. 2014, Van Der Heide et al. 2017). For example, in a recent report, Klok and colleagues have investigated RGD-peptide functionalized 2-(2-methoxyethoxy)ethyl methacrylate (MEO$_2$MA) based poly(MEO$_2$MA-co-HEMA-co-PEGMA) copolymer brushes as an alternative, promising thermoresponsive polymer platform for modulating cell adhesion and detachment (Desseaux and Klok 2014). RGD-functionalized poly(MEO$_2$MA-co-HEMA-co-PEGMA) copolymer brushes is prepared via surface-initiated ATRP (SI-ATRP) and the as-fabricated films enabled integrin-mediated adhesion of 3T3 fibroblasts at 37°C (above the LCST) and allowed the release of proliferated cells by cooling to 23°C. The use of cell-adhesive RGD containing peptide ligands, which can be thermo-reversibly masked or unmasked, is interesting since it allows the use of serum-free cell culture media (similar to RGDS-modified PNiPAAm cell culture dishes, described earlier). This is attractive as it not only reduces the immunological side effects, but also imparts additional opportunities to select specific cell types for integrin-mediated adhesion and therefore, facilitate control over the stimulation and differentiation of cells.

Poly(2-oxazoline)-based (co)polymers are also considered a viable candidate for the fabrication of temperature-responsive supports for the cell sheet technology due to their sharp thermoresponsive transitions. Dworak, Kawecki and colleagues immobilized amine end-functional poly(2-isopropyl-2-oxazoline) (P^{i}PrOx) and poly((2-ethyl-2-oxazoline)-*co*-(2-nonyl-2-oxazoline)) (PEtOx-*co*-PNonOx) onto glass surfaces via the simple grafting-to approach (Dworak et al. 2014). These as-fabricated thermoresponsive (co)poly(2-oxazoline) surfaces appeared to be effective supports for the adhesion as well as proliferation of dermal fibroblasts. The detachment of intact dermal fibroblast sheets could be manipulated simply by a variation of the temperature, like PNiPAAm-based dishes, without requiring mechanical or enzymatic methods for cell detachment. The PAOx-based surfaces did not have any adverse effect on cultured dermal fibroblasts, as established from the gene expression profile and genotoxicity experiments. Therefore, such thermoresponsive PAOx-based surfaces could be a potential candidate as a substrate to generate dermal fibroblast sheets, which are required for the treatment of wounds and in skin tissue engineering. In a recent report, Gröll et al. established an easy, user-friendly, and low-cost technique for creating thermoresponsive poly(2-n-propyl-oxazoline) (P^{n}PrOx) surfaces, applicable for cell-sheet technology and tissue-engineering purposes (Ryma et al. 2019). In this report, different standard cell culture dishes were repeatedly coated with 0.1 wt% aqueous solutions of P^{n}PrOx and dried in an oven to create a fully covered and thermoresponsive surface. The process is indeed very simple and inexpensive as it does not require any grafting techniques. Different cell types (including endothelial cells, mesenchymal stem cells, and fibroblasts) were able to adhere and proliferate (until confluency) on P^{n}PrOx surface in a similar manner to non-treated culture dishes. By reducing the temperature below the LCST of P^{n}PrOx, the polymer becomes hydrophilic and the detached cell sheets can be harvested for further processing. The cellular junctions between single cells within the sheet could also be detected using immunostainings, indicating that strong and intact intracellular contacts are preserved in the harvested sheets.

Conclusion and future directions

The past several decades have noticed a rapid advancement in the development of stimuli-responsive polymer materials for biomedical applications. Among the available physical and chemical stimuli, temperature is one of the simplest stimuli, which mostly occurs naturally or can be easily applied artificially from outside the system, and the changes usually start/stop when the temperature is switched on/off. This allows the precise spatial and temporal control of the thermal response. Therefore, in comparison to other physiologically relevant stimuli, such as pH, redox potential gradients among others, temperature stimuli provide higher degree of freedom in pathophysiological conditions of the target and, therefore, thermoresponsive polymers offer great advantages in biomedical sectors such as programmed site-specific drug delivery, gene delivery, and tissue engineering. Moreover, precise operating temperatures can be achieved by modulating the thermoresponsive polymer properties through variation of polymer architecture, comonomer addition, composition, and presence of functional groups among others. Also, adequately devised polymeric systems sensitive to both temperature and pH have great potential for enhanced and site-specific intracellular drug/gene delivery to effectively treat cancers as certain malignancies can cause a slight increase in temperature and decrease in extracellular pH around the tumour site (as tumour cells have a strong metabolism). In this present chapter, we highlighted how thermoresponsive polymers can be exploited for advanced biomedical applications in drug delivery, gene delivery and tissue engineering. In the examples of thermoresponsive micelles for drug delivery applications, the literature reports often show that the drug release can be improved upon hyperthermia. However, with few exceptions, most of the literature studies reported so far have revealed the biomedical application potential of these thermoresponsive polymer systems *in vitro* only. Therefore, additional studies focusing on the evaluation of *in vivo* efficacy of these systems are needed to be performed on different therapeutic conditions (e.g., tumour) to reduce the present large gap in between the *in vitro* results and *in vivo* performance. In terms of critical temperature, some studies reported polymer systems with a LCST at 37°C for a nanocarrier that should release its payload (drug/gene) in a controlled fashion. Yet at this temperature, there is no way of controlling the drug release *in vivo* and these studies usually suggest further work to alter their molecular design to tune the critical temperature towards mild hyperthermia. In addition, changing the molecular design of the thermoresponsive block copolymers micelles with different targeting moieties/ agents can alter the drug release and cellular uptake efficacies of these micelles and could be highly interesting in terms of treating multidrug resistance (MDR) cells or cancer stem cells and therefore warrants further investigations. Generally, "smart" carriers from responsive polymer systems are most effective in reaching their cellular targets when high molecular weight polymers are used. Such high molecular weight polymers are not readily excreted through renal excretion after delivering the drug, resulting in the significant accumulation in the human body and that would lead to serious side effects. That could be another important reason why they have not been usually tested in clinical trials. Clinical trial or medical translation is an expensive process, and it is not economically feasible to proceed with clinical trials if there is a real potential for toxicity. In this context, side-effects of thermoresponsive materials on the body also needs to be considered when investigating for biorelevant applications. Most of the thermoresponsive polymers mentioned in this chapter have not undergone extensive biocompatibility investigations. Until such studies are measured, application of these novel thermoresponsive materials for biorelevant applications is likely to be limited to the laboratory. There is also severe lack of understanding of how the parameters, for example, polymer architecture, molar mass, copolymer composition, hydrophilic-hydrophobic balance (HLB), size and size distribution of the self-assembled "smart" nanocarriers will affect the loading efficiency, cellular uptake, and drug release. A thorough understanding of these factors is necessary to be able to design and fabricate optimal nano-carrier systems for enhanced therapeutic applications. Finally, elaborated *in vivo* evaluation such as pharmacokinetic, pharmacodynamics, bio-distribution and degradation pathways of the materials need to be performed to maximize or enhance the clinical potential of a thermoresponsive nanocarriers and for this, the system needs to be fabricated adequately in the presence of efficient collaboration with research groups and clinicians. A major challenge that remains in the non-viral gene delivery is to quantitatively predict the optimal design of polymer-based gene carriers (polyplexes). Major limiting factors in the cationic polymer-based gene transfection are cytotoxicity, biocompatibility, limited stability of polyplexes during blood circulation targeting ability, and finally, the unpacking of gene intracellularly from the vector. The tightly formed polyplexes mostly enter the cells easily; however, it doesn't lead to good transfection efficiency if the nucleic acid remains tightly bound to carrier. Gene unpacking is a major limiting step in the process of polymer-based non-viral gene transfection and therefore, it is necessary to understand precisely about how loaded genes are released from non-viral vectors. Moreover, the design of biodegradable "smart" vectors capable of releasing their loaded gene payloads in a timely and spatially controlled manner is indeed an option for the development of next-generation gene-delivery vectors.

Over the past few decades, Okano and colleagues has extensively investigated thermoresponsive PNiPAAm as material for thermoresponsive surfaces for tissue engineering applications. The literature review presented in this chapter have also confirmed that PNiPAAm is the most investigated and typically favoured thermoresponsive polymer surface for cell-sheet fabrication and cell culture dishes prepared from electron beam-mediated thermoresponsive PNiPAAm surfaces are now available commercially and already applied clinically in few cases as tissue substitutes such as cornea replacement among others. However, they are not extensively investigated, especially with focus on clinical applications. Also, there is a certain limitation as electron beam irradiation is an expensive process, and temperature-triggered efficient cell-sheet fabrication is only possible in a small range of thickness of the prepared PNiPAAm coated surfaces. This led researchers to investigate for alternative surface preparation techniques. In this context, thermoresponsive polyelectrolyte multilayer films via layer-by-layer (LbL) approach

(Liao et al. 2010) have appeared as potential alternative surface modification method, that allows the precise control over the thickness of surface coatings. However, these are relatively new techniques and further in-depth investigation is required for clinical translation. One further approach of tuning PNiPAAm properties is co-polymerization with different other functional monomer and it has been widely used to fabricate functional thermoresponsive surface with tunable LCST and different adhesion kinetics. The combination of thermoresponsive PNiPAAm with more bioactive polymer could also be interesting in this context. We have also demonstrated the potential tissue engineering application possibilities of other interesting thermoresponsive polymers such as poly(oligo(ethylene glycol) methacrylate)s (POEGMA) and pseudo-peptidic poly(2-oxazoline)s (PAOx) among others. However, in comparison to PNiPAAm, these polymers are not extensively studied, especially with focus on clinical applications, thus highlighting opportunities for further investigations.

References

Abraham, T.N., V. Raj, T. Prasad, P.A. Kumar, K. Sreenivasan and T. Kumary. 2010. A Novel thermoresponsive graft copolymer containing phosphorylated HEMA for generating detachable cell layers. J. Appl. Polym. Sci. 115(1): 52–62.

Abulateefeh, S.R., S.G. Spain, J.W. Aylott, W.C. Chan, M.C. Garnett and C. Alexander. 2011. Thermoresponsive polymer colloids for drug delivery and cancer therapy. Macromol. Biosci. 11(12): 1722–1734.

Adams, N. and U.S. Schubert. 2007. Poly(2-oxazolines) in biological and biomedical application contexts. Adv. Drug Deliv. Rev. 59(15): 1504–1520.

Agarwal, S., Y. Zhang, S. Maji and A. Greiner. 2012. PDMAEMA based gene delivery materials. Mater Today 15(9): 388–393.

Agut, W., A. Brulet, C. Schatz, D. Taton and S. Lecommandoux. 2010. pH and temperature responsive polymeric micelles and polymersomes by self-assembly of poly[2-(dimethylamino) ethyl methacrylate]-b-poly(glutamic acid) double hydrophilic block copolymers. Langmuir 26(13): 10546–10554.

Akimoto, J., M. Nakayama, K. Sakai and T. Okano. 2009. Temperature-induced intracellular uptake of thermoresponsive polymeric micelles. Biomacromolecules 10(6): 1331–1336.

Akimoto, J., M. Nakayama and T. Okano. 2014. Temperature-responsive polymeric micelles for optimizing drug targeting to solid tumors. J. Control Release 193: 2–8.

Akiyama, Y., A. Kikuchi, M. Yamato and T. Okano. 2004. Ultrathin poly(N-isopropylacrylamide) grafted layer on polystyrene surfaces for cell adhesion/detachment control. Langmuir 20(13): 5506–5511.

Al Fatease, A., V. Shah, D.X. Nguyen, B. Cote, N. LeBlanc, D.A. Rao et al. 2019. Chemosensitization and mitigation of Adriamycin-induced cardiotoxicity using combinational polymeric micelles for co-delivery of quercetin/resveratrol and resveratrol/curcumin in ovarian cancer. Nanomed: Nanotechnol. Biol. Med. 19: 39–48.

Alexandridis, P. and T.A. Hatton. 1995. Poly(ethylene oxide)-poly(propylene oxide)-poly(ethylene oxide) block copolymer surfactants in aqueous solutions and at interfaces: thermodynamics, structure, dynamics, and modeling. Colloids Surf A: Physicochem Eng. 96(1-2): 1–46.

Aluri, S., S.M. Janib and J.A. Mackay. 2009. Environmentally responsive peptides as anticancer drug carriers. Adv. Drug Deliv. Rev. 61(11): 940–952.

Anderson, C.R., C. Abecunas, M. Warrener, A. Laschewsky and E. Wischerhoff. 2017. Effects of methacrylate-based thermoresponsive polymer brush composition on fibroblast adhesion and morphology. Cell Mol. Bioeng. 10: 75–88.

Andrew Mackay, J. and A. Chilkoti. 2008. Temperature sensitive peptides: engineering hyperthermia-directed therapeutics. Int. J. Hyperthermia. 24(6): 483–495.

Annabi, N., S.M. Mithieux, E.A. Boughton, A.J. Ruys, A.S. Weiss and F. Dehghani. 2009. Synthesis of highly porous crosslinked elastin hydrogels and their interaction with fibroblasts *in vitro*. Biomaterials 30(27): 4550–4557.

Aoki, T., K. Nakamura, K. Sanui, A. Kikuchi, T. Okano, Y. Sakurai et al. 1999. Adenosine-induced changes of the phase transition of poly(6-(acryloyloxymethyl) uracil) aqueous solution. Polym J. 31(11): 1185–1188.

Arisaka, Y., J. Kobayashi, M. Yamato, Y. Akiyama and T. Okano. 2013. Switching of cell growth/ detachment on heparin-functionalized thermoresponsive surface for rapid cell sheet fabrication and manipulation. Biomaterials 34(17): 4214–4222.

Arndt, K.-F., T. Schmidt and R. Reichelt. 2001. Thermo-sensitive poly(methyl vinyl ether) micro-gel formed by high energy radiation. Polymer 42(16): 6785–6791.

Aseyev, V., H. Tenhu and F.M. Winnik. 2011. Non-ionic thermoresponsive polymers in water. Adv. Polym. Sci. 242: 29–89.

Avila-Salas, F. and E.F. Duran-Lara. 2020. An overview of injectable thermo-responsive hydrogels and advances in their biomedical applications. Curr. Med. Chem. 27(34): 5773–5789.

Bagheri, M., S. Shateri, H. Niknejad and A.A. Entezami. 2014. Thermosensitive biotinylated hydroxypropyl cellulose-based polymer micelles as a nano-carrier for cancer-targeted drug delivery. J. Polym. Res. 21: 1–15.

Bayer, C.L. and N.A. Peppas. 2008. Advances in recognitive, conductive and responsive delivery systems. J. Control Release 132(3): 216–221.

Beija, M., J.-D. Marty and M. Destarac. 2011. Thermoresponsive poly(N-vinyl caprolactam)-coated gold nanoparticles: sharp reversible response and easy tunability. Chem. Commun. 47(10): 2826–2828.

Bellingham, C.M., M.A. Lillie, J.M. Gosline, G.M. Wright, B.C. Starcher, A.J. Bailey et al. 2003. Recombinant human elastin polypeptides self-assemble into biomaterials with elastin-like properties. Biopolymers: Original Research on Biomolecules 70(4): 445–455.

Belting, M. and P. Petersson. 1999. Intracellular accumulation of secreted proteoglycans inhibits cationic lipid-mediated gene transfer: co-transfer of glycosaminoglycans to the nucleus. J. Biol. Chem. 274(27): 19375–19382.

Bhatia, S.N., G.H. Underhill, K.S. Zaret and I.J. Fox. 2014. Cell and tissue engineering for liver disease. Sci. Transl. Med. 6(245): 1–49.

Bhattacharjee, A., K. Kumar, A. Arora and D.S. Katti. 2016. Fabrication and characterization of Pluronic modified poly(hydroxybutyrate) fibers for potential wound dressing applications. Mater Sci. Eng C. 63: 266–273.

Bigini, P., M. Gobbi, M. Bonati, A. Clavenna, M. Zucchetti, S. Garattini et al. 2021. The role and impact of polyethylene glycol on anaphylactic reactions to COVID-19 nano-vaccines. Nat. Nanotechnol. 16(11): 1169–1171.

Bloksma, M.M., D.J. Bakker, C. Weber, R. Hoogenboom and U.S. Schubert. 2010. The effect of Hofmeister Salts on the LCST transition of Poly(2-oxazoline)s with varying hydrophilicity. Macromol. Rapid Commun. 31(8): 724–728.

Bloksma, M.M., R.M. Paulus, H.P. van Kuringen, F. van der Woerdt, H.M. Lambermont-Thijs, U.S. Schubert et al. 2012. Thermoresponsive Poly(2-oxazine)s. Macromol. Rapid Commun. 33(1): 92–96.

Bogomolova, A., M. Hruby, J. Panek, M. Rabyk, S. Turner, S. Bals et al. 2013. Small-angle X-ray scattering and light scattering study of hybrid nanoparticles composed of thermoresponsive triblock copolymer F127 and thermoresponsive statistical polyoxazolines with hydrophobic moieties. J. Appl. Crystallogr. 46(6): 1690–1698.

Bordat, A., T. Boissenot, J. Nicolas and N. Tsapis. 2019. Thermoresponsive polymer nanocarriers for biomedical applications. Adv. Drug Deliv. Rev. 138: 167–192.

Braunecker, W.A. and K. Matyjaszewski. 2007. Controlled/living radical polymerization: Features, developments, and perspectives. Prog. Polym. Sci. 32(1): 93–146.

Brissault, B., A. Kichler, C. Guis, C. Leborgne, O. Danos and H. Cheradame. 2003. Synthesis of linear polyethylenimine derivatives for DNA transfection. Bioconjugate Chem. 14(3): 581–587.

Brun-Graeppi, A.K.A.S., C. Richard, M. Bessodes, D. Scherman and O.-W. Merten. 2010. Thermoresponsive surfaces for cell culture and enzyme-free cell detachment. Prog. Polym. Sci. 35(11): 1311–1324.

Cabane, E., X. Zhang, K. Langowska, C.G. Palivan and W. Meier. 2012. Stimuli-responsive polymers and their applications in nanomedicine. Biointerphases 7(1): 9.

Calejo, M.T., S.A. Sande and B. Nyström. 2013. Thermoresponsive polymers as gene and drug delivery vectors: architecture and mechanism of action. Expert Opin Drug Deliv. 10(12): 1669–1686.

Calejo, M.T., A.M.S. Cardoso, A.-L. Kjoniksen, K. Zhu, C.M. Morais, S.A. Sande et al. 2013. Temperature-responsive cationic block copolymers as nanocarriers for gene delivery. Int. J. Pharm. 448(1): 105–114.

Canavan, H.E., X. Cheng, D.J. Graham, B.D. Ratner and D.G. Castner. 2005. Cell sheet detachment affects the extracellular matrix: a surface science study comparing thermal liftoff, enzymatic, and mechanical methods. J. Biomed. Mater Res. Part A. 75(1): 1–13.

Chen, Y.-C., L.-C. Liao, P.-L. Lu, C.-L. Lo, H.-C. Tsai, C.-Y. Huang et al. 2012a. The accumulation of dual pH and temperature responsive micelles in tumors. Biomaterials 33(18): 4576–4588.

Chen, Q., K. Osada, T. Ishii, M. Oba, S. Uchida, T.A. Tockary et al. 2012b. Homo-catiomer integration into PEGylated polyplex micelle from block-catiomer for systemic anti-angiogenic gene therapy for fibrotic pancreatic tumors. Biomaterials 33(18): 4722–4730.

Cheng, C.-C., Y.-C. Yen and F.-C. Chang. 2011. Hierarchical structures formed from self-complementary sextuple hydrogen-bonding arrays. RSC Adv. 1(7): 1190–1194.

Cheng, C.-C., F.-C. Chang, W.-Y. Kao, S.-M. Hwang, L.-C. Liao, Y.-J. Chang et al. 2016. Highly efficient drug delivery systems based on functional supramolecular polymers: *In vitro* evaluation. Acta Biomater. 33: 194–202.

Chilkoti, A., M.R. Dreher and D.E. Meyer. 2002. Design of thermally responsive, recombinant polypeptide carriers for targeted drug delivery. Adv. Drug Deliv. Rev. 54(8): 1093–1111.

Cohn, D., A. Sosnik and A. Levy. 2003. Improved reverse thermo-responsive polymeric systems. Biomaterials 24(21): 3707–3714.

Cortez-Lemus, N.A. and A. Licea-Claverie. 2016. Poly(N-vinylcaprolactam), a comprehensive review on a thermoresponsive polymer becoming popular. Prog Polym. Sci. 53: 1–51.

Crespy, D. and R.M. Rossi. 2007. Temperature-responsive polymers with LCST in the physiological range and their applications in textiles. Polym Int. 56(12): 1461–1468.

Cui, H., M. Nowicki, J.P. Fisher and L.G. Zhang. 2017. 3D bioprinting for organ regeneration. Adv. Healthc Mater. 6(1): 1601118.

Cunliffe, D., C. de las Heras Alarcón, V. Peters, J.R. Smith and C. Alexander. 2003. Thermoresponsive surface-grafted poly(N−isopropylacrylamide) copolymers: effect of phase transitions on protein and bacterial attachment. Langmuir. 19(7): 2888–2899.

Dai, S., P. Ravi and K.C. Tam. 2009. Thermo- and photo-responsive polymeric systems. Soft Matter. 5(13): 2513–2533.

de la Rosa, V.R., E. Bauwens, B.D. Monnery, B.G. De Geest and R. Hoogenboom. 2014. Fast and accurate partial hydrolysis of poly(2-ethyl-2-oxazoline) into tailored linear polyethylenimine copolymers. Polym. Chem. 5(17): 4957–4964.

De las Heras Alarcón, C., S. Pennadam and C. Alexander. 2005. Stimuli responsive polymers for biomedical applications. Chem. Soc. Rev. 34(3): 276–285.

Demirel, A.L., M. Meyer and H. Schlaad. 2007. Formation of polyamide nanofibers by directional crystallization in aqueous solution. Angew Chem. Int. Ed. 119(45): 8776–8778.

Deshmukh, S.A., G. Kamath, K.J. Suthar, D.C. Mancini and S.K. Sankaranarayanan. 2014. Non-equilibrium effects evidenced by vibrational spectra during the coil-to-globule transition in poly (N-isopropylacrylamide) subjected to an ultrafast heating–cooling cycle. Soft Matter. 10(10): 1462–1480.

Desseaux, S. and H.-A. Klok. 2014. Temperature-controlled masking/unmasking of cell-adhesive cues with poly(ethylene glycol) methacrylate based brushes. Biomacromolecules 15(10): 3859–3865.

Doberenz, F., K. Zeng, C. Willems, K. Zhang and T. Groth. 2020. Thermoresponsive polymers and their biomedical application in tissue engineering—a review. J. Mater Chem. B. 8(4): 607–628.

Dreher, M.R., D. Raucher, N. Balu, O.M. Colvin, S.M. Ludeman and A. Chilkoti. 2003. Evaluation of an elastin-like polypeptide–doxorubicin conjugate for cancer therapy. J. Control Release 91(1-2): 31–43.

Dumortier, G., J.L. Grossiord, F. Agnely and J.C. Chaumeil. 2006. A review of poloxamer 407 pharmaceutical and pharmacological characteristics. Pharm. Res. 23: 2709–2728.

Dworak, A., A. Utrata-Wesolek, N. Oleszko, W. Walach, B. Trzebicka, J. Aniol et al. 2014. Poly(2-substituted-2-oxazoline) surfaces for dermal fibroblasts adhesion and detachment. J. Mater Sci. Mater. Med. 25: 1149–1163.

Ebara, M., M. Yamato, M. Hirose, T. Aoyagi, A. Kikuchi, K. Sakai et al. 2003. Copolymerization of 2-carboxyisopropylacrylamide with N-isopropylacrylamide accelerates cell detachment from grafted surfaces by reducing temperature. Biomacromolecules 4(2): 344–349.

Ebara, M., M. Yamato, T. Aoyagi, A. Kikuchi, K. Sakai and T. Okano. 2004a. Temperature-responsive cell culture surfaces enable "on-off" affinity control between cell integrins and RGDS ligands. Biomacromolecules 5(2): 505–510.

Ebara, M., M. Yamato, T. Aoyagi, A. Kikuchi, K. Sakai and T. Okano. 2004b. Immobilization of cell-adhesive peptides to temperature-responsive surfaces facilitates both serum-free cell adhesion and noninvasive cell harvest. Tissue Eng. 10(7-8): 1125–1135.

Ebara, M., M. Yamato, T. Aoyagi, A. Kikuchi, K. Sakai and T. Okano. 2008a. A novel approach to observing synergy effects of PHSRN on integrin–RGD binding using intelligent surfaces. Adv. Mater. 20(16): 3034–3038.

Ebara, M., M. Yamato, T. Aoyagi, A. Kikuchi, K. Sakai and T. Okano. 2008b. The effect of extensible PEG tethers on shielding between grafted thermo-responsive polymer chains and integrin–RGD binding. Biomaterials 29(27): 3650–3655.

Echeverria, C., D. Lopez and C. Mijangos. 2009. UCST responsive microgels of poly(acrylamide–acrylic acid) copolymers: Structure and viscoelastic properties. Macromolecules 42(22): 9118–9123.

Eeckman, F., A.J. Moes and K. Amighi. 2004. Synthesis and characterization of thermosensitive copolymers for oral controlled drug delivery. Eur. Polym. J. 40(4): 873–881.

El-Aassar, M., G. El Fawal, N.M. El-Deeb, H.S. Hassan and X. Mo. 2016. Electrospun polyvinyl alcohol/pluronic F127 blended nanofibers containing titanium dioxide for antibacterial wound dressing. Appl. Biochem. Biotechnol. 178: 1488–1502.

Elloumi-Hannachi, I., M. Yamato and T. Okano. 2010. Cell sheet engineering: a unique nanotechnology for scaffold-free tissue reconstruction with clinical applications in regenerative medicine. J. Intern. Med. 267(1): 54–70.

Emamzadeh, M., D. Desmaële, P. Couvreur and G. Pasparakis. 2018. Dual controlled delivery of squalenoyl-gemcitabine and paclitaxel using thermo-responsive polymeric micelles for pancreatic cancer. J. Mater. Chem. B. 6(15): 2230–2239.

Emamzadeh, M., M. Emamzadeh and G. Pasparakis. 2019. Dual controlled delivery of gemcitabine and cisplatin using polymer-modified thermosensitive liposomes for pancreatic cancer. ACS Appl. Bio Mater. 2(3): 1298–1309.

Feil, H., Y.H. Bae, J. Feijen and S.W. Kim. 1993. Effect of comonomer hydrophilicity and ionization on the lower critical solution temperature of N-isopropylacrylamide copolymers. Macromolecules 26(10): 2496–2500.

Feng, G., H. Chen, J. Li, Q. Huang, M.J. Gupte, H. Liu et al. 2015. Gene therapy for nucleus pulposus regeneration by heme oxygenase-1 plasmid DNA carried by mixed polyplex micelles with thermo-responsive heterogeneous coronas. Biomaterials 52: 1–13.

Feng, S., E. Zheng, H. Liu, C. Wang and F. Liu. 2008. Synthesis and characterization of poly(N-vinyl caprolactam) and its graft copolymers of dextran and dextrose. Polymer. 49: 1081–1082.

Fernandes, J.C., X. Qiu, F.M. Winnik, M. Benderdour, X. Zhang, K. Dai et al. 2013. Linear polyethylenimine produced by partial acid hydrolysis of poly(2-ethyl-2-oxazoline) for DNA and siRNA delivery *in vitro*. Int. J. Nanomedicine 11: 4091–4102.

Flemming, P., A.S. Munch, A. Fery and P. Uhlmann. 2021. Constrained thermoresponsive polymers—new insights into fundamentals and applications. Beilstein J. Org. Chem. 17: 2123–2163.

Fliervoet, L.A., C.F. van Nostrum, W.E. Hennink and T. Vermonden. 2019. Balancing hydrophobic and electrostatic interactions in thermosensitive polyplexes for nucleic acid delivery. Multifunct Mater. 2(2): 024002.

Fujishige, S., K. Kubota and I. Ando. 1989. Phase transition of aqueous solutions of poly(N-isopropylacrylamide) and poly(N-isopropylmethacrylamide). J. Phys. Chem. 93(8): 3311–3313.

Furgeson, D.Y., M.R. Dreher and A. Chilkoti. 2006. Structural optimization of a "smart" doxorubicin–polypeptide conjugate for thermally targeted delivery to solid tumors. J. Control Release. 110(2): 362–369.

Furyk, S., Y. Zhang, D. Ortiz-Acosta, P.S. Cremer and D.E. Bergbreiter. 2006. Effects of end group polarity and molecular weight on the lower critical solution temperature of poly(N-isopropylacrylamide). J. Polym. Sci, Part A: Polym. Chem. 44(4): 1492–1501.

Gandhi, A., A. Paul, S.O. Sen and K.K. Sen. 2015. Studies on thermoresponsive polymers: Phase behaviour, drug delivery and biomedical applications. Asian J. Pharm. Sci. 10(2): 99–107.

Gao, Y., M. Wei, X. Li, W. Xu, A. Ahiabu, J. Perdiz et al. 2017. Stimuli-responsive polymers: Fundamental considerations and applications. Macromol. Res. 25(6): 513–527.

Gauthier, M.A., M.I. Gibson and H.-A. Klok. 2009. Synthesis of functional polymers by post-polymerization modification. Angew Chem. Int. Ed. 48(1): 48–58.

Glassner, M., M. Vergaelen and R. Hoogenboom. 2018. Poly(2-oxazoline)s: A comprehensive overview of polymer structures and their physical properties. Polym. Int. 67(1): 32–45.

Guo, X. and L. Huang. 2012. Recent advances in nonviral vectors for gene delivery. Acc Chem. Res. 45(7): 971–979.

Guo, X., D. Li, G. Yang, C. Shi, Z. Tang, J. Wang et al. 2014. Thermo-triggered drug release from actively targeting polymer micelles. ACS Appl. Mater. Interfaces 6(11): 8549–8559.

Gutteridge, J.M. 1984. Lipid peroxidation and possible hydroxyl radical formation stimulated by the self-reduction of a doxorubicin-iron (III) complex. Biochem. Pharmacol. 33(11): 1725–1728.

Haladjova, E., M. Smolicek, I. Ugrinova, D. Momekova, P. Shestakova, Z. Kroneková et al. 2020a. DNA delivery systems based on copolymers of poly(2-methyl-2-oxazoline) and polyethyleneimine: Effect of polyoxazoline moieties on the endo-lysosomal escape. J. Appl. Polym. Sci. 137(45): 49400.

Haladjova, E., S. Rangelov and C. Tsvetanov. 2020b. Thermoresponsive polyoxazolines as vectors for transfection of nucleic acids. Polymers 12(11): 2609.

Halperin, A., M. Kröger and F.M. Winnik. 2015. Poly(N-isopropylacrylamide) phase diagrams: fifty years of research. Angew Chem. Int. Ed. 54(51): 15342–15367.

Hatakeyama, H., A. Kikuchi, M. Yamato and T. Okano. 2005. Influence of insulin immobilization to thermoresponsive culture surfaces on cell proliferation and thermally induced cell detachment. Biomaterials 26(25): 5167–5176.

Hatakeyama, H., A. Kikuchi, M. Yamato and T. Okano. 2006. Bio-functionalized thermoresponsive interfaces facilitating cell adhesion and proliferation. Biomaterials 27(29): 5069–5078.

Hatakeyama, H., A. Kikuchi, M. Yamato and T. Okano. 2007. Patterned biofunctional designs of thermoresponsive surfaces for spatiotemporally controlled cell adhesion, growth, and thermally induced detachment. Biomaterials 28(25): 3632–3643.

Heyda, J. and J. Dzubiella. 2014. Thermodynamic description of Hofmeister effects on the LCST of thermosensitive polymers. J. Phys Chem. B. 118(37): 10979–10988.

Hiruta, Y., Y. Kanda, N. Katsuyama and H. Kanazawa. 2017. Dual temperature-and pH-responsive polymeric micelle for selective and efficient two-step doxorubicin delivery. RSC Adv. 7(47): 29540–29549.

Hocine, S. and M.-H. Li. 2013. Thermoresponsive self-assembled polymer colloids in water. Soft Matter. 9(25): 5839–5861.

Hoffman, A.S., P.S. Stayton, V. Bulmus, G. Chen, J. Chen, C. Cheung et al. 2000. Really smart bioconjugates of smart polymers and receptor proteins. J. Biomed. Mater. Res. 52(4): 577–586.

Hoffman, A.S. 2008. The origins and evolution of "controlled" drug delivery systems. J. Control Release. 132(3): 153–163.

Hoffman, A.S. 2013. Stimuli-responsive polymers: Biomedical applications and challenges for clinical translation. Adv. Drug Delivery Rev. 65(1): 10–16.

Hogan, K.J. and A.G. Mikos. 2020. Biodegradable thermoresponsive polymers: Applications in drug delivery and tissue engineering. Polymer. 211: 123063.

Hong, W., D. Chen, L. Jia, J. Gu, H. Hu, X. Zhao et al. 2014. Thermo- and pH-responsive copolymers based on PLGA-PEG-PLGA and poly(L-histidine): synthesis and *in vitro* characterization of copolymer micelles. Acta Biomater. 10(3): 1259–1271.

Hoogenboom, R., M.W. Fijten, H.M. Thijs, B.M. van Lankvelt and U.S. Schubert. 2005. Microwave-assisted synthesis and properties of a series of poly(2-alkyl-2-oxazoline)s. Des Monomers Polym. 8(6): 659–671.

Hoogenboom, R., H.M. Thijs, M.J. Jochems, B.M. van Lankvelt, M.W. Fijten and U.S. Schubert. 2008. Tuning the LCST of poly (2-oxazoline)s by varying composition and molecular weight: alternatives to poly (N-isopropylacrylamide)? Chem. Commun. (44): 5758–5760.

Hoogenboom, R. 2009. Poly(2-oxazoline) s: a polymer class with numerous potential applications. Angew Chem. Int. Ed. 48(43): 7978–7994.

Hoogenboom, R. and H. Schlaad. 2017. Thermoresponsive poly(2-oxazoline)s, polypeptoids, and polypeptides. Polym. Chem. 8(1): 24–40.

Hruby, M., S.K. Filippov, J. Panek, M. Novakova, H. Mackova, J. Kucka et al. 2010. Polyoxazoline thermoresponsive micelles as radionuclide delivery systems. Macromol. Biosci. 10(8): 916–924.

Hsiue, G.-H., H.-Z. Chiang, C.-H. Wang and T.-M. Juang. 2006. Nonviral gene carriers based on diblock copolymers of poly (2-ethyl-2-oxazoline) and linear polyethylenimine. Bioconjugate Chem. 17(3): 781–786.

Hu, Y., V. Darcos, S. Monge and S. Li. 2015. Thermo-responsive drug release from self-assembled micelles of brush-like PLA/PEG analogues block copolymers. Int. J. Pharm. 491(1-2): 152–161.

Huglin, M.B. and M.A. Radwan. 1991. Unperturbed dimensions of a zwitterionic polymethacrylate. Polym. Int. 26(2): 97–104.

Iansante, V., R. Mitry, C. Filippi, E. Fitzpatrick and A. Dhawan. 2018. Human hepatocyte transplantation for liver disease: current status and future perspectives. Pediatr. Res. 83(1): 232–240.

Idziak, I., D. Avoce, D. Lessard, D. Gravel and X. Zhu. 1999. Thermosensitivity of aqueous solutions of poly(N, N-diethylacrylamide). Macromolecules 32(4): 1260–1263.

Indermun, S., M. Govender, P. Kumar, Y.E. Choonara and V. Pillay. 2018. Stimuli-responsive polymers as smart drug delivery systems: Classifications based on carrier type and triggered-release mechanism. pp. 43–58. *In*: Makhlouf, A.S.H. and N.Y. Abu-Thabit (eds.). Stimuli Responsive Polymeric Nanocarriers for Drug Delivery Applications, Volume 1: Woodhead Publishing.

Jana, S., Y. Biswas, M. Anas, A. Saha and T.K. Mandal. 2018. Poly [oligo (2-ethyl-2-oxazoline) acrylate]-based poly (ionic liquid) random copolymers with coexistent and tunable lower critical solution temperature-and upper critical solution temperature-type phase transitions. Langmuir. 34(42): 12653–12663.

Jana, S. and M. Uchman. 2020. Poly(2-oxazoline)-based stimulus-responsive (Co) polymers: An overview of their design, solution properties, surface-chemistries and applications. Prog. Polym. Sci. 106: 101252.

Jana, S. and R. Hoogenboom. 2022. Poly(2-oxazoline)s: a comprehensive overview of polymer structures and their physical properties—an update. Polym. Int. 71(8): 935–949.

Jeong, J.H., S.H. Song, D.W. Lim, H. Lee and T.G. Park. 2001. DNA transfection using linear poly (ethylenimine) prepared by controlled acid hydrolysis of poly(2-ethyl-2-oxazoline). J. Control Release. 73(2-3): 391–399.

Jing, Y. and P. Wu. 2013. Study on the thermoresponsive two phase transition processes of hydroxypropyl cellulose concentrated aqueous solution: from a microscopic perspective. Cellulose 20: 67–81.

Joseph, N., T. Prasad, V. Raj, P. Anil Kumar, K. Sreenivasan and T. Kumary. 2010. A cytocompatible poly(n-isopropylacrylamide-co-glycidylmethacrylate) coated surface as new substrate for corneal tissue engineering. J. Bioact. Compat. Polym. 25(1): 58–74.

Kanto, R., R. Yonenuma, M. Yamamoto, H. Furusawa, S. Yano, M. Haruki et al. 2021. Mixed polyplex micelles with thermoresponsive and lysine-based zwitterionic shells derived from two poly(vinyl amine)-based block copolymers. Langmuir. 37(10): 3001–3014.

Katsumoto, Y. and N. Kubosaki. 2008. Tacticity effects on the phase diagram for poly(N-isopropylacrylamide) in water. Macromolecules 41(15): 5955–5956.

Katsumoto, Y., A. Tsuchiizu, X. Qiu and F.M. Winnik. 2012. Dissecting the mechanism of the heat-induced phase separation and crystallization of poly(2-isopropyl-2-oxazoline) in water through vibrational spectroscopy and molecular orbital calculations. Macromolecules 45(8): 3531–3541.

Kay, M.A. 2011. State-of-the-art gene-based therapies: the road ahead. Nat. Rev. Genet. 12(5): 316–328.

Kermagoret, A., C.-A. Fustin, M. Bourguignon, C. Detrembleur, C. Jerome and A. Debuigne. 2013. One-pot controlled synthesis of double thermoresponsive N-vinylcaprolactam-based copolymers with tunable LCSTs. Polym. Chem. 4(8): 2575–2583.

Khademhosseini, A., J.P. Vacanti and R. Langer. 2009. Progress in tissue engineering. Sci. Am. 300(5): 64–71.

Khademhosseini, A. and R. Langer. 2016. A decade of progress in tissue engineering. Nat. Protoc. 11(10): 1775–1781.

Kikuchi, A. and T. Okano. 2002. Intelligent thermoresponsive polymeric stationary phases for aqueous chromatography of biological compounds. Prog. Polym. Sci. 27(6): 1165–1193.

Kim, K., K. Ohashi, R. Utoh, K. Kano and T. Okano. 2012. Preserved liver-specific functions of hepatocytes in 3D co-culture with endothelial cell sheets. Biomaterials 33(5): 1406–1413.

Kim, Y.-J. and Y.T. Matsunaga. 2017. Thermo-responsive polymers and their application as smart biomaterials. J. Mater Chem. B. 5(23): 4307–4321.

Klouda, L. 2015. Thermoresponsive hydrogels in biomedical applications: A seven-year update. Eur. J. Pharm Biopharm. 97: 338–349.

Knop, K., R. Hoogenboom, D. Fischer and U.S. Schubert. 2010. Poly(ethylene glycol) in drug delivery: pros and cons as well as potential alternatives. Angew Chem. Int. Ed. 49(36): 6288–6308.

Kong, Y.W. and E.C. Dreaden. 2022. PEG: Will it come back to you? polyethelyne glycol immunogenicity, COVID vaccines, and the case for new PEG derivatives and alternatives. Front Bioeng. Biotechnol. 10: 879988.

Konradi, R., C. Acikgoz and M. Textor. 2012. Polyoxazolines for nonfouling surface coatings-a direct comparison to the gold standard PEG. Macromol. Rapid Commun. 33(19): 1663–1676.

Kumar, P.A., K. Sreenivasan and T. Kumary. 2007. Alternate method for grafting thermoresponsive polymer for transferring *in vitro* cell sheet structures. J. Appl. Polym. Sci. 105(4): 2245–2251.

Kumashiro, Y., M. Yamato and T. Okano. 2010. Cell attachment–detachment control on temperature-responsive thin surfaces for novel tissue engineering. Ann. Biomed. Eng. 38: 1977–1988.

Kurisawa, M., M. Yokoyama and T. Okano. 2000. Gene expression control by temperature with thermo-responsive polymeric gene carriers. J. Control Release. 69(1): 127–137.

Kwon, O.H., A. Kikuchi, M. Yamato, Y. Sakurai and T. Okano. 2000. Rapid cell sheet detachment from Poly(N-isopropylacrylamide)-grafted porous cell culture membranes. J. Biomed. Mater. Res. 50(1): 82–89.

Kwon, O.H., A. Kikuchi, M. Yamato and T. Okano. 2003. Accelerated cell sheet recovery by co-grafting of PEG with PIPAAm onto porous cell culture membranes. Biomaterials 24(7): 1223–1232.

Lambermont-Thijs, H.M., F.S. van der Woerdt, A. Baumgaertel, L. Bonami, F.E. Du Prez, U.S. Schubert et al. 2010. Linear poly(ethylene imine)s by acidic hydrolysis of poly(2-oxazoline) s: kinetic screening, thermal properties, and temperature-induced solubility transitions. Macromolecules 43(2): 927–933.

Langer, R. and J.T. Vacanti. 1993. Tissue engineering. Science 260: 920–926.

Langer, R. and J. Vacanti. 2016. Advances in tissue engineering. J. Pediatr. Surg. 51(1): 8–12.

Lee, Y. and K. Kataoka. 2012. Delivery of nucleic acid drugs. Adv. Polym. Sci. 95–134.

Li, J., Q. Chen, Z. Zha, H. Li, K. Toh, A. Dirisala et al. 2015. Ternary polyplex micelles with PEG shells and intermediate barrier to complexed DNA cores for efficient systemic gene delivery. J. Control Release. 209: 77–87.

Li, M., M.J. Mondrinos, M.R. Gandhi, F.K. Ko, A.S. Weiss and P.I. Lelkes. 2005. Electrospun protein fibers as matrices for tissue engineering. Biomaterials 26(30): 5999–6008.

Li, W., J. Li, J. Gao, B. Li, Y. Xia, Y. Meng et al. 2011. The fine-tuning of thermosensitive and degradable polymer micelles for enhancing intracellular uptake and drug release in tumors. Biomaterials 32(15): 3832–3844.

Li, Y., J. Li, B. Chen, Q. Chen, G. Zhang, S. Liu et al. 2014. Polyplex micelles with thermoresponsive heterogeneous coronas for prolonged blood retention and promoted gene transfection. Biomacromolecules 15(8): 2914–2923.

Li, Z., B. Hao, Y. Tang, H. Li, T.-C. Lee, A. Feng et al. 2020. Effect of end-groups on sulfobetaine homopolymers with the tunable upper critical solution temperature (UCST). Eur Polym J. 132: 109704.

Liang, X., F. Liu, V. Kozlovskaya, Z. Palchak and E. Kharlampieva. 2015. Thermoresponsive micelles from double LCST-poly(3-methyl-N-vinylcaprolactam) block copolymers for cancer therapy. ACS Macro Lett. 4(3): 308–311.

Liao, T., M.D. Moussallem, J. Kim, J.B. Schlenoff and T. Ma. 2010. N-isopropylacrylamide-based thermoresponsive polyelectrolyte multilayer films for human mesenchymal stem cell expansion. Biotechnol. Prog. 26(6): 1705–1713.

Lim, Y.M., J.P. Jeun, J.H. Lee, Y.M. Lee and Y.C. Nho. 2007. Cell sheet detachment from poly(N-vinylcaprolactam-co-N-isopropylacrylamide) grafted onto tissue culture polystyrene dishes. J. Ind. Eng. Chem. 13(1): 21–26.

Liu, J., C. Detrembleur, M.-C. De Pauw-Gillet, S. Mornet, E. Duguet and C. Jérôme. 2014. Gold nanorods coated with a thermo-responsive poly (ethylene glycol)-b-poly(N-vinylcaprolactam) corona as drug delivery systems for remotely near infrared-triggered release. Polym. Chem. 5(3): 799–813.

Liu, N., B. Li, C. Gong, Y. Liu, Y. Wang and G. Wu. 2015. A pH- and thermo-responsive poly(amino acid)-based drug delivery system. Colloids Surf B: Biointerfaces 136: 562–569.

Liu, R., M. Fraylich and B.R. Saunders. 2009. Thermoresponsive copolymers: from fundamental studies to applications. Colloid Polym. Sci. 287: 627–643.

Liu, Y., K. Li, J. Pan, B. Liu and S.-S. Feng. 2010. Folic acid conjugated nanoparticles of mixed lipid monolayer shell and biodegradable polymer core for targeted delivery of Docetaxel. Biomaterials 31(2): 330–338.

Lo, C.-L., K.-M. Lin and G.-H. Hsiue. 2005. Preparation and characterization of intelligent core-shell nanoparticles based on poly(D, L-lactide)-g-poly(N-isopropyl acrylamide-co-methacrylic acid). J. Control Release. 104(3): 477–488.

Lu, Y., K. Zhou, Y. Ding, G. Zhang and C. Wu. 2010. Origin of hysteresis observed in association and dissociation of polymer chains in water. Phys. Chem. Chem. Phys. 12(13): 3188–3194.

Lutz, J.-F. and A. Hoth. 2006. Preparation of ideal PEG analogues with a tunable thermosensitivity by controlled radical copolymerization of 2-(2-methoxyethoxy) ethyl methacrylate and oligo(ethylene glycol) methacrylate. Macromolecules 39(2): 893–896.

Lutz, J.-F., O. Akdemir and A. Hoth. 2006. Point by point comparison of two thermosensitive polymers exhibiting a similar LCST: is the age of poly(NIPAM) over? J. Am. Chem. Soc. 128(40): 13046–13047.

Lutz, J.F. 2008. Polymerization of oligo(ethylene glycol)(meth) acrylates: Toward new generations of smart biocompatible materials. J. Polym. Sci. Part A: Polym. Chem. 46(11): 3459–3470.

Lutz, J.-F., A. Hoth and K. Schade. 2009. Design of oligo(ethylene glycol)-based thermoresponsive polymers: an optimization study. Des Monomers Polym. 12(4): 343–353.

Lutz, J.-F. 2011. Thermo-Switchable Materials Prepared Using the OEGMA-Platform. Adv. Mater. 23(19): 2237–2243.

Ma, Y., S. Hou, B. Ji, Y. Yao and X. Feng. 2010. A novel temperature-responsive polymer as a gene vector. Macromol. Biosci. 10(2): 202–210.

Madathil, B.K., P.R. Anil Kumar and T.V. Kumary. 2014. N-isopropylacrylamide-co-glycidylmethacrylate as a thermoresponsive substrate for corneal endothelial cell sheet engineering. Biomed. Res. Int. 2014: 1–7.

Manouras, T., E. Koufakis, S.H. Anastasiadis and M. Vamvakaki. 2017. A facile route towards PDMAEMA homopolymer amphiphiles. Soft Matter. 13(20): 3777–3782.

Mao, Z., L. Ma, J. Yan, M. Yan, C. Gao and J. Shen. 2007. The gene transfection efficiency of thermoresponsive N, N, N-trimethyl chitosan chloride-g-poly(N-isopropylacrylamide) copolymer. Biomaterials 28(30): 4488–4500.

Matsumoto, M., T. Tada, T.-A. Asoh, T. Shoji, T. Nishiyama, H. Horibe et al. 2018. Dynamics of the phase separation in a thermoresponsive polymer: Accelerated phase separation of stereocontrolled poly(N, N-diethylacrylamide) in water. Langmuir. 34(45): 13690–13696.

Matyjaszewski, K. and J. Xia. 2001. Atom transfer radical polymerization. Chem. Rev. 101(9): 2921–2990.

McDaniel, J.R., D.C. Radford and A. Chilkoti. 2013. A unified model for *de novo* design of elastin-like polypeptides with tunable inverse transition temperatures. Biomacromolecules 14(8): 2866–2872.

Mees, M.A. and R. Hoogenboom. 2018. Full and partial hydrolysis of poly(2-oxazoline)s and the subsequent post-polymerization modification of the resulting polyethylenimine (co)polymers. Polym. Chem. 9(40): 4968–4978.

Mendes, P.M. 2008. Stimuli-responsive surfaces for bio-applications. Chem. Soc. Rev. 37(11): 2512–2529.

Mertoglu, M., S. Garnier, A. Laschewsky, K. Skrabania and J. Storsberg. 2005. Stimuli responsive amphiphilic block copolymers for aqueous media synthesised via reversible addition fragmentation chain transfer polymerisation (RAFT). Polymer. 46(18): 7726–7740.

Meyer, D.E., B. Shin, G. Kong, M. Dewhirst and A. Chilkoti. 2001. Drug targeting using thermally responsive polymers and local hyperthermia. J. Control Release. 74(1-3): 213–224.

Mishra, V., S.-H. Jung, H.M. Jeong and H.-i. Lee. 2014. Thermoresponsive ureido-derivatized polymers: The effect of quaternization on UCST properties. Polym. Chem. 5(7): 2411–2416.

Mizutani, A., A. Kikuchi, M. Yamato, H. Kanazawa and T. Okano. 2008. Preparation of thermoresponsive polymer brush surfaces and their interaction with cells. Biomaterials 29(13): 2073–2081.

Moad, G., E. Rizzardo and S.H. Thang. 2005. Living radical polymerization by the RAFT process. Aust. J. Chem. 58(6): 379–410.

Moad, G., E. Rizzardo and S.H. Thang. 2008. Toward living radical polymerization. Acc Chem. Res. 41(9): 1133–1142.

Moerkerke, R., F. Meeussen, R. Koningsveld, H. Berghmans, W. Mondelaers, E. Schacht et al. 1998. Phase transitions in swollen networks. 3. Swelling behavior of radiation cross-linked poly(vinyl methyl ether) in water. Macromolecules 31(7): 2223–2229.

Mohammed, M.N., K.B. Yusoh and J.H.B.H. Shariffuddin. 2018. Poly(N-vinyl caprolactam) thermoresponsive polymer in novel drug delivery systems: A review. Mater Express. 8(1): 21–34.

Moran, M.T., W.M. Carroll, A. Gorelov and Y. Rochev. 2007. Intact endothelial cell sheet harvesting from thermoresponsive surfaces coated with cell adhesion promoters. J. R Soc. Interface 4(17): 1151–1157.

Municoy, S., M.I. Alvarez Echazu, P.E. Antezana, J.M. Galdopórpora, C. Olivetti, A.M. Mebert et al. 2020. Stimuli-responsive materials for tissue engineering and drug delivery. Int. J. Mol. Sci. 21(13): 4724.

Nagase, K., J. Kobayashi and T. Okano. 2009. Temperature-responsive intelligent interfaces for biomolecular separation and cell sheet engineering. J. R Soc. Interface 6(suppl_3): S293–S309.

Nagase, K., Y. Hatakeyama, T. Shimizu, K. Matsuura, M. Yamato, N. Takeda et al. 2015. Thermoresponsive cationic copolymer brushes for mesenchymal stem cell separation. Biomacromolecules 16(2): 532–540.

Nagase, K., M. Yamato, H. Kanazawa and T. Okano. 2018. Poly(N-isopropylacrylamide)-based thermoresponsive surfaces provide new types of biomedical applications. Biomaterials 153: 27–48.

Nagase, K., A. Ota, T. Hirotani, S. Yamada, A.M. Akimoto and H. Kanazawa. 2020a. Thermoresponsive cationic block copolymer brushes for temperature-modulated stem cell separation. Macromol. Rapid Commun. 41(19): 2000308.

Nagase, K., N. Uchikawa, T. Hirotani, A.M. Akimoto and H. Kanazawa. 2020b. Thermoresponsive anionic copolymer brush-grafted surfaces for cell separation. Colloids Surf B: Biointerfaces 185: 110565.

Nagase, K., H. Wakayama, J. Matsuda, N. Kojima and H. Kanazawa. 2023. Thermoresponsive mixed polymer brush to effectively control the adhesion and separation of stem cells by altering temperature. Mater Today Bio. 20: 100627.

Ng, W.S., L.A. Connal, E. Forbes and G.V. Franks. 2018. A review of temperature-responsive polymers as novel reagents for solid-liquid separation and froth flotation of minerals. Miner Eng. 123: 144–159.

Nicolas, J., Y. Guillaneuf, C. Lefay, D. Bertin, D. Gigmes and B. Charleux. 2013. Nitroxide-mediated polymerization. Prog Polym. Sci. 38(1): 63–235.

Nishida, K., J.D. Kang, L.G. Gilbertson, S.H. Moon, J.K. Suh, M.T. Vogt et al. 1999. Modulation of the biologic activity of the rabbit intervertebral disc by gene therapy: an *in vivo* study of adenovirus-mediated transfer of the human transforming growth factor beta 1 encoding gene. Spine. 24(23): 2419–2425.

Nithya, J., P.R. Anil Kumar and T.V. Kumary. 2011. Tunable stimuli-responsive polymers for cell sheet engineering. *In*: Daniel, E. (ed.). Regenerative Medicine and Tissue Engineering. Rijeka: IntechOpen. p. Ch. 23.

Nitschke, M., S. Gramm, T. Gotze, M. Valtink, J. Drichel, B. Voit et al. 2007. Thermo-responsive poly(NiPAAm-co-DEGMA) substrates for gentle harvest of human corneal endothelial cell sheets. J. Biomed. Mater. Res. Part A. 80(4): 1003–1010.

Nuyken, O. and S.D. Pask. 2013. Ring-opening polymerization-an introductory review. Polymers 5(2): 361–403.

Oda, H., Y. Yoshida, A. Kawamura and A. Kakinuma. 2008. Cell shape, cell–cell contact, cell–extracellular matrix contact and cell polarity are all required for the maximum induction of CYP2B1 and CYP2B2 gene expression by phenobarbital in adult rat cultured hepatocytes. Biochem. Pharmacol. 75(5): 1209–1217.

Okano, T., N. Yamada, M. Okuhara, H. Sakai and Y. Sakurai. 1995. Mechanism of cell detachment from temperature-modulated, hydrophilic-hydrophobic polymer surfaces. Biomaterials 16(4): 297–303.

Osawa, S., K. Osada, S. Hiki, A. Dirisala, T. Ishii and K. Kataoka. 2016. Polyplex micelles with double-protective compartments of hydrophilic shell and thermoswitchable palisade of poly(oxazoline)-based block copolymers for promoted gene transfection. Biomacromolecules 17(1): 354–361.

Oupicky, D., T. Reschel, C. Konak and L. Oupicka. 2003. Temperature-controlled behavior of self-assembly gene delivery vectors based on complexes of DNA with poly(L-lysine)-g raft-poly(N-isopropylacrylamide). Macromolecules 36(18): 6863–6872.

Pack, D.W., A.S. Hoffman, S. Pun and P.S. Stayton. 2005. Design and development of polymers for gene delivery. Nat. Rev. Drug Discov. 4(7): 581–593.

Panek, J., S.K. Filippov, M. Hruby, M. Rabyk, A. Bogomolova, J. Kucka et al. 2012. Thermoresponsive nanoparticles based on poly(2-alkyl-2-Oxazolines) and pluronic F127. Macromol. Rapid Commun. 33(19): 1683–1689.

Panja, S., G. Dey, R. Bharti, K. Kumari, T. Maiti, M. Mandal et al. 2016. Tailor-made temperature-sensitive micelle for targeted and on-demand release of anticancer drugs. ACS Appl. Mater. Interfaces 8(19): 12063–12074.

Park, S., H.Y. Cho, J.A. Yoon, Y. Kwak, A. Srinivasan, J.O. Hollinger et al. 2010. Photo-cross-linkable thermoresponsive star polymers designed for control of cell-surface interactions. Biomacromolecules 11(10): 2647–2652.

Pasparakis, G. and C. Tsitsilianis. 2020. LCST polymers: Thermoresponsive nanostructured assemblies towards bioapplications. Polymer. 211: 123146.

Patel, N.G. and G. Zhang. 2013. Responsive systems for cell sheet detachment. Organogenesis 9(2): 93–100.

Perrier, S. 2017. 50th Anniversary Perspective: RAFT Polymerization—A User Guide. Macromolecules. 50(19): 7433–7447.

Pham, Q.P., U. Sharma and A.G. Mikos. 2006. Electrospinning of polymeric nanofibers for tissue engineering applications: a review. Tissue Eng. 12(5): 1197–1211.

Place, E.S., J.H. George, C.K. Williams and M.M. Stevens. 2009. Synthetic polymer scaffolds for tissue engineering. Chem. Soc. Rev. 38(4): 1139–1151.

Place, E.S., N.D. Evans and M.M. Stevens. 2009. Complexity in biomaterials for tissue engineering. Nat. Mater. 8(6): 457–470.

Plunkett, K.N., X. Zhu, J.S. Moore and D.E. Leckband. 2006. PNIPAM chain collapse depends on the molecular weight and grafting density. Langmuir. 22(9): 4259–4266.

Prabaharan, M., J.J. Grailer, D.A. Steeber and S. Gong. 2008. Stimuli-responsive chitosan-graft-poly(N-vinylcaprolactam) as a promising material for controlled hydrophobic drug delivery. Macromol. Biosci. 8(9): 843–851.

Praveen, W., B.K. Madathil, R.S. Raj, T. Kumary and P.A. Kumar. 2017. A flexible thermoresponsive cell culture substrate for direct transfer of keratinocyte cell sheets. Biomed. Mater. 12(6): 065012.

Qiao, P., Q. Niu, Z. Wang and D. Cao. 2010. Synthesis of thermosensitive micelles based on poly(N-isopropylacrylamide) and poly(l-alanine) for controlled release of adriamycin. Chem. Eng. J. 159(1-3): 257–263.

Rackaitis, M., K. Strawhecker and E. Manias. 2002. Water-soluble polymers with tunable temperature sensitivity: Solution behavior. J. Polym. Sci. Part B Polym. Phys. 40(19): 2339–2342.

Reed, J.A., A.E. Lucero, S. Hu, L.K. Ista, M.T. Bore, G.P. López et al. 2010. A low-cost, rapid deposition method for "smart" films: applications in mammalian cell release. ACS Appl. Mater Interfaces. 2(4): 1048–1051.

Rezaei, S.J.T., M.R. Nabid, H. Niknejad and A.A. Entezami. 2012. Folate-decorated thermoresponsive micelles based on star-shaped amphiphilic block copolymers for efficient intracellular release of anticancer drugs. Int. J. Pharm. 437(1-2): 70–79.

Rossegger, E., V. Schenk and F. Wiesbrock. 2013. Design strategies for functionalized poly(2-oxazoline) s and derived materials. Polymers 5(3): 956–1011.

Roth, P.J., F.D. Jochum, F.R. Forst, R. Zentel and P. Theato. 2010. Influence of end groups on the stimulus-responsive behavior of poly[oligo (ethylene glycol) methacrylate] in water. Macromolecules 43(10): 4638–4645.

Roy, D., W.L. Brooks and B.S. Sumerlin. 2013. New directions in thermoresponsive polymers. Chem. Soc. Rev. 42(17): 7214–7243.

Ryma, M., J. Blohbaum, R. Singh, A. Sancho, J. Matuszak, I. Cicha et al. 2019. Easy-to-prepare coating of standard cell culture dishes for cell-sheet engineering using aqueous solutions of poly (2-n-propyl-oxazoline). ACS Biomater. Sci. Eng. 5(3): 1509–1517.

Rzaev, Z.M., S. Dincer and E. Pişkin. 2007. Functional copolymers of N-isopropylacrylamide for bioengineering applications. Prog. Polym. Sci. 32(5): 534–595.

Samal, S.K., M. Dash, S. Van Vlierberghe, D.L. Kaplan, E. Chiellini, C. Van Blitterswijk et al. 2012. Cationic polymers and their therapeutic potential. Chem. Soc. Rev. 41(21): 7147–7194.

Sambe, L., V.R. de La Rosa, K. Belal, F. Stoffelbach, J. Lyskawa, F. Delattre et al. 2014. Programmable polymer-based supramolecular temperature sensor with a memory function. Angew Chem. Int. Ed. 126(20): 5144–5148.

Sarwan, T., P. Kumar, Y.E. Choonara and V. Pillay. 2020. Hybrid thermo-responsive polymer systems and their biomedical applications. Front Mater. 7: 73.

Schild, H.G. and D.A. Tirrell. 1990. Microcalorimetric detection of lower critical solution temperatures in aqueous polymer solutions. J. Phys. Chem. 94(10): 4352–4356.

Schild, H.G. 1992. Poly(N-isopropylacrylamide): experiment, theory and application. Prog. Polym. Sci. 17(2): 163–249.

Schulz, D., D. Peiffer, P. Agarwal, J. Larabee, J. Kaladas, L. Soni et al. 1986. Phase behaviour and solution properties of sulphobetaine polymers. Polymer 27(11): 1734–1742.

Sedlacek, O., P. Cernoch, J. Kucka, R. Konefal, P. Stepanek, M. Vetrik et al. 2016. Thermoresponsive polymers for nuclear medicine: which polymer is the best? Langmuir. 32(24): 6115–6122.

Sekine, H., T. Shimizu, S. Kosaka, E. Kobayashi and T. Okano. 2006. Cardiomyocyte bridging between hearts and bioengineered myocardial tissues with mesenchymal transition of mesothelial cells. J. Heart Lung Transplant. 25(3): 324–332.

Senaratne, W., L. Andruzzi and C.K. Ober. 2005. Self-assembled monolayers and polymer brushes in biotechnology: current applications and future perspectives. Biomacromolecules 6(5): 2427–2448.

Seuring, J. and S. Agarwal. 2010. Non-ionic homo-and copolymers with H-donor and H-acceptor units with an UCST in water. Macromol. Chem. Phys. 211(19): 2109–2117.

Seuring, J. and S. Agarwal. 2012. Polymers with upper critical solution temperature in aqueous solution. Macromol. Rapid Commun. 33(22): 1898–1920.

Shakya, A.K., R. Holmdahl, K.S. Nandakumar and A. Kumar. 2014. Polymeric cryogels are biocompatible, and their biodegradation is independent of oxidative radicals. J. Biomed. Mater. Res. Part A. 102(10): 3409–3418.

Shimada, N., H. Ino, K. Maie, M. Nakayama, A. Kano and A. Maruyama. 2011. Ureido-derivatized polymers based on both poly (allylurea) and poly (L-citrulline) exhibit UCST-type phase transition behavior under physiologically relevant conditions. Biomacromolecules 12(10): 3418–3422.

Shimizu, K., H. Fujita and E. Nagamori. 2010. Oxygen plasma-treated thermoresponsive polymer surfaces for cell sheet engineering. Biotechnol. Bioeng. 106(2): 303–310.

Shimizu, T., H. Sekine, Y. Isoi, M. Yamato, A. Kikuchi and T. Okano. 2006. Long-term survival and growth of pulsatile myocardial tissue grafts engineered by the layering of cardiomyocyte sheets. Tissue Eng. 12(3): 499–507.

Smith, E., J. Yang, L. McGann, W. Sebald and H. Uludag. 2005. RGD-grafted thermoreversible polymers to facilitate attachment of BMP-2 responsive C2C12 cells. Biomaterials 26(35): 7329–7338.

Soltantabar, P., E.L. Calubaquib, E. Mostafavi, M.C. Biewer and M.C. Stefan. 2020. Enhancement of loading efficiency by coloading of doxorubicin and quercetin in thermoresponsive polymeric micelles. Biomacromolecules 21(4): 1427–1436.

Song, L., Y. Yan, M. Marzano and Y. Li. 2019. Studying Heterotypic Cell–cell interactions in the human brain using pluripotent stem cell models for neurodegeneration. Cells 8(4): 299.

Soppimath, K.S., D.W. Tan and Y.Y. Yang. 2005. pH-triggered thermally responsive polymer core–shell nanoparticles for drug delivery. Adv. Mater. 17(3): 318–323.

Sosnik, A., D. Cohn, J.S. Román and G.A. Abraham. 2003. Crosslinkable PEO-PPO-PEO-based reverse thermo-responsive gels as potentially injectable materials. J. Biomater. Sci, Polym. Ed. 14(3): 227–239.

Sponchioni, M., R. Ferrari, L. Morosi and D. Moscatelli. 2016. Influence of the polymer structure over self-assembly and thermo-responsive properties: The case of PEG-b-PCL grafted copolymers via a combination of RAFT and ROP. J. Polym. Sci. Part A: Polym. Chem. 54(18): 2919–2931.

Sponchioni, M., U.C. Palmiero and D. Moscatelli. 2019. Thermo-responsive polymers: Applications of smart materials in drug delivery and tissue engineering. Mater. Sci. Eng. C. 102: 589–605.

Stuart, M.A.C., W.T. Huck, J. Genzer, M. Müller, C. Ober, M. Stamm et al. 2010. Emerging applications of stimuli-responsive polymer materials. Nat. Mater. 9(2): 101–113.

Sun, F., Y. Wang, Y. Wei, G. Cheng and G. Ma. 2014. Thermo-triggered drug delivery from polymeric micelles of poly(N-isopropylacrylamide-co-acrylamide)-b-poly(n-butyl methacrylate) for tumor targeting. J. Bioact. Compat. Polym. 29(4): 301–317.

Sun, X.-L., P.-C. Tsai, R. Bhat, E. Bonder, B. Michniak-Kohn and A. Pietrangelo. 2015. Thermoresponsive block copolymer micelles with tunable pyrrolidone-based polymer cores: structure/property correlations and application as drug carriers. J. Mater. Chem. B. 3(5): 814–823.

Sung, T.-C., H.C. Su, Q.-D. Ling, S.S. Kumar, Y. Chang, S.-T. Hsu et al. 2020. Efficient differentiation of human pluripotent stem cells into cardiomyocytes on cell sorting thermoresponsive surface. Biomaterials 253: 120060.

Tai, W., R. Mahato and K. Cheng. 2010. The role of HER2 in cancer therapy and targeted drug delivery. J. Control Release. 146(3): 264–275.

Takezawa, T., Y. Mori and K. Yoshizato. 1990. Cell culture on a thermo-responsive polymer surface. Bio/technology 8(9): 854–856.

Tan, I., F. Roohi and M.-M. Titirici. 2012. Thermoresponsive polymers in liquid chromatography. Anal. Methods 4(1): 34–43.

Tang, Z., Y. Akiyama and T. Okano. 2012. Temperature-responsive polymer modified surface for cell sheet engineering. Polymers 4(3): 1478–1498.

Teotia, A.K., H. Sami and A. Kumar. 2015. Thermo-responsive polymers: structure and design of smart materials. pp. 3–43. *In*: Zhang, Z. (ed.). Switchable and Responsive Surfaces and Materials for Biomedical Applications. Oxford: Woodhead Publishing.

Tokarev, I. and S. Minko. 2009. Stimuli-responsive hydrogel thin films. Soft Matter. 5(3): 511–524.

Trzebicka, B., R. Szweda, D. Kosowski, D. Szweda, Ł. Otulakowski, E. Haladjova et al. 2017. Thermoresponsive polymer-peptide/protein conjugates. Prog. Polym. Sci. 68: 35–76.

Tsuda, Y., A. Kikuchi, M. Yamato, Y. Sakurai, M. Umezu and T. Okano. 2004. Control of cell adhesion and detachment using temperature and thermoresponsive copolymer grafted culture surfaces. J. Biomed Mater. Res. Part A. 69(1): 70–78.

Tsuda, Y., A. Kikuchi, M. Yamato, A. Nakao, Y. Sakurai, M. Umezu et al. 2005. The use of patterned dual thermoresponsive surfaces for the collective recovery as co-cultured cell sheets. Biomaterials 26(14): 1885–1893.

Tsuda, Y., A. Kikuchi, M. Yamato, G. Chen and T. Okano. 2006. Heterotypic cell interactions on a dually patterned surface. Biochem. Biophys. Res. Commun. 348(3): 937–944.

Turk, M., S. Dincer, I.G. Yulug and E. Piskin. 2004. *In vitro* transfection of HeLa cells with temperature sensitive polycationic copolymers. J. Control Release. 96(2): 325–340.

Twaites, B.R., C. de las Heras Alarcon, M. Lavigne, A. Saulnier, S.S. Pennadam, D. Cunliffe et al. 2005. Thermoresponsive polymers as gene delivery vectors: cell viability, DNA transport and transfection studies. J. Control Release. 108(2-3): 472–483.

Uhlig, K., B. Boysen, A. Lankenau, M. Jaeger, E. Wischerhoff, J.-F. Lutz et al. 2012. On the influence of the architecture of poly (ethylene glycol)-based thermoresponsive polymers on cell adhesion. Biomicrofluidics 6(2).

Urry, D.W., T.M. Parker, M.C. Reid and D.C. Gowda. 1991. Biocompatibility of the bioelastic materials, poly(GVGVP) and its γ-irradiation cross-linked matrix: summary of generic biological test results. J. Bioact. Compat. Polym. 6(3): 263–282.

Urry, D.W. 1992. Free energy transduction in polypeptides and proteins based on inverse temperature transitions. Prog. Biophys. Mol. Biol. 57(1): 23–57.

Urry, D.W. 1997. Physical chemistry of biological free energy transduction as demonstrated by elastic protein-based polymers. J. Phys. Chem. B; p. 11007–11028.

Van Der Heide, D., B. Verbraeken, R. Hoogenboom, T. Dargaville and D. Hickey. 2017. Porous poly(2-oxazoline) scaffolds for developing 3D primary human tissue culture. Biomater Tissue Technol. 1(1): 1–5.

Vancoillie, G., D. Frank and R. Hoogenboom. 2014. Thermoresponsive poly(oligo ethylene glycol acrylates). Prog. Polym. Sci. 39(6): 1074–1095.

Vanparijs, N., L. Nuhn and B.G. De Geest. 2017. Transiently thermoresponsive polymers and their applications in biomedicine. Chem. Soc. Rev. 46(4): 1193–1239.

Varghese, V.M., V. Raj, K. Sreenivasan and T. Kumary. 2010. *In vitro* cytocompatibility evaluation of a thermoresponsive NIPAAm-MMA copolymeric surface using L929 cells. J. Mater. Sci: Mater. Med. 21: 1631–1639.

Vert, M. 2007. Polymeric biomaterials: Strategies of the past vs. strategies of the future. Prog. Polym. Sci. 32(8-9): 755–761.

Viegas, T.X., M.D. Bentley, J.M. Harris, Z. Fang, K. Yoon, B. Dizman et al. 2011. Polyoxazoline: chemistry, properties, and applications in drug delivery. Bioconjugate Chem. 22(5): 976–986.

Vihola, H., A. Laukkanen, L. Valtola, H. Tenhu and J. Hirvonen. 2005. Cytotoxicity of thermosensitive polymers poly(N-isopropylacrylamide), poly(N-vinylcaprolactam) and amphiphilically modified poly(N-vinylcaprolactam). Biomaterials 26(16): 3055–3064.

Wang, X., S. Li, Z. Wan, Z. Quan and Q. Tan. 2014. Investigation of thermo-sensitive amphiphilic micelles as drug carriers for chemotherapy in cholangiocarcinoma *in vitro* and *in vivo*. Int. J. Pharm. 463(1): 81–88.

Ward, M.A. and T.K. Georgiou. 2011. Thermoresponsive polymers for biomedical applications. Polymers 3(3): 1215–1242.

Washington, K.E., R.N. Kularatne, M.C. Biewer and M.C. Stefan. 2018. Combination loading of doxorubicin and resveratrol in polymeric micelles for increased loading efficiency and efficacy. ACS Biomater. Sci. Eng. 4(3): 997–1004.

Weber, C., R. Hoogenboom and U.S. Schubert. 2012. Temperature responsive bio-compatible polymers based on poly(ethylene oxide) and poly(2-oxazoline)s. Prog. Polym. Sci. 37(5): 686–714.

Wei, M., Y. Gao, X. Li and M.J. Serpe. 2017. Stimuli-responsive polymers and their applications. Polym. Chem. 8(1): 127–143.

Wong, S.Y., J.M. Pelet and D. Putnam. 2007. Polymer systems for gene delivery-past, present, and future. Prog. Polym. Sci. 32(8-9): 799–837.

Wu, Q., X. Tang, X. Liu, Y. Hou, H. Li, C. Yang et al. 2016. Thermo/pH dual responsive mixed-shell polymeric micelles based on the complementary multiple hydrogen bonds for drug delivery. Chem. Asian J. 11(1): 112–119.

Xia, X., S. Tang, X. Lu and Z. Hu. 2003. Formation and volume phase transition of hydroxypropyl cellulose microgels in salt solution. Macromolecules 36(10): 3695–3698.

Xu, X., N. Bizmark, K.S. Christie, S.S. Datta, Z.J. Ren and R.D. Priestley. 2022. Thermoresponsive polymers for water treatment and collection. Macromolecules 55(6): 1894–1909.

Yamada, N., T. Okano, H. Sakai, F. Karikusa, Y. Sawasaki and Y. Sakurai. 1990. Thermo-responsive polymeric surfaces; control of attachment and detachment of cultured cells. Makromol. Chem. Rapid Commun. 11(11): 571–576.

Yamato, M., O.H. Kwon, M. Hirose, A. Kikuchi and T. Okano. 2001. Novel patterned cell coculture utilizing thermally responsive grafted polymer surfaces. J. Biomed. Mater. Res. 55(1): 137–140.

Yan, J., W. Ji, E. Chen, Z. Li and D. Liang. 2008. Association and aggregation behavior of poly(ethylene oxide)-b-poly(N-isopropylacrylamide) in aqueous solution. Macromolecules 41(13): 4908–4913.

Yan, N., J. Zhang, Y. Yuan, G.-T. Chen, P.J. Dyson, Z.-C. Li et al. 2010. Thermoresponsive polymers based on poly-vinylpyrrolidone: applications in nanoparticle catalysis. Chem. Commun. 46(10): 1631–1633.

Yang, J., M. Yamato, K. Nishida, T. Ohki, M. Kanzaki, H. Sekine et al. 2006. Cell delivery in regenerative medicine: the cell sheet engineering approach. J. Control Release. 116(2): 193–203.

Yang, J., P. Zhang, L. Tang, P. Sun, W. Liu, P. Sun et al. 2010. Temperature-tuned DNA condensation and gene transfection by PEI-g-(PMEO2MA-b-PHEMA) copolymer-based nonviral vectors. Biomaterials 31(1): 144–155.

Yang, L., X. Fan, J. Zhang and J. Ju. 2020. Preparation and characterization of thermoresponsive poly(N-isopropylacrylamide) for cell culture applications. Polymers 12(2): 389.

Yokoyama, M., M. Kurisawa and T. Okano. 2001. Influential factors on temperature-controlled gene expression using thermoresponsive polymeric gene carriers. J. Artif Organs. 4: 138–145.

Yokoyama, M. 2002. Gene delivery using temperature-responsive polymeric carriers. Drug Discov Today. 7(7): 426–432.

You, Y.-Z., D.S. Manickam, Q.-H. Zhou and D. Oupicky. 2007. Reducible poly(2-dimethylaminoethyl methacrylate): synthesis, cytotoxicity, and gene delivery activity. J. Control Release 122(3): 217–225.

Zhang, Q. and R. Hoogenboom. 2015. Polymers with upper critical solution temperature behavior in alcohol/water solvent mixtures. Prog. Polym. Sci. 48: 122–142.

Zhang, Q., C. Weber, U.S. Schubert and R. Hoogenboom. 2017. Thermoresponsive polymers with lower critical solution temperature: from fundamental aspects and measuring techniques to recommended turbidimetry conditions. Mater Horiz. 4(2): 109–116.

Zhang, Q., M.J. Serpe and S.M. Mugo. 2017. Stimuli responsive polymer-based 3D optical crystals for sensing. Polymers 9(11): 436.

Zhang, Q., Y. Zhang, Y. Wan, W. Carvalho, L. Hu and M.J. Serpe. 2021. Stimuli-responsive polymers for sensing and reacting to environmental conditions. Prog. Polym. Sci. 116: 101386.

Zhao, C., L. Dolmans and X. Zhu. 2019. Thermoresponsive behavior of poly(acrylic acid-co-acrylonitrile) with a UCST. Macromolecules 52(12): 4441–4446.

Zhao, J., R. Hoogenboom, G. Van Assche and B. Van Mele. 2010. Demixing and remixing kinetics of poly(2-isopropyl-2-oxazoline) (PIPOZ) aqueous solutions studied by modulated temperature differential scanning calorimetry. Macromolecules 43(16): 6853–6860.

Chapter 2

Engineering Smart Nanomaterials

Afroz Karim, Jyotish Kumar, Ummy Habiba Sweety,
Sofia A. Delgado and *Mahesh Narayan**

Introduction

The etymology of the prefix 'nano,' originating from the Greek term signifying 'dwarf' or minute, underscores its representation of one billionth of a meter (10^9 m). One can make distinction between nanoscience and nanotechnology, the former being concerned with scrutinizing structures and molecules within the nanometer range of 1 to 100 nm, while the latter pertains to the application of this knowledge in practical contexts such as devices (Mansoori and Soelaiman 2005). The epoch before the advent of nanomaterials was characterized by a paradigm reliant on conventional materials, governed by their macroscopic properties. However, the pursuit of more precise control, tailored functionality, and innovative breakthroughs steered the scientific community toward a realm of enhanced possibilities. This transformative transition was galvanized by the discovery of nanomaterials, marking a watershed moment in the early 1980s. These materials, defined by their nanoscale dimensions, have novel quantum effects, unraveling unforeseen prospects. The 1985 revelation of fullerenes and the subsequent unveiling of carbon nanotubes in 1991 crystallized the emergence of nanomaterials (Aqel et al. 2012, Hirsch 2010). This catalytic period was augmented by technological breakthroughs like atomic force microscopy and transmission electron microscopy, equipping scientists with the tools to manipulate materials at the nanoscale (Binnig 1990).

The era after the discovery of nanomaterials has been one of profound transformation across scientific domains, none more so than in biomedicine.

Department of Chemistry and Biochemistry, The University of Texas at El Paso (UTEP), El Paso, Texas 79968, United States.
* Corresponding author: mnarayan@utep.edu

Nanomaterials harnessed their unique physicochemical properties derived from their diminutive size and distinctive surface characteristics, addressing challenges that had long eluded conventional materials. In the biomedical realm, nanomaterials orchestrated targeted drug delivery, augmented imaging contrast agents, and enabled groundbreaking diagnostic modalities. This paradigm shift heralded enhanced therapeutic efficacy, diminished side effects, and the potential to traverse physiological barriers, revolutionizing medical interventions (Roco 2011).

In recent years, the field of nanotechnology has witnessed remarkable advancements, paving the way for the development of novel materials with unprecedented functionalities. Among these, smart nanomaterials have garnered significant attention due to their unique ability to respond to specific external stimuli, enabling precise control over their properties and behavior. These materials hold immense promise for revolutionizing various scientific and technological domains, particularly in biomedicine (Thangudu 2020). As far as our understanding extends, the pioneering articulation of "intelligent materials" occurred when Toshinori Takagi presented the concept in April 1990. He defined these materials as substances capable of adjusting to environmental shifts at their most optimal state, thus unveiling their inherent functions in response to these alterations (Takagi 1990). At that time, understanding the complete scope and practicality of this idea was difficult. But there was a feeling of excitement in the air because it pointed to new areas waiting to be discovered in the fields of science and innovation. This important statement had the power to inspire a fresh start, bringing in new ideas that might change the way we look at and use materials.

In this chapter, we delve into the comprehensive landscape of smart nanomaterials, encompassing their diverse types, versatile applications in gene and drug delivery, and the synthesis of biodegradable polymers for enhanced biocompatibility as depicted in Fig. 1.

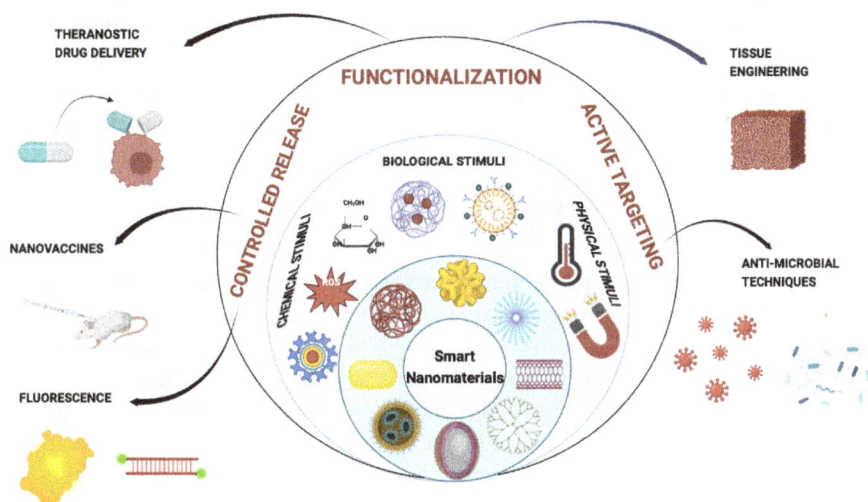

Fig. 1. Smart nanomaterials and their diverse biomedical applications (Created with BioRender.com).

Smart nanomaterials

Many smart materials of macroscopic dimensions are also known and share the smart behavior described below for smart nanomaterials. Smart nanomaterials are a class of materials that can respond to various external stimuli by altering their properties in a controlled and specific manner. These stimuli can include factors like temperature, pH, light, magnetic fields, and even the presence of specific molecules. The unique feature of smart nanomaterials lies in their ability to adapt and change in response to these triggers, making them highly versatile and valuable in a wide range of applications (Yoshida and Lahann 2008). The mechanism of action for these smart nanomaterials relies on the specific properties of the material and the stimulus. For instance, temperature-responsive nanogels change their swelling behavior due to temperature-induced changes in polymer chain interactions. pH-sensitive liposomes alter their membrane structure in response to pH changes, leading to the release of encapsulated drugs. Light-responsive nanoparticles can heat up upon light absorption, leading to local temperature changes that can have therapeutic effects. Magnetic nanoparticles are guided by external magnetic fields to specific locations, where they can exert their therapeutic or imaging functions. This adaptability of these smart nanomaterials allows them to be tailored for specific applications in fields such as medicine, imaging, and diagnostics (Aflori 2021). Herein, we elucidate several classifications of these smart nanomaterials along with their distinctive attributes.

Types of nanomaterials

Physically responsive nanomaterial

Physically responsive nanomaterials are a class of nanoscale materials that exhibit reversible changes in their physical properties in response to external stimuli such as temperature, light, pressure, magnetic fields, or electric fields. These materials can undergo alterations in their structural, optical, electrical, or mechanical characteristics, making them highly versatile for a wide range of applications in fields such as medicine, electronics, sensing, energy, and more. The ability to manipulate and control these properties on the nanoscale has led to the development of innovative and adaptive materials with enhanced functionalities (Blum et al. 2015). Examples of physico-sensitive nanomaterials and their applications are elaborated in the subsequent sections.

Temperature - sensitive nanomaterial

Temperature-responsive nanomaterials are a class of materials that exhibit reversible changes in their properties in response to variations in temperature. These materials are designed to undergo distinct transformations, such as changes in solubility, shape, volume, or aggregation state, within a specific temperature range. Thermo-sensitive nanomaterials have garnered significant attention due to their potential applications in various fields, particularly in biomedicine, drug delivery, and responsive coatings (Sahle et al. 2018). Thermosensitive polymers, a subset of temperature-responsive materials, have gained prominence for their ability to undergo reversible changes in their physical properties in response to temperature variations (Arafa et al. 2018).

The responsiveness of thermosensitive polymers stems from the balance between hydrophilic and hydrophobic interactions within the polymer structure. Commonly, these polymers contain hydrophobic segments that aggregate at lower temperatures and hydrophilic segments that promote solubility at higher temperatures. This interplay results in distinct phase transitions, typically characterized by a lower critical solution temperature (LCST) or an upper critical solution temperature (UCST), depending on the nature of the polymer (Roy et al. 2013). The LCST behavior is rooted in the balance between hydrophilic and hydrophobic interactions within the polymer structure. In aqueous solutions, below the LCST, polymer chains exhibit favorable interactions with water molecules due to their hydrophilic moieties. As the temperature rises beyond the LCST, the entropic gain from the release of water molecules surpasses the enthalpic contributions of hydrophilic interactions, causing the polymer chains to aggregate. This aggregation leads to phase separation and the formation of a coacervate or a polymer-rich phase that separates from the surrounding solvent (Zhang et al. 2017). LCST polymers can be categorized into two main types based on their phase transition behavior: some of these polymers exhibit a single LCST, where the phase transition occurs at a specific temperature for a given solvent system. Poly(N-isopropylacrylamide) (PNIPAM) is a classic example of a simple LCST polymer. Below its LCST, PNIPAM chains are solvated by water, while above the LCST, they collapse and aggregate, leading to phase separation (Chen et al. 2018). In gradient LCST polymers, the phase transition behavior can be tuned by varying the polymer composition or molecular weight. These polymers display a gradual transition range rather than a sharp LCST. This property allows for fine-tuning of the polymer's responsiveness to temperature changes. Copolymers like poly (N-isopropylacrylamide-co-acrylic acid) exhibit gradient LCST behavior due to the presence of both hydrophobic and hydrophilic groups (Shin et al. 2003).

Tamaki and Kojima (2020) investigated dendrimers with LCST and UCST behaviors, which means that they undergo phase transitions and solubility changes at specific temperature ranges. These dendrimers are modified with phenylalanine groups, which introduce pH responsiveness alongside their thermosensitivity. The researchers conducted experiments to understand how the dendrimers' LCST and UCST behaviors were influenced by pH and temperature variations. They employ various techniques to characterize the dendrimers' structural changes, solubility, and aggregation patterns under different conditions. The results demonstrate that the dendrimers exhibited a transition from LCST to UCST behavior with changes in pH, which is a unique and interesting phenomenon (Tamaki and Kojima 2020).

Chen et al. introduced a hydrogel composed of three components: poly(N-isopropylacrylamide) (PNIPAM), poly(2-(dimethylamino)ethyl methacrylate) (PDMA), and poly (acrylic acid) (PAA). This composition imparts both thermo-responsive and pH-responsive properties to the hydrogel. The hydrogel exhibits responsiveness to changes in both temperature and pH. This allows it to undergo reversible sol-gel transitions, changing its structure from a liquid-like state to a gel-like state and vice versa under specific conditions. The hydrogel demonstrates the ability to form diverse nanostructures, likely including micelles and vesicles, depending on environmental factors such as temperature and pH. These nanostructures can serve as potential drug carriers (Chen et al. 2018).

Electrical and electrochemical stimuli-responsive nanomaterials

In the realm of nanotechnology, a compelling avenue of research has emerged through the exploration of electrical and electrochemical stimuli-responsive nanomaterials. This subfield represents a convergence of nanoscience and stimuli-triggered behavior, offering remarkable possibilities for controlled and dynamic material responses. These stimuli-responsive materials exhibit changes in their properties, such as electrical conductivity, charge distribution, and structural arrangements, driven by specific electrical potentials or electrochemical reactions. They are designed by integrating specific responsive components into their structure. These components can include conductive polymers, redox-active molecules, or electroactive groups. The selection of materials depends on the desired response and application. Stimuli-responsive nanomaterials can be engineered to encapsulate therapeutic agents and release them in a controlled manner upon exposure to specific electrical or electrochemical signals. This capability enables precise targeting of diseased cells or tissues, minimizing off-target effects and enhancing therapeutic efficacy. These materials could potentially revolutionize chemotherapy, reducing systemic toxicity while maximizing drug delivery to cancerous cells (Abidian et al. 2006).

Electrical and electrochemical stimuli-responsive nanomaterials encompass a diverse spectrum of materials characterized by their dynamic responsiveness to specific electrical or electrochemical signals. These materials can be classified into several distinct categories based on their structural composition, mechanisms of response, and functional applications. One prominent class comprises conductive polymers, such as polypyrrole, polyaniline, and polythiophene. These polymers exhibit reversible redox reactions upon exposure to electrical stimuli, resulting in changes in their oxidation state and conductivity. Due to their unique properties, they find utility in applications spanning biosensing, drug delivery, and neural stimulation (Svirskis et al. 2010).

Redox-active nanoparticles form another significant category. These nanoparticles incorporate species like metal ions (e.g., gold, silver) or metal oxide nanoparticles (e.g., Fe_3O_4) that undergo reversible changes in oxidation state or charge distribution in response to electrochemical stimuli. These alterations can lead to shifts in optical, electrical, or magnetic properties, making these nanoparticles valuable in diagnostics, imaging, and targeted therapeutic strategies (Feng et al. 2014).

Electroactive hydrogels represent a distinct class, characterized by their capacity to change volume, shape, or mechanical attributes upon application of electrical potentials. These hydrogels play crucial roles in controlled drug release, the fabrication of tissue engineering scaffolds, and the development of artificial muscles, showcasing their versatility in various biomedical applications. Electroactive polyelectrolytes constitute a class of materials combining polyelectrolyte properties with electroactivity. Responsive to changes in ionic strength, pH, and electrical potentials, these polymers are employed in domains such as drug delivery, sensors, and actuators, capitalizing on their ability to adapt to varying environmental conditions (Benselfelt et al. 2023).

Conductive nanocomposites represent a diverse group, amalgamating conductive elements like carbon nanotubes, graphene, or conductive polymers with other materials to enhance electrical and electrochemical responsiveness. These composite materials find utility in diverse applications, ranging from biosensors to energy storage solutions (Zhao et al. 2012).

Another category encompasses electroactive nanofibers, which can be functionalized with electroactive constituents to create materials that respond to electrical or electrochemical cues. Their adaptable characteristics make them suitable for applications like wound healing, drug delivery, and tissue engineering, where tailored responsiveness is vital (Zhang et al. 2021).

Nanoporous electrodes, characterized by their well-defined pores, can undergo alterations in surface area and porosity when subjected to electrical potentials. Such materials are pertinent to applications such as supercapacitors and sensing platforms due to their unique electrochemical response (Zhao et al. 2014).

Bioelectrodes, often fashioned from conductive polymers or nanomaterials, serve as interfaces with biological systems and have the capability to modulate cellular behavior through electrical and electrochemical signals. Their integration within neuroprosthetics, biosensors, and regenerative medicine underscores their role in shaping innovative biomedical technologies (Mishra et al. 2014).

In essence, electrical and electrochemical stimuli-responsive nanomaterials encompass a diverse array of material classes, each imbued with specific properties and functionalities that can be harnessed across a wide spectrum of biomedical and technological applications.

Light-responsive nanomaterials

Light-sensitive nanomaterials, known as photoresponsive or photonic nanomaterials, constitute a captivating class of materials characterized by their ability to undergo dynamic changes in response to light stimuli. These materials have garnered substantial attention owing to their multifaceted applications across diverse fields. Notably, they have emerged as indispensable in offering the ability to pinpoint drug delivery sites, gauge targeting efficacy, and visualize tumors through their adaptability to an array of parameters (Zhang et al. 2022). These nanomaterials encompass various categories, including photochromic materials that exhibit reversible color changes, photothermal agents that convert light energy into heat for therapies, and photoluminescent nanoparticles pivotal in imaging and biosensing. In the realm of intelligent materials, they hold promise as modulators of structural and mechanical properties upon light exposure, thereby advancing areas like drug delivery, tissue engineering, and soft robotics. Particularly intriguing are the photoconductive nanomaterials capable of altering electrical conductivity under light illumination, with involvements spanning photodetectors to solar cells. Plasmonic nanoparticles, manifesting in materials like gold and silver, offer enhanced light-matter interactions with applications in imaging, sensing, and photothermal therapies. Additionally, nanophotonic structures, such as photonic crystals and metamaterials, offer unparalleled optical properties due to their intricate periodic arrangement, enabling light manipulation and advanced optical devices. A distinct attribute of these materials lies in their capacity to tailor responses by adjusting input parameters such as intensity, wavelength, beam diameter, and exposure time. This intrinsic versatility positions light-sensitive smart materials at the forefront of pioneering technologies, harnessing light-matter interactions to innovate diverse applications and redefine possibilities across scientific and technological landscapes (Bertrand and Gohy 2017). In addressing the intricate challenges associated with precise drug release and the realm of light-mediated theranostics, the utilization of intelligent nanomaterials emerges as a promising alternative, with nanopolymers taking center stage. These nanopolymers offer a strategic avenue to exercise meticulous control over drug delivery to specific target sites, facilitating the attainment of precise drug concentrations within specified time frames. Among the diverse array of light-sensitive polymers, a notable subset incorporates chromophores such as azobenzene groups, spiropyran groups, and nitrobenzyl groups. These chromophores play a pivotal role in rendering these polymers light responsive. While numerous photomaterials have demonstrated remarkable efficacy in biomedical applications, a significant limitation lies in their activation confined to the UV/VIS spectral region. This constraint curtails tissue penetration and raises concerns about the potential harm inflicted on adjacent tissues. To surmount these challenges, the development of photoresponsive smart nanomaterials emerges as a transformative strategy. These materials exhibit a unique blend of attributes, including the ability to generate reactive oxygen species (such as singlet oxygen, peroxide, and hydroxyl radicals), robust targeting capabilities, water solubility, and crucially, activation by near-infrared light (NIR). This near-infrared activation circumvents limitations posed by traditional UV/VIS activation, facilitating deeper tissue penetration while minimizing collateral damage to non-target tissues. In essence, the evolution of photoresponsive smart nanomaterials presents a transformative approach to circumvent prevailing limitations, showcasing their potential to revolutionize therapeutic interventions by enabling controlled drug release and precise phototherapeutic actions in complex biological environments (Sun et al. 2019).

Sun et al. (2017) highlights the development of an innovative amphiphilic ruthenium polymetallodrugs that exhibits dual phototherapeutic effects, combining photodynamic therapy (PDT) and photochemotherapy (PCT). The study demonstrates the material's potential for effective tumor inhibition *in vivo*, showcasing its promise as a multifunctional nanomedicine for advanced cancer treatment strategies. The polymetallodrug possesses amphiphilic properties, enabling it to self-assemble into nanoparticles in aqueous environments, enhancing its delivery and accumulation in tumor tissues. The researchers demonstrate its effectiveness in producing reactive oxygen species upon light activation, causing cellular damage *in vitro*. In addition, the polymetallodrug is shown to generate cytotoxic species that enhance it phototherapeutic efficacy in a combined PDT and PCT approach (Sun et al. 2017).

Mena-Giraldo et al. (2020) explored the utilization of photosensitive nanocarriers as a strategy for targeted intracellular cargo delivery. The nanocarriers are engineered to respond to light stimuli, enabling controlled cargo release within cells. These nanocarriers are designed to encapsulate cargo and respond to specific light wavelengths, triggering cargo release through photoinduced mechanisms. This was achieved through the strategic modification of chitosan, a biocompatible polymer, with ultraviolet-photosensitive azobenzene molecules. Notably, their investigation delved into the profound influence of ultraviolet (UV) light exposure on the controlled release of cargo, guided meticulously by nanobioconjugates. The research ambitiously embraced a range of cargo models, namely Nile red and dofetilide, effectively establishing a comprehensive encapsulation/release paradigm. Of particular interest was their utilization of photoresponsive polymeric nanocarriers, ingeniously functionalized with a cardiac transmembrane peptide. Through judicious implementation of UV light irradiation, the therapeutic efficacy of their devised regime was notably augmented. This augmentation materialized as an elevation in the intracellular delivery concentration, concomitantly mitigating the extent of cargo reactions and overall cargo amount. The paper thus shines a spotlight on a meticulously designed strategy, harnessing photoresponsive nanocarriers to revolutionize controlled drug delivery, further underscoring the substantial impact of tailored light stimuli in amplifying therapeutic precision (Mena-Giraldo et al. 2020).

Magnetic-responsive nanomaterials

Magnetic-responsive nanomaterials or magnetic nanomaterials are a class of materials that can be manipulated or controlled using external magnetic fields. These materials typically consist of small particles, often on the nanoscale, that exhibit magnetic properties due to the presence of ferromagnetic, ferrimagnetic, or superparamagnetic materials. Magnetic responsiveness enables these nanomaterials to interact with and respond to magnetic fields, offering unique functionalities that can be harnessed for various applications in biomedicine, electronics, sensing, and more. Magnetic nanoparticles are usually composed of materials like iron, cobalt, or nickel and can range in size from a few nanometers to hundreds of nanometers (Thevenot et al. 2013). Their small size imparts them with superparamagnetic behavior, meaning they do not retain permanent magnetization but can be magnetized in the presence of an external magnetic field. In biomedicine, magnetic-responsive nanomaterials have gained significant attention for their potential in targeted drug delivery, hyperthermia-based cancer treatment, magnetic resonance imaging (MRI) contrast agents, and as carriers for biomolecules like proteins and genes. Their ability to be guided by external magnetic fields allows for site-specific drug delivery and enhanced imaging contrast, making them valuable tools in diagnosing and treating various diseases (Ghosh et al. 2019).

The study conducted by Zhang et al. presents a novel nanogel system for targeted breast cancer treatment and magnetic resonance imaging (MRI) by integrating doxorubicin (DOX) and iron oxide nanoparticles (Fe_3O_4) within a polymeric matrix. This system utilizes the specific binding ability of trastuzumab (Herceptin), a monoclonal antibody targeting human epidermal growth factor receptor 2 (HER2), which is often overexpressed in breast cancer cells. The nanogel is composed of a copolymer poly (N-isopropylacrylamide-co-acrylic acid-co-methacryloyl poly (ethylene glycol)) (P(NIPAM-AA-MAPEG)) that undergoes volume phase transition in response to temperature changes. In this approach, the DOX-loaded Fe_3O_4 nanoparticles are encapsulated within the P(NIPAM-AA-MAPEG) nanogel. The conjugation of Herceptin to the surface of the nanogel allows for targeted delivery to HER2-positive breast cancer cells. The system demonstrates several key features: enhanced cellular internalization due to HER2-targeting, pH-responsive DOX release within cancer cells, and the potential for hyperthermia-based cancer treatment due to the magnetic properties of Fe_3O_4 nanoparticles. Additionally, the nanogel system serves as a contrast agent for MRI due to the presence of Fe_3O_4, enabling non-invasive imaging of tumor sites. The combination of targeted drug delivery and diagnostic imaging in a single nanogel system provides a multifunctional platform for breast cancer therapy. The study showcases the potential for synergistic effects by integrating different therapeutic modalities within a single nanomaterial (Zhang et al. 2022).

Nanomaterials responsive to chemicals

Nanomaterials responsive to chemicals are a specialized category of nanomaterials that exhibit changes in their properties or behavior in response to specific chemical cues or changes in their surrounding chemical environment. These nanomaterials are designed to interact with certain molecules, ions, or chemical conditions, triggering a measurable and often reversible response at the nanoscale. Chemical responsive nanomaterials can manifest various types of responses, such as changes in color, fluorescence, conductivity, or morphology, among others. These responses enable their application in fields such as sensing, diagnostics, drug delivery, and environmental monitoring (Shah et al. 2020).

pH-responsive nanomaterials

pH-responsive nanomaterials are a class of nanoscale materials that can change their properties or behavior in response to variations in the pH level of their surrounding environment. These nanomaterials are designed to undergo specific changes, such as swelling, degradation, or release of encapsulated molecules, when exposed to different pH conditions. pH-responsive nanomaterials are widely used in drug delivery systems. Many diseased tissues, such as tumors, have an acidic microenvironment. pH-responsive nanocarriers can remain stable at neutral pH (found in healthy tissues) and release their payload selectively in the acidic conditions of disease sites. This ensures that drugs are delivered precisely where they are needed, minimizing side effects on healthy tissues (Verma et al. 2016). pH-responsive polymers can be categorized into distinct groups, each exhibiting specific behaviors and responses to pH changes. One notable category encompasses pH-dependent swelling polymers, which display alterations in volume (swelling) in response to shifts in pH levels. Poly (acrylic acid) (PAA), poly (methacrylic acid) (PMAA), and poly (itaconic acid) (PIA) are illustrative examples that exemplify this group's characteristics (Aguilar and San Román 2019). Another significant class involves pH-responsive hydrogels, three-dimensional networks of polymers with the ability to absorb and retain water. These hydrogels undergo swelling transformations in response to variations in pH, with applications spanning drug delivery, tissue engineering, and wound healing (Rao et al. 2021). pH-responsive micelles constitute another compelling classification, formed through self-assembly of amphiphilic polymers in solvents. These micelles demonstrate the potential to adjust structure, stability, or drug release characteristics as pH changes occur, finding utility in targeted therapies and drug delivery strategies (Wu et al. 2010). A crucial category comprises pH-sensitive polyelectrolytes, distinguished by the presence of ionizable groups within their molecular structure. This unique characteristic allows these polymers to undergo changes in their charge and solubility in direct response to fluctuations in pH levels. By adapting their charge and solubility profiles, these polymers can facilitate the controlled release of therapeutic agents, enhance cellular uptake of genetic material, and modulate interactions within complex biological systems (Sui and Schlenoff 2004). Volodkin et al. (2003) present a model system for controlled protein release using pH-sensitive polyelectrolyte microparticles. The authors focus on designing microparticles that can respond to changes in pH by releasing encapsulated proteins. The pH-sensitive behavior is achieved by incorporating ionizable groups, enabling the microparticles to alter their charge and solubility in response to pH variations. The study explores the fabrication process of these microparticles and investigates their release behavior under different pH conditions. The researchers employ a technique known as layer-by-layer (LBL) assembly to create the microparticles. This technique involves the sequential deposition of polyelectrolyte layers onto a template material. The template can be solid particles or droplets suspended in a liquid. The layering process is facilitated by the electrostatic interactions between the charged polyelectrolytes, leading to the formation of multilayered coatings on the template. Once the multilayer coating is established, the template is typically removed, leaving behind hollow microcapsules with a shell composed of alternating layers of polyelectrolytes. These pH-sensitive polyelectrolyte shells are what give the microparticles their responsive behavior to changes in pH (Volodkin et al. 2003).

Additionally, another noteworthy classification in the realm of pH-responsive nanomaterials is pH-responsive membranes. These specialized materials hold the remarkable capability to adjust their permeability or selectivity in direct correlation with shifts in pH. This inherent responsiveness allows pH-sensitive membranes to act as dynamic barriers or filters, permitting the passage of specific molecules or ions under defined pH conditions (Purkait et al. 2018). The study by Nguyen et al. (2015) presents a novel soluble peptide characterized by its pH-responsive membrane insertion behavior. The peptide is prepared through solid-phase peptide synthesis, a widely used method for synthesizing peptides with precise sequences. The conjugation process involves attaching the fluorescence label to the peptide using appropriate chemical reactions. This modified peptide is then employed in experiments to probe its pH-responsive membrane insertion behavior. The peptide's ability to insert itself into lipid membranes in response to changes in pH is investigated through a series of biophysical and spectroscopic techniques (Nguyen et al. 2015).

Redox – responsive nanomaterials

Redox-responsive nanomaterials are a class of materials that exhibit changes in their properties or behaviors in response to changes in the redox (reduction-oxidation) state of their environment. Redox reactions involve the transfer of electrons between different chemical species, leading to changes in the oxidation states of the involved components. Redox-responsive nanomaterials are designed to exploit these electron transfer processes for specific applications, often in the realm of controlled release, drug delivery, sensing, and catalysis. These materials are engineered to respond to variations in redox conditions by undergoing reversible changes in structure, charge, or other properties. One common approach to achieving redox responsiveness is through the incorporation of redox-active functional groups or moieties, such as disulfide bonds, quinones, or transition metal complexes, into the material's structure. Under specific redox conditions, these groups can undergo changes that trigger structural modifications, disassembly, or release of encapsulated cargo (Hsu and Almutairi 2021).

Polymers containing disulfide bonds (S-S) stand out as a prime example of redox-sensitive linkages. These polymers can undergo cleavage and reformation in direct response to changes in the redox environment. Oxidative settings foster the creation of disulfide bonds, thereby stabilizing the polymer structure, whereas reducing conditions prompt the cleavage of disulfide bonds, leading to polymer degradation. This characteristic makes these polymers notably effective in drug delivery systems, where they can release therapeutic payloads in response to the reducing milieu found within cellular environments (Zhang et al. 2018). The study by Wen et al. presents a novel supramolecular gene carrier system that demonstrates high efficiency and multifunctionality. This carrier system is constructed through self-assembly of redox-sensitive and zwitterionic polymer blocks. The redox-sensitive polymer blocks are designed to respond to changes in the cellular redox environment, facilitating controlled release of the encapsulated genetic material. The zwitterionic polymer blocks contribute to the stability and biocompatibility of the carrier, minimizing nonspecific interactions with biological components. The resulting supramolecular gene carrier exhibits efficient gene delivery, protection of genetic cargo, and controlled release. The study showcases the potential of combining redox-responsive and zwitterionic components in designing effective and versatile gene carrier systems for therapeutic purposes (Wen et al. 2014). Another noteworthy category involves polymer-coated metal nanoparticles, wherein modifications in redox conditions can cause the polymer to bind to or detach from the nanoparticle surface, thereby influencing the properties of the resultant nanocomposite. This adaptability has far-reaching applications in catalysis, sensing, and drug delivery (Ejderyan et al. 2022). Qiu et al. introduced a hybrid nanoparticle system composed of redox-responsive polymer prodrugs and silver nanoparticles (AgNPs) for drug delivery applications. This system is built upon the foundation of redox-responsive polymers, specifically P[(2-((2-((camptothecin)-oxy)ethyl)disulfanyl)ethylmethacrylate)-co-(2-(D-galactose) methylmethacrylate)] (P(MACPTS-co-MAGP)), along with the incorporation of silver nanoparticles (AgNPs). The primary objective is to effectively deliver the anti-cancer drug camptothecin (CPT) while simultaneously enabling monitoring of the drug release process. The key innovation lies in the recovery of CPT's fluorescence, which serves as an indicator for drug release. This sophisticated drug carrier system holds significant potential in enhancing cancer treatment strategies by facilitating controlled drug release and offering a mechanism for real-time monitoring of the drug delivery process (Qiu et al. 2018).

Nanomaterials responsive to biological stimulus

These nanomaterials are a class of advanced materials designed to interact with and adapt to specific biological signals within a biological environment. These signals can encompass a range of physiological factors such as pH, temperature, enzymes, and specific biomolecules present in biological systems. The primary objective of these materials is to harness and respond to these signals in a controlled and tailored manner, allowing for dynamic and responsive behavior in the presence of biological stimuli. The design and engineering of these materials often involve the incorporation of specific molecular or chemical components that can recognize and interact with the biological cues. This interaction triggers a cascade of changes, leading to alterations in the material's structure, surface properties, solubility, or other characteristics. As a result, biological responsive nanomaterials offer a promising avenue to enhance the precision, efficiency, and effectiveness. The hallmark of biological responsive nanomaterials lies in their ability to translate these biological signals into measurable changes in their properties, functions, or behaviors. This adaptability enables the materials to perform various tasks such as targeted drug delivery, diagnostic imaging, controlled release of therapeutic agents, and tissue engineering, while minimizing adverse effects on healthy tissues. As a result, biological responsive nanomaterials offer a promising avenue to enhance the precision, efficiency, and effectiveness of biomedical applications by harnessing the dynamic nature of biological environments (Lu and Liu 2007). For example, Kang et al. (2008) utilized polymeric carriers for targeted gene delivery in cancer cells by exploiting abnormal protein kinase Cα (PKCα) activation. The aim was to create a selective gene delivery system for cancer cells with elevated PKCα activity. Their approach involved designing polymeric carriers equipped with a PKCα-responsive peptide linker, which would cleave upon active PKCα presence, enabling controlled therapeutic gene release within cancer cells. The study successfully synthesized and characterized these carriers, which demonstrated efficient gene delivery to cancer cells with high PKCα activity, while sparing normal cells with lower PKCα levels. This exemplifies the use of biological responsive nanomaterials, particularly polymeric carriers tailored to respond to intracellular signals (PKCα activation), for precise gene delivery in potential cancer therapy (Kang et al. 2008).

Enzyme-responsive nanomaterial

Enzyme-responsive nanomaterials are a specialized class of nanomaterials designed to interact with and respond to specific enzymes present in biological systems. Enzymes are highly specific biomolecules that catalyze biochemical reactions within cells and tissues.

Enzyme-responsive nanomaterials possess distinct advantages that set them apart from other types of stimuli-responsive nanomaterials. Enzymes are highly specific, and their biological relevance makes them well-suited for biomedical applications. They can seamlessly integrate into physiological processes, minimizing the potential for side effects or immune responses. Unlike some other nanomaterials, which might be recognized as foreign entities by the immune system or disrupt normal cellular functions, enzyme-responsive nanomaterials can function harmoniously within the intricate biological milieu. This seamless integration is owed to the fact that enzymes, the central actors in biochemical reactions, naturally occur within living systems.

Enzymes, as biological catalysts, possess remarkable capabilities that make them particularly attractive for triggering specific reactions. One of the key benefits is the ability of enzymes to catalyze chemical reactions under conditions that are generally gentle and compatible with biological systems. These conditions often involve mild temperatures, neutral pH values, and buffered aqueous solutions. This stands in contrast to many conventional chemical reactions, which might require harsher conditions that could potentially disrupt the delicate balance of biological environments. Moreover, enzymes demonstrate an extraordinary level of substrate selectivity, meaning they have a remarkable capacity to interact specifically with their target molecules. This selectivity is a result of the intricate interactions between enzymes and their substrates. Such high selectivity translates into the potential for executing precise and sophisticated chemical transformations that mimic the intricate processes found in biological systems. This biologically inspired specificity offers a valuable tool for designing reactions that are tailored to yield desired products while minimizing unwanted byproducts (Ulijn 2006).

Enzymes are often dysregulated in various diseases, including cancers and inflammatory disorders. Enzyme-responsive nanomaterials can capitalize on these disease-related enzyme alterations to enable targeted drug delivery. Enzyme activity can be monitored in real time, providing valuable information about disease progression, treatment efficacy, or other physiological changes (Hu et al. 2014).

The work by Datta et al. (2017) presents the development of enzyme-responsive nucleotide functionalized silver nanoparticles (AgNPs) with notable antimicrobial and anticancer properties. The study focuses on designing nanomaterials that exhibit a response to specific enzymes and assessing their effectiveness against microbial infections and cancer cells. In the paper, the authors describe a new and innovative technique they have developed for the metallization of nucleotides. This method involves using silver nanoparticles and adenosine triphosphate (ATP), a molecule found in cells that serves as a source of energy. By interacting silver nanoparticles with ATP, they successfully coat the nanoparticles with this nucleotide. The resulting nanoparticles are highly stable, showcasing remarkable durability in a typical physiological environment. What makes these nanoparticles even more intriguing is their responsiveness to enzymes. When exposed to specific enzymes, these adenosine triphosphate-coated silver nanoparticles exhibit a unique behavior. This responsiveness has practical implications, as it enables controlled reactions within biological systems (Datta et al. 2017).

Cai et al. (2018) presents the development of a multifunctional polymeric conjugate with enzyme-sensitive and biodegradable properties for theranostic applications in nanomedicine. The researchers designed an innovative amphiphilic triblock copolymer, pHPMA-Gd-PTX-Cy5.5, composed of N-(2-hydroxypropyl methyl) acrylamide (HPMA), gadolinium, paclitaxel (PTX), and Cyanine5.5

(Cy5.5). The copolymer was synthesized through a two-step RAFT polymerization process and formed self-assembled nanoparticles of approximately 85 nm in diameter. An enzyme-sensitive tetrapeptide linker (GFLG) was incorporated into the polymer's backbone to enhance its responsiveness to enzymatic degradation. This allowed the high molecular weight conjugate (92 kDa) to be broken down into smaller fragments (44 kDa) upon exposure to specific enzymes, facilitating controlled drug release. The conjugate exhibited both drug delivery capabilities through the incorporation of PTX and imaging functionality through Cy5.5. The researchers demonstrated that the multifunctional conjugate possessed promising potential as a theranostic nanomedicine, enabling simultaneous cancer therapy and imaging within a single nanoscale system. The development of this enzyme-sensitive and biodegradable polymeric conjugate represents a significant advancement in the field of nanomedicine, offering a versatile platform for improved cancer treatment strategies (Cai et al. 2018).

Glucose responsive nanomaterials

Glucose-responsive nanomaterials are a class of innovative materials that exhibit a direct response to changes in glucose concentration. These materials have garnered significant interest due to their potential applications in diabetes management and drug delivery. Glucose-responsive nanomaterials are designed to interact with glucose molecules and undergo specific changes in their properties, such as swelling, contraction, or release of cargo, in response to varying glucose levels (Mishra et al. 2008). Polymers have garnered extensive attention, particularly in the context of glucose detection and insulin delivery fields. This interest arises from their unique properties that make them suitable for addressing the challenges associated with diabetes management. In the realm of glucose detection, polymers can be engineered to create sensitive and selective glucose sensors. These sensors utilize the interactions between glucose molecules and the polymer matrix to produce detectable signals, such as changes in color, conductivity, or fluorescence. This ability to transduce glucose concentration into measurable signals is vital for accurate and timely glucose monitoring. Additionally, polymers play a crucial role in insulin delivery strategies. In diabetes treatment, maintaining optimal insulin levels is essential for controlling blood glucose. Polymers can be formulated into various delivery systems, such as hydrogels or nanoparticles, that encapsulate and release insulin in response to glucose levels. These systems can take advantage of the glucose-responsive behavior of specific polymers. For instance, certain polymers exhibit swelling or changes in permeability in the presence of glucose, enabling controlled and on-demand insulin release. This approach offers the potential to mimic the body's natural insulin release mechanism more closely, leading to improved glucose regulation and reduced risk of hypoglycemia (Kim and Park 2001, Volpatti et al. 2019).

The study by Hadiya et al. (2021) focuses on developing a novel approach for insulin delivery using nanoparticles that combine natural and synthetic polymers. The researchers aimed to enhance the efficiency and effectiveness of insulin therapy for diabetes management. In this work, a hybrid system was designed by incorporating chitosan, a natural polymer, and poly (lactic-co-glycolic acid) (PLGA), a synthetic polymer, to create nanoparticles capable of encapsulating and delivering insulin. The researchers employed a two-step method involving ionotropic gelation and solvent evaporation techniques to fabricate the nanoparticles. Chitosan, known for its biocompatibility and mucoadhesive properties, was used to form the core of the nanoparticles and encapsulate insulin. PLGA, on the other hand, served as the outer shell, providing stability and control over insulin release. This hybrid system aimed to capitalize on the advantages of both natural and synthetic polymers to achieve enhanced insulin delivery (Hadiya et al. 2021).

The study by Tang and Chen (2020) focuses on the development of hydrogel-based colloidal photonic crystal devices for glucose sensing applications. The researchers aim to leverage the unique optical properties of colloidal photonic crystals embedded in hydrogels to create a sensitive and selective glucose sensing platform. The study involves the fabrication of hydrogel-based colloidal photonic crystal devices using a combination of templating and polymerization techniques. The colloidal crystal structure within the hydrogel matrix responds to changes in the surrounding glucose concentration by inducing shifts in its optical properties, such as color changes. This phenomenon, known as the "stop-band shift," serves as the basis for glucose detection. The researchers systematically investigated the performance of the developed sensing platform by assessing the changes in the photonic crystal's optical properties in response to varying glucose concentrations. They experimentally validated the sensitivity and specificity of the device for glucose sensing, demonstrating its potential as a reliable tool for monitoring glucose levels (Tang and Chen 2020).

Dual and multi-responsive nanomaterials

A dual or multi-responsive nanomaterial refers to a type of nanoscale material that can respond to multiple external stimuli or triggers in a controlled manner. These nanomaterials possess the ability to exhibit distinct changes in their properties or behavior when exposed to two or more different cues, such as changes in temperature, pH, light, magnetic fields, or specific chemical signals. The concept of dual or multi-responsiveness in nanomaterials adds an extra layer of complexity and versatility to their functionalities. By incorporating multiple responsive elements, these nanomaterials can adapt to different environmental conditions or signals, allowing them to perform intricate tasks or functions that respond to a combination of factors.

The development of multi-stimuli-sensitive polymeric nanoparticles addresses several challenges in conventional drug delivery. Their ability to respond to multiple cues enables them to navigate complex biological environments, such as the acidic and enzymatic conditions found in tumor tissues. This results in improved site-specific accumulation and controlled release of therapeutic payloads. Additionally, the versatility of these nanoparticles allows for tailored responses to specific conditions, enabling on-demand drug release and minimizing off-target effects. In these systems, the polymer-based nanoparticles serve as carriers for encapsulating drugs, genes, or imaging agents. Through careful design and modification of the polymer structure, the nanoparticles can be engineered to respond to distinct stimuli in a synchronized manner. For instance, pH-sensitive polymers may trigger drug release in the acidic tumor microenvironment, while temperature-sensitive polymers may respond to localized heating methods (Cheng et al. 2013).

Chen and Du (2013) present a novel approach to drug delivery using polymer vesicles that respond to both ultrasound and pH conditions. These polymer vesicles are designed to enhance the targeted and controlled release of anticancer drugs, addressing the challenges of conventional chemotherapy. In this study, the ultrasound serves as an external trigger, allowing the release of encapsulated drugs from the vesicles when exposed to ultrasound waves. On the other hand, the vesicles exhibit pH sensitivity, which is particularly relevant in cancer therapy due to the typically acidic microenvironment of tumors. The study demonstrates the effectiveness of these dual-responsive polymer vesicles for encapsulating and delivering anticancer drugs. Under normal physiological conditions, the vesicles remain stable and retain the drug payload. However, upon exposure to ultrasound waves or a shift in pH, the vesicles release the drugs, targeting cancer cells more precisely and minimizing harm to healthy tissues (Chen and Du 2013).

The flexibility and versatility of dual multi-responsive nanomaterials are notable strengths. Their capacity to adapt to complex physiological conditions, where multiple stimuli may be present, underscores their potential to provide adaptive and context-dependent responses. By incorporating responses to multiple triggers, these nanomaterials can offer a more nuanced and comprehensive approach to addressing intricate biological systems. This adaptability is particularly valuable in the dynamic and heterogeneous environments encountered within the body. Synergistic effects represent a compelling advantage of dual multi-responsive nanomaterials. By responding to multiple triggers simultaneously, these materials can capitalize on interactions between stimuli, resulting in amplified therapeutic effects or more efficient responses. This aspect opens possibilities for optimizing drug delivery profiles, enhancing imaging capabilities, and overall improving the efficacy of therapeutic interventions. These nanomaterials also minimize false positives, ensuring a high level of specificity in their responses. The integration of multiple triggers reduces the likelihood of unintended activation, leading to a higher degree of accuracy and reliability in the nanomaterial's behavior. This specificity is crucial in achieving the desired therapeutic outcomes while minimizing any potential adverse effects (Cheng et al. 2013).

Smart nanomaterials for biomedical applications

Recent developments in nanotechnology have raised the demand for new nanomaterials with versatile properties and multiple functionalities. Nanomaterials have a high surface area to volume ratio, small size, adaptable physicochemical properties, tunable plasmonic properties, excellent photo and magnetic properties and functional quantum properties. Nevertheless, the use of nanomaterials for biomedical applications still has many challenges such as low target capacity, less cellular uptake, fast renal clearance and multiple side effects. To overcome these challenges, a new class of "Smart nanomaterials" must be developed with better physicochemical and biological properties. Some of the important biomedical applications from recent smart nanomaterial developments are mentioned as follows.

Smart nanomaterials for anticancer drug and Si RNA delivery

The use of several nanomaterials for drug delivery has been exploited in the past several years and they are used exponentially in modern-day drug delivery systems, for example long-established tablets to advanced nanoformulations. The nanoparticle possesses a unique property of high surface area to volume ratio as a result of their small sizes as compared to macro-sized particles. Recently, nanoparticles have been extensively used in targeted anticancer drug delivery in several types of cancer treatment. The targeted and precise delivery of drugs requires extensive knowledge and a clear understanding of their physicochemical behavior in biological systems along with an understanding of limitations and reasons for the failure of the newly developed drug delivery nanosystems. To enhance drug availability and reduce side effects, designing of novel smart nanoparticle systems must be developed. Many smart nanoparticles have been developed recently for cancer treatment, mainly liposomes or polymer-based nanoparticles. Many hydrophobic drugs are easily carried to the hydrophobic tumor sites by the encapsulation in liposomes. Recently, liposomes with rigid bilayers were more efficient in delivering loaded drugs than with more flexible bilayers (Lee et al. 2015). Similarly, PEGylated liposomes release hydrophobic drugs at a much higher rate than the control. The fast release of drugs is not restricted to liposomes but is also seen in other nanomaterials such as polymeric micelles facilitated by interaction with blood proteins (Chen et al. 2008). The drug-loaded liposomal or polymeric nanocarriers release their content by a diffusional process accompanied by interaction with blood proteins (Chen et al. 2008). For all intravenous drug formulations, it is very important to maintain the rate of drug release such that they release the drug slowly in blood but faster when activated by corresponding environmental factors (Dokoumetzidis and Macheras 2006). A smart drug delivery system refers to a delivery system that is capable of releasing drug molecules more efficiently when exposed to certain physicochemical conditions such as temperature, light, magnetic field, pH, ions, and biological conditions such as enzymes, small molecules and antibodies (Lee et al. 2015). Another interesting observation is corona formation: when drug/drug carriers interact with the biological fluid, it instantly forms a biocorona by adsorbing biological molecules around them. After corona formation, drug nanoparticles mandate the drug delivery process (Gupta and Roy 2020).

The nanocarrier system was developed by Dong et al. (2015) for the delivery of chemotherapeutic drug and siRNA cancer treatment. In this study, they synthesized a multifunctional PHD/PPF/siRNA complex by one step assembly method using prefunctionalized polymers (Dong et al. 2013). Another attractive option developed by Seo et al. (2015) developed a better approach for co-delivery method using multifunctional nanocarriers for controlled release in cancer treatment (Seo et al. 2015). In that study, they have synthesized co-delivery system for gene and anticancer drug delivery using polyethyleneimine, folic acid and pluronic acid. The smart delivery system shows a controlled release in hypothermic environmental condition. This multifunctional nanocarrier system promises a targeted delivery in cancer treatment (Seo et al. 2015).

Recently, Kim et al. (2014) developed a smart triblock copolymer micelles drug delivery system for Si-RNA delivery. They have synthesized a triblock polymer using nonionic and hydrophilic poly (ethylene glycol) for smooth delivery of Si-RNA from triblock copolymer micelles within the cells and enhanced the performance of polymeric micelles for efficient Si-RNA delivery. This smart delivery system helped to overcome the irreversible aggregate formation (Kim et al. 2014).

Smart nanomaterials for neurodegenerative disorders

With increase in aging population in recent years, there is an urgent need for the development of treatment strategies for neurodegenerative disorders (ND) such as Parkinson's disease (PD), Alzheimer's disease (AD), Huntington's disease and amyotrophic lateral sclerosis (ALS) (Masoudi Asil et al. 2020). The main challenge in their treatment is failure to cross blood brain barrier (BBB). Current development in nanotechnology provides an opportunity to develop smart nanomaterials for overcome this barrier and deliver drugs to CNS (Lamptey et al. 2022). Nanoengineered particles exhibit the property to cross BBB. The nanomaterial systems have been developed from polymeric, magnetic and inorganic nanoparticles for improved drug delivery for the treatment of CNS disorders (Poovaiah et al. 2018).

In recent years, nanomaterials such as nanoparticles, nanotubes, quantum dots and nanofibers have been exploited for various biomedical applications such as drug delivery, bioimaging and biosensors. Engineered nanomaterials designed for neurodegenerative disorders are called as neuronanomedicine (Soni et al. 2016). Such nanoscale smart devices can interact with biological system at molecular level and can trigger physiological response in target cells and tissues. Some nanoparticles are designed to enhance the BBB penetration and for targeting specific domains within cells (Ramanathan et al. 2018). The target is to reach intracellular molecules and protein aggregates such as amyloid plaques in AD.

Gold nanoparticles (AuNPs) are commonly used nanomaterials for biomedical applications due to its multifunctional characteristics such as shape and stability, surface plasmon resonance (SPR), drug and gene delivery applications, antimicrobial, anticancer, sensing application and imaging. AuNPs combined with exosomes derived membrane have been proved effective delivery to CNS (Ajnai et al. 2014). Additionally, targeted exosome conjugated AuNPs were accumulated in mouse brain by bioimaging. Synthetic AuNPs with the brain targeted exosome can be a highly effective and novel strategy for brain targeting (Sanati et al. 2019). Similarly, multifunctional Aβ inhibitor designed using AuNPs such as AuNPs@POMD-pep (POMD: polyoxometalate with Wells–Dawson structure, pep: peptide) have shown synergistic inhibition of Aβ aggregation, minimizing Aβ induced cytotoxicity and decreasing Aβ-mediated peroxidase activity (Gao et al. 2015). Another similar study reported that specific peptide conjugated AuNPs have shown promising results in dissolving Aβ aggregates. Interestingly, modified β-sheet breaker peptide (CLPFFD) conjugated with AuNPs has been a very effective method for destroying aggregates of Aβ. Introduction of this peptide increased the permeability of BBB by interacting with interferin receptors of endothelial cells in brain vasculature (Masoudi Asil et al. 2020).

Recently, Yang et al. (2020) developed a bioinspired hybrid-nanoscaffold-based therapeutic intervention in CNS inhibitory microenvironment in case of brain injuries. In this study, researchers have developed a 3 D porous nanoscaffold using self-assembly method for delivery of anti-inflammatory molecules to suppress the neuroinhibitory microenvironment. When they tailored the structural and biomedical properties of nanoscaffold, the axonal growth was escalated in human pluripotent stem cell (hiPSC) in an *in vitro* model of neuroinflammation (Yang et al. 2020).

Conclusion

In conclusion, the realm of smart nanomaterials has unveiled a new paradigm in the field of biomedical applications. Throughout this chapter, we have explored the remarkable diversity of stimuli-responsive nanomaterials that hold immense potential for revolutionizing healthcare. The spectrum of stimuli, including light, electrical, pH, magnetic, and glucose responsiveness, has empowered the design of nanomaterials that adapt and respond to their environment with unparalleled precision. Smart nanomaterials have transformed the landscape of anticancer therapeutics, allowing for personalized treatment approaches through targeted drug and siRNA delivery. Their potential in addressing neurodegenerative disorders provides hope for interventions that can slow or halt disease progression. Through the convergence of cutting-edge science, engineering, and medicine, smart nanomaterials have paved the way for innovative strategies to tackle some of the most pressing medical challenges of our time.

In comparison to conventional medicines, the advent of smart nanomaterials represents a paradigm shift that offers unparalleled advantages in biomedical applications. The ability of these materials to respond dynamically to specific triggers allows for targeted interventions with minimized off-target effects. This precision-driven approach not only enhances therapeutic efficacy but also reduces adverse side effects, addressing one of the longstanding challenges of traditional treatments. As we look ahead, the potential of smart nanomaterials in biomedical applications is truly groundbreaking. Their multifunctional nature, coupled with their capacity to respond to various stimuli, positions them as formidable tools for personalized medicine. The prospect of tailoring treatments to individual patient profiles holds the potential to revolutionize disease management and significantly improve patient outcomes.

In an era where precision and customization are at the forefront of medical advancements, smart nanomaterials stand as beacons of innovation. The convergence of nanotechnology, medicine, and engineering has opened doors to innovative therapies, diagnosis, and monitoring strategies that were once deemed unattainable. The journey of smart nanomaterials in biomedicine has just begun, promising a future where science fiction becomes reality, and where the limitations of conventional approaches are transcended by the promise of targeted, effective, and patient-centric interventions.

Acknowledgement

The authors would like to thank the National Institutes of Health's National Institute of General Medical Sciences (NIH/NIGMS) for funding this research under Award Number 1R16GM145575-01.

References

Abidian, M.R., D. Kim and D.C. Martin. 2006. Conducting-polymer nanotubes for controlled drug release. Advanced Materials 18: 405–409.

Aflori, M. 2021. Smart nanomaterials for biomedical applications—a review. Nanomaterials 11: 396.

Aguilar, M.R. and J. San Román. 2019. Smart Polymers and their Applications. Woodhead Publishing.

Ajnai, G., A. Chiu, T. Kan, C.C. Cheng, T.H. Tsai and J. Chang. 2014. Trends of gold nanoparticle-based drug delivery system in cancer therapy. J. Exp. Clin. Med. 6: 172–178. https://doi.org/10.1016/J.JECM.2014.10.015.

Aqel, A., K.M.M. Abou El-Nour, R.A.A. Ammar and A. Al-Warthan. 2012. Carbon nanotubes, science and technology part (I) structure, synthesis and characterisation. Arabian Journal of Chemistry 5: 1–23.

Arafa, M.G., R.F. El-Kased and M.M. Elmazar. 2018. Thermoresponsive gels containing gold nanoparticles as smart antibacterial and wound healing agents. Sci. Rep. 8: 13674.

Benselfelt, T., J. Shakya, P. Rothemund, S.B. Lindström, A. Piper, T.E. Winkler, A. Hajian, L. Wågberg, C. Keplinger and M.M. Hamedi. 2023. Electrochemically controlled hydrogels with electrotunable permeability and uniaxial actuation. Advanced Materials 2303255.

Bertrand, O. and J.-F. Gohy. 2017. Photo-responsive polymers: synthesis and applications. Polym. Chem. 8: 52–73.

Binnig, G.K. 1990. Atomic force microscope and method for imaging surfaces with atomic resolution.

Blum, A.P., J.K. Kammeyer, A.M. Rush, C.E. Callmann, M.E. Hahn and N.C. Gianneschi. 2015. Stimuli-responsive nanomaterials for biomedical applications. J. Am. Chem. Soc. 137: 2140–2154.

Cai, H., X. Wang, H. Zhang, L. Sun, D. Pan, Q. Gong, Z. Gu and K. Luo. 2018. Enzyme-sensitive biodegradable and multifunctional polymeric conjugate as theranostic nanomedicine. Appl. Mater. Today 11: 207–218.

Chen, H., S. Kim, L. Li, S. Wang, K. Park and J.X. Cheng. 2008. Release of hydrophobic molecules from polymer micelles into cell membranes revealed by Forster resonance energy transfer imaging. Proc. Natl. Acad. Sci. U S A 105: 6596–6601. https://doi.org/10.1073/PNAS.0707046105.

Chen, W. and J. Du. 2013. Ultrasound and pH dually responsive polymer vesicles for anticancer drug delivery. Sci. Rep. 3: 2162.

Chen, Y., Y. Gao, L.P. da Silva, R.P. Pirraco, M. Ma, L. Yang, R.L. Reis and J. Chen. 2018. A thermo-/pH-responsive hydrogel (PNIPAM-PDMA-PAA) with diverse nanostructures and gel behaviors as a general drug carrier for drug release. Polym. Chem. 9: 4063–4072.

Cheng, R., F. Meng, C. Deng, H.-A. Klok and Z. Zhong. 2013. Dual and multi-stimuli responsive polymeric nanoparticles for programmed site-specific drug delivery. Biomaterials 34: 3647–3657.

Datta, L.P., A. Chatterjee, K. Acharya, P. De and M. Das. 2017. Enzyme responsive nucleotide functionalized silver nanoparticles with effective antimicrobial and anticancer activity. New Journal of Chemistry 41: 1538–1548.

Dokoumetzidis, A. and P. Macheras. 2006. A century of dissolution research: from Noyes and Whitney to the biopharmaceutics classification system. Int. J. Pharm. 321: 1–11. https://doi.org/10.1016/J.IJPHARM.2006.07.011.

Ejderyan, N., R. Sanyal and A. Sanyal. 2022. Stimuli-responsive polymer-coated iron oxide nanoparticles as drug delivery platforms. pp. 133–169. In: Stimuli-Responsive Nanocarriers. Elsevier.

Feng, X., X. Sui, M.A. Hempenius and G.J. Vancso. 2014. Electrografting of stimuli-responsive, redox active organometallic polymers to gold from ionic liquids. J. Am. Chem. Soc. 136: 7865–7868.

Gao, N., H. Sun, K. Dong, J. Ren and X. Qu. 2015. Gold-nanoparticle-based multifunctional amyloid-β inhibitor against Alzheimer's disease. Chemistry 21: 829–835. https://doi.org/10.1002/CHEM.201404562.

Ghosh, S., P.D. Patil and R.D. Kitture. 2019. Physically responsive nanostructures in breast cancer theranostics. pp. 1–2. *In*: External Field and Radiation Stimulated Breast Cancer Nanotheranostics. IOP Publishing Bristol, UK.

Hadiya, S., R. Radwan, M. Zakaria, T. El-Sherif, M.A. Hamad and M. Elsabahy. 2021. Nanoparticles integrating natural and synthetic polymers for *in vivo* insulin delivery. Pharm. Dev. Technol. 26: 30–40.

Hirsch, A. 2010. The era of carbon allotropes. Nat. Mater 9: 868–871.

Hsu, P.-H. and A. Almutairi. 2021. Recent progress of redox-responsive polymeric nanomaterials for controlled release. J. Mater. Chem. B 9: 2179–2188.

Hu, Q., P.S. Katti and Z. Gu. 2014. Enzyme-responsive nanomaterials for controlled drug delivery. Nanoscale 6: 12273–12286.

Kim, H.J., K. Miyata, T. Nomoto, M. Zheng, A. Kim, X. Liu, H. Cabral, R.J. Christie, N. Nishiyama and K. Kataoka. 2014. siRNA delivery from triblock copolymer micelles with spatially-ordered compartments of PEG shell, siRNA-loaded intermediate layer, and hydrophobic core. Biomaterials 35: 4548–4556. https://doi.org/10.1016/J.BIOMATERIALS.2014.02.016.

Kim, J.J. and K. Park. 2001. Modulated insulin delivery from glucose-sensitive hydrogel dosage forms. Journal of Controlled Release 77: 39–47.

Lamptey, R.N.L., B. Chaulagain, R. Trivedi, A. Gothwal, B. Layek and J. Singh. 2022. A review of the common neurodegenerative disorders: current therapeutic approaches and the potential role of nanotherapeutics. Int. J. Mol. Sci. 23. https://doi.org/10.3390/ijms23031851.

Lee, B.K., Y.H. Yun and K. Park. 2015. Smart nanoparticles for drug delivery: boundaries and opportunities. Chem. Eng. Sci. 125: 158. https://doi.org/10.1016/J.CES.2014.06.042.

Lu, Y. and J. Liu. 2007. Smart nanomaterials inspired by biology: dynamic assembly of error-free nanomaterials in response to multiple chemical and biological stimuli. Acc Chem. Res. 40: 315–323.

Mansoori, G.A. and T.A.F. Soelaiman. 2005. Nanotechnology—An Introduction for the Standards Community. ASTM International.

Masoudi Asil, S., J. Ahlawat, G. Guillama Barroso and M. Narayan. 2020. Nanomaterial based drug delivery systems for the treatment of neurodegenerative diseases. Biomater. Sci. 8: 4109. https://doi.org/10.1039/D0BM00809E.

Mena-Giraldo, P., S. Pérez-Buitrago, M. Londoño-Berrío, I.C. Ortiz-Trujillo, L.M. Hoyos-Palacio and J. Orozco. 2020. Photosensitive nanocarriers for specific delivery of cargo into cells. Sci. Rep. 10: 2110.

Mishra, M., H. Kumar, R.K. Singh and K. Tripathi. 2008. Diabetes and nanomaterials. Digest J. Nanomater. Biostruct. 3: 109–113.

Mishra, S.K., V. Sharma, D. Kumar and Rajesh. 2014. Biofunctionalized gold nanoparticle-conducting polymer nanocomposite based bioelectrode for CRP detection. Appl. Biochem. Biotechnol. 174: 984–997.

Nguyen, V.P., D.S. Alves, H.L. Scott, F.L. Davis and F.N. Barrera. 2015. A novel soluble peptide with pH-responsive membrane insertion. Biochemistry 54: 6567–6575.

Poovaiah, N., Z. Davoudi, H. Peng, B. Schlichtmann, S. Mallapragada, B. Narasimhan and Q. Wang. 2018. Treatment of neurodegenerative disorders through the blood-brain barrier using nanocarriers. Nanoscale 10: 16962–16983. https://doi.org/10.1039/c8nr04073g.

Purkait, M.K., M.K. Sinha, P. Mondal and R. Singh. 2018. pH-responsive membranes. pp. 39–66. *In*: Interface Science and Technology. Elsevier.

Qiu, L., L. Zhao, C. Xing and Y. Zhan. 2018. Redox-responsive polymer prodrug/AgNPs hybrid nanoparticles for drug delivery. Chinese Chemical Letters 29: 301–304.

Ramanathan, S., G. Archunan, M. Sivakumar, S.T. Selvan, A.L. Fred, S. Kumar, B. Gulyás and P. Padmanabhan. 2018. Theranostic applications of nanoparticles in neurodegenerative disorders. Int. J. Nanomedicine 13: 5561–5576. https://doi.org/10.2147/IJN.S149022.

Rao, T., P. Chvs, M. Yamini and C.H. Prasad. 2021. Hydrogels the three dimensional networks: a review. Int. J. Curr. Pharm. Res. 13: 12–17.

Roco, M.C. 2011. The long view of nanotechnology development: the National Nanotechnology Initiative at 10 years. Journal of Nanoparticle Research.

Roy, D., W.L.A. Brooks and B.S. Sumerlin. 2013. New directions in thermoresponsive polymers. Chem. Soc. Rev. 42: 7214–7243.

Sahle, F.F., M. Gulfam and T.L. Lowe. 2018. Design strategies for physical-stimuli-responsive programmable nanotherapeutics. Drug Discov Today 23: 992–1006.

Sanati, M., F. Khodagholi, S. Aminyavari, F. Ghasemi, M. Gholami, A. Kebriaeezadeh, O. Sabzevari, M.J. Hajipour, M. Imani, M. Mahmoudi and M. Sharifzadeh. 2019. Impact of gold nanoparticles on amyloid β-induced alzheimer's disease in a rat animal model: involvement of STIM proteins. ACS Chem. Neurosci. 10: 2299–2309. https://doi.org/10.1021/ACSCHEMNEURO.8B00622.

Seo, S.J., S.Y. Lee, S.J. Choi and H.W. Kim. 2015. Tumor-targeting co-delivery of drug and gene from temperature-triggered micelles. Macromol. Biosci. 15: 1198–1204. https://doi.org/10.1002/MABI.201500137.

Shah, R.A., E.M. Frazar and J.Z. Hilt. 2020. Recent developments in stimuli responsive nanomaterials and their bionanotechnology applications. Curr. Opin. Chem. Eng. 30: 103–111.

Shin, B.C., S.S. Kim, J.K. Ko, J. Jegal and B.M. Lee. 2003. Gradual phase transition of poly (N-isopropylacrylamide-co-acrylic acid) gel induced by electric current. Eur. Polym. J. 39: 579–584.

Soni, S., R.K. Ruhela and B. Medhi. 2016. Nanomedicine in Central Nervous System (CNS) disorders: A present and future prospective. Adv. Pharm. Bull 6: 319. https://doi.org/10.15171/APB.2016.044.

Sui, Z. and J.B. Schlenoff. 2004. Phase separations in pH-responsive polyelectrolyte multilayers: charge extrusion versus charge expulsion. Langmuir 20: 6026–6031.

Sun, S., S. Liang, W.-C. Xu, G. Xu and S. Wu. 2019. Photoresponsive polymers with multi-azobenzene groups. Polym. Chem. 10: 4389–4401.

Sun, W., S. Li, B. Häupler, J. Liu, S. Jin, W. Steffen, U.S. Schubert, H. Butt, X. Liang and S. Wu. 2017. An amphiphilic ruthenium polymetallodrug for combined photodynamic therapy and photochemotherapy *in vivo*. Advanced Materials 29: 1603702.

Svirskis, D., J. Travas-Sejdic, A. Rodgers and S. Garg. 2010. Electrochemically controlled drug delivery based on intrinsically conducting polymers. Journal of Controlled Release 146: 6–15.

Takagi, T. 1990. A concept of intelligent materials. J. Intell. Mater. Syst. Struct. 1: 149–156.

Tamaki, M. and C. Kojima. 2020. pH-Switchable LCST/UCST-type thermosensitive behaviors of phenylalanine-modified zwitterionic dendrimers. RSC Adv. 10: 10452–10460.

Tang, W. and C. Chen. 2020. Hydrogel-based colloidal photonic crystal devices for glucose sensing. Polymers (Basel) 12: 625.

Thangudu, S. 2020. Next generation nanomaterials: Smart nanomaterials, significance, and biomedical applications. Applications of Nanomaterials in Human Health 287–312.

Thevenot, J., H. Oliveira, O. Sandre and S. Lecommandoux. 2013. Magnetic responsive polymer composite materials. Chem. Soc. Rev. 42: 7099–7116.

Verma, A., S. Sharma, P.K. Gupta, A. Singh, B.V. Teja, P. Dwivedi, G.K. Gupta, R. Trivedi and P.R. Mishra. 2016. Vitamin B12 functionalized layer by layer calcium phosphate nanoparticles: A mucoadhesive and pH responsive carrier for improved oral delivery of insulin. Acta Biomater. 31: 288–300.

Volodkin, D.V, N.G. Balabushevitch, G.B. Sukhorukov and N.I. Larionova. 2003. Model system for controlled protein release: pH-sensitive polyelectrolyte microparticles. STP Pharma Sciences 13: 163–170.

Volpatti, L.R., M.A. Matranga, A.B. Cortinas, D. Delcassian, K.B. Daniel, R. Langer and D.G. Anderson. 2019. Glucose-responsive nanoparticles for rapid and extended self-regulated insulin delivery. ACS Nano 14: 488–497.

Wen, Y., Z. Zhang and J. Li. 2014. Highly efficient multifunctional supramolecular gene carrier system self-assembled from redox-sensitive and zwitterionic polymer blocks. Adv. Funct. Mater. 24: 3874–3884.

Wu, X.L., J.H. Kim, H. Koo, S.M. Bae, H. Shin, M.S. Kim, B.-H. Lee, R.-W. Park, I.-S. Kim and K. Choi. 2010. Tumor-targeting peptide conjugated pH-responsive micelles as a potential drug carrier for cancer therapy. Bioconjug. Chem. 21: 208–213.

Yang, L., B.M. Conley, S.R. Cerqueira, T. Pongkulapa, S. Wang, J.K. Lee and K.B. Lee. 2020. Effective modulation of CNS inhibitory microenvironment using bioinspired hybrid-nanoscaffold-based therapeutic interventions. Advanced Materials 32: 2002578. https://doi.org/10.1002/ADMA.202002578.

Yoshida, M. and J. Lahann. 2008. Smart nanomaterials. ACS Nano 2: 1101–1107.

Zhang, P., J. Wu, F. Xiao, D. Zhao and Y. Luan. 2018. Disulfide bond based polymeric drug carriers for cancer chemotherapy and relevant redox environments in mammals. Med. Res. Rev. 38: 1485–1510.

Zhang, Q., C. Weber, U.S. Schubert and R. Hoogenboom. 2017. Thermoresponsive polymers with lower critical solution temperature: from fundamental aspects and measuring techniques to recommended turbidimetry conditions. Mater. Horiz 4: 109–116.

Zhang, X., L. Li, J. Ouyang, L. Zhang, J. Xue, H. Zhang and W. Tao. 2021. Electroactive electrospun nanofibers for tissue engineering. Nano Today 39: 101196.

Zhang, Xu, S. Wang, G. Cheng, P. Yu and J. Chang. 2022. Light-responsive nanomaterials for cancer therapy. Engineering 13: 18–30.

Zhang, Xiaojuan, P. Wei, Z. Wang, Y. Zhao, W. Xiao, Y. Bian, D. Liang, Q. Lin, W. Song and W. Jiang. 2022. Herceptin-conjugated DOX-Fe_3O_4/P (NIPAM-AA-MAPEG) nanogel system for HER2-targeted breast cancer treatment and magnetic resonance imaging. ACS Appl. Mater. Interfaces 14: 15956–15969.

Zhao, H., C. Wang, R. Vellacheri, M. Zhou, Y. Xu, Q. Fu, M. Wu, F. Grote and Y. Lei. 2014. Self-supported metallic nanopore arrays with highly oriented nanoporous structures as ideally nanostructured electrodes for supercapacitor applications. Advanced Materials 26: 7654–7659.

Zhao, X., Y. Liu, J. Lu, J. Zhou and J. Li. 2012. Temperature-responsive polymer/carbon nanotube hybrids: smart conductive nanocomposite films for modulating the bioelectrocatalysis of NADH. Chemistry–A European Journal 18: 3687–3694.

Chapter 3

Exploiting Smart Hydrogels in Biotechnological Applications of Microfluidics

Rossana E. Madrid,[1,]* *Carla B. Goy*[1,2] and *Katia Gianni*[3]

Microfluidics

Microfluidic refers to the science and technology of fluids at the microscale (Whitesides 2006). The technological branch of microfluidic refers to little microsystems that allow working with very small volumes of fluids -micro or nanoliters- (Whitesides 2006), having precise control over the spatiotemporal dynamics of the microenvironment (Mross et al. 2015, Sackmann et al. 2014). Some advantages of this technology involve integration, automation, the parallelism of processes (Mark et al. 2009), and high sensitivity (Ren et al. 2001). Also, the small dimensions of the channels achieve a substantial reduction in the sample volume and the use and cost of reagents, while the integration of several processes inside one chip reduces user intervention and time (Mross et al. 2015). The traditional concept of microfluidic devices considers them as little laboratories on a chip (*lab-on-a-chip*), with the ability to perform several analytical processes with high efficiency and repeatability. The final goal of these systems is to perform real-time

---

[1] Laboratorio de Medios e Interfases (LAMEIN), DBI, FACET, National University of Tucumán, and Superior Institute of Biological Research (INSIBIO), CONICET, Av. Independencia 1800, C.P. 4000, San Miguel de Tucumán, Argentina.

[2] Instituto de investigaciones de Bioingeniería (IIBI), Facultad de Ingeniería, Universidad del Norte Santo Tomás de Aquino, Av. Juan Domingo Perón 2085, C.P. 4107, Yerba Buena, Tucumán.

[3] Planta Piloto de Procesos Industriales Microbiológicos (PROIMI), CONICET, Pje Caseros C.P. 4000, San Miguel de Tucumán, Argentina.

Emails: carlabgoy@gmail.com; ka20_04@hotmail.com

* Corresponding author: rmadrid@herrera.unt.edu.ar

sample analysis in the location where the sample is taken (Point of Care Testing -POCT-) without requiring specialized equipment and personnel. Nowadays, the use of these devices is ubiquitous, including in fields such as biology, medicine, pharmacy, chemistry, agroindustry, and environment (Babikian et al. 2011).

Regarding the concept of microfluidic as a science, it is a subdiscipline of fluid mechanics, which implies that the equations that describe the behavior of the fluid at the macroscale are identical to that at the microscale. A system is considered microfluidic when its dimensions are from tens to hundreds of micrometers (Whitesides 2006); such dimensions involve special physics of the fluids (Squires and Quake 2005).

Some relevant characteristics of fluid physics at the microscale are laminar flow (as a consequence of low Reynolds number), mixing only by diffusion, high surface tension, and dominant capillarity effect, among others. Laminar flow implies that the lines of flow are ordered and that the fluid moves in parallel sheets without intermingling, where each fluid particle follows a smooth path, called a streamline. This kind of flow is a consequence of the relationship between the inertial and viscous forces over a fluid, which is expressed by the Reynolds number (Re) -dimensionless-. When the viscous forces are predominant, Re is low (for practical purposes < 2000) and the movement of flow particles will be laminar, as in microfluidic devices (Re typically of the order of 10^{-2}–10). This kind of flow implies that the mixing of fluid inside the chips will be predominantly by diffusion (Novotný and Foret 2017). This can be a good situation when it is necessary, for example, to obtain a chemical separation, but mixing can be challenging (Novotný and Foret 2017). When mixing occurs only by diffusion, it requires times in the order of minutes or more (Squires and Quake 2005).

The surface tension can be interpreted as an elastic membrane in a liquid surface that imposes a resistance to penetrate the liquid. Because of this phenomenon, some insects can stand over some liquids without sinking. This concept is associated with the amount of energy necessary to increase the fluid surface by the unit of area and it is a consequence of the intermolecular forces that exist at the interface. Liquids whose molecules have strong intermolecular attractive forces will have high surface tension (Mohseni 2015). At the microscale, surface tension dominates over classical mechanics and it is imposed over Newtonian forces (gravity and inertia) (Kim 2000).

Regarding the dominant effect of capillarity in microfluidic technology, it is a consequence of large surface-to-volume ratios (Novotný and Foret 2017), and it is a great advantage of these systems since it brings an alternative to produce flow inside the microchannels without the need for external pump systems (Ren et al. 2001), just by capillarity.

Some recent and important examples of the applications of microfluidics devices are given below. Ito et al. (Ito et al. 2021) created a pump-free microfluidic hemofiltration device that uses the difference between the arterial end venous pressure to produce backfiltration of the blood. For this, the device must be connected between an artery and a vein and can remove water and waste improving the patient's quality of life; the filtrate is brought to the bladder and discarded as urine (Fig. 1A).

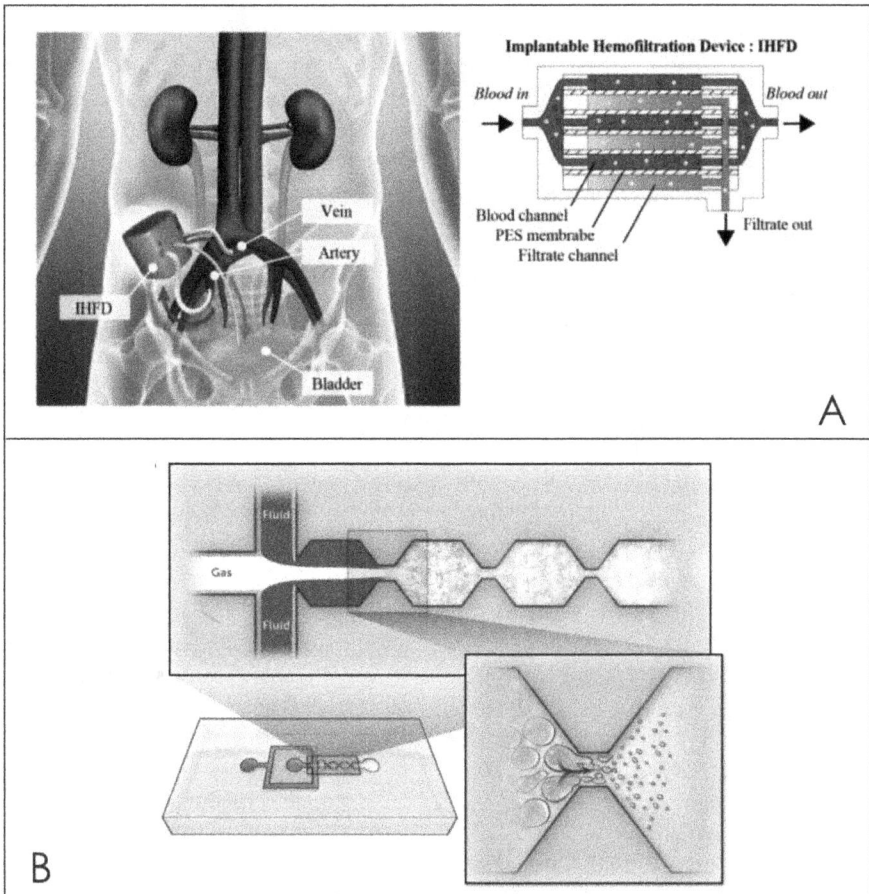

Fig. 1. A. Implantable hemofiltration device connected between an artery and a vein to produce backfiltration and to eliminate water and waste from blood (left). (Right) Detailed image of the system (image from (Ito et al. 2021) with free Creative Commons License). B. Scheme of the device configuration for microbubbles production (image from (Vutha et al. 2022) with free Creative Commons License).

Vutha et al. created a device for real-time on-demand intravenous oxygen delivery (Vutha et al. 2022). The device is able to administer oxygen gas directly to the bloodstream in real-time, which has great significance in hypoxemia conditions. This is achieved by means of the creation of oxygen bubbles (smaller than a red blood cell) covered by a membrane; microbubbles are able to traverse the lung without causing obstruction and provide arterial oxygenation (Fig. 1B). Other good examples include organ-on-a-chip microfluidics to be used in cancer research (Regmi et al. 2022), microfluidic biosensors for infectious and non-infectious diseases (Bhardwaj et al. 2022), among many others.

Fabrication techniques

In its beginning (early 1990's), the fabrication of microfluidic devices took advantage of the techniques used in the microelectronics industry (photolithography, etching, and deposition processes) (Becker and Locascio 2002, Nielsen et al. 2020, Scott et al. 2021). The problem with this methodology is the necessity of high-level equipment and clean rooms, which involves high fabrication costs and is not reachable for most potential users (Nielsen et al. 2020). Because of this, over the last few years, other fabrication approaches emerged using low-cost materials and even being equipment free.

Nowadays, the most common way to obtain a microfluidic device includes the fabrication of a mold with a pattern of channels, over which a polymer is cast, cured, and de-molded. Hence, the polymer contains a negative replica of the channel pattern. Finally, the polymer with the channels adheres face down over a support layer (that can be of a lot of different materials, such as polymer, glass, plastic, etc.) in order to seal the chip. This procedure is known as *soft-lithography* and is mostly carried out using the polydimethylsiloxane (PDMS) polymer. Some advantages of polymers are low cost (when compared to silicon or glass), optical transparency, gas permeability, and electrical insulation, among others.

Originally, the mold was fabricated using the standard photolithographic process used for the fabrication of microelectronic devices, and silicon or glass as a substrate (Becker and Locascio 2002). New recent techniques use the ubiquitous 3D impressions (Amin et al. 2016, Rogers et al. 2015, Waheed et al. 2016) as mold or printed circuit boards with the transferred channel pattern etched (Chang and Hui 2019, Jain et al. 2017, Li et al. 2013, Tiwari et al. 2020). These new approaches reduce costs related to infrastructure and equipment, which makes them more accessible to users (Amin et al. 2016).

One alternative to the fabrication of microfluidic devices by replication is direct manufacturing, which involves the removal of material where channels and structures are desired in the final chip. In this case, the mold is not needed. However, this methodology requires high-cost equipment and is time-consuming. Paper-based microfluidic devices are another alternative extensively used for analytical purposes; these sensors are economical, easy to fabricate, have the ability to store reagents (Noviana et al. 2021), and the flow of fluids is driven by capillarity without the need for active pumping. In paper microfluidics, hydrophobic barriers must be created to guide the fluid flow.

Hydrogels and their highlights

Although microfluidic science has driven technological developments in a lot of different areas, enhanced capabilities can be obtained when they are combined with hydrogels (Goy et al. 2019). Hydrogels are tri-dimensional (3D) polymers that can absorb a large amount of water and swell, being insoluble in water (Katmiwati and Nakanishi 2014). The first characteristic is due to the hydrophilic functional groups in the polymer, while its resistance to dissolution is due to the cross-links between the chains of the network. Hydrogels not only absorb water but can also absorb other solvents, and have high permeability to oxygen and other water-soluble metabolites. These polymers have several highlights that make them very interesting for biotechnological applications like biocompatibility (Bulpitt and Aeschlimann 1999, Lee and Mooney 2001, Molinaro et al. 2002, Sechriest et al. 2000), transparency, inert behavior, a structure similar to that of the extracellular matrix (Ibrahim and Richardson 2017), the ability to provide a protective environment to biological entities (Choi et al. 2008, Ibrahim and Richardson 2017, Koh and Pishko 2005, Logun et al. 2016), controllable degradation rates and tunable stiffness. Also, some hydrogels can respond to external stimuli (temperature, pH, light, electric and magnetic field, among others (Ahmed 2015, Qiu and Park 2001)); these are known as stimulus-sensitive hydrogels and are very useful to act as active elements since they can swell and shrink according to the conditions of the environment (Franke et al. 2017). Natural hydrogels include gelatin, collagen, silk, agarose, etc., and examples of synthetic ones are poly(vinyl alcohol) (PVA), polyethylene(glycol) (PEG), poly(hydroxyethyl methacrylate) (pHEMA) and poly(n-isopropyl acrylamide) (pNIPAA). The last two are also good examples of stimuli-sensitive polymers that react to the electric field and to the temperature, respectively (Koetting et al. 2015). Also, pNIPAA is the preferred stimuli-sensitive hydrogel to act as an active element inside microfluidic chips, due to the ease of using temperature as stimuli (Goy et al. 2019).

Three compounds are needed for the synthesis of hydrogels, which are the monomers, the initiator, and the crosslinking agent. Various methods for synthesis have been reported, such as chemical crosslinking, physical crosslinking, polymerization grafting, and radiation cross-linking. Each one of these methods produces different dimensions, configurations, and properties of the hydrogels. Physical crosslinks are reversible and result from physical or interlocking chain interactions such as ionic interactions, hydrogen bonding, hydrophobic interactions, and crystallite formation. In contrast, chemically cross-linked networks exhibit permanent covalent bonds (Beck et al. 2023). Material properties such as softness, biocompatibility, and elasticity as well as sensitivity to stimuli can be tailored through chemistry, compounds, and resulting structures.

The combination of microfluidics and hydrogels has been widely used in the last few years (Goy et al. 2019). Together, they have been exploited to obtain cell culture devices, biosensing platforms, separation, modeling, and gradient generation devices, among other applications. The already mentioned advantages of both technologies make them extremely attractive for biotechnological applications. In order to obtain the combination of microfluidics and hydrogel, the polymers can be embedded inside a microfluidic device made of a polymeric material, glass or paper, etc., or microfluidic channels can be created inside hydrogel bulks.

The most popular method to embed the hydrogel inside the microchannels consists in *in-situ* photopolymerization of the hydrogel using UV light and, sometimes, a photomask. One advantage of this strategy is a high spatial resolution for the polymerization of very complex hydrogel structures (Koh et al. 2016, Piao et al. 2015, Zhang et al. 2016).

The second approach to obtain the combination is the creation of microchannels inside a hydrogel bulk, which can be carried out by using: (a) micro-molding, a process similar to the one followed to fabricate microfluidic chips (detailed above) (Mu et al. 2013, Nie et al. 2018, Toepke et al. 2013); (b) use of removable structures as sacrificial materials to create the channels (Bellan et al. 2012, Jafarkhani et al. 2018, Jiang et al. 2018, Kruss et al. 2012, Shin and Hyun 2017, Štumberger and Vihar 2018, Zhao et al. 2016); (c) or hydrogel degradation in order to create a microchannel using, for example, a laser with image-guided control to degrade the gel (Heintz et al. 2016).

Microfluidics and hydrogel combination and its contribution to biotechnology

Biotechnology is the field of exploration and exploitation of technology and biological life forms to create new products and processes. Biotechnology is being applied in various fields to produce more solutions to tackle the challenges of humankind. The four major areas of biotechnology application are medical (named red biotechnology, for example, tissue engineering applications, therapeutics devices), agriculture (green biotechnology), industrial (white biotechnology), and environmental sectors (grey biotechnology) (Ortseifen et al. 2020).

In this chapter, we use this classification of biotechnology areas in order to analyze the contribution of the microfluidics and hydrogel combination (MHC) to each of them. First, the state of the art was explored using PubMed and Scopus databases and the search engine "microfluidics+hydrogel+biotechnology" (the time period considered was from 01/01/2010 to 28/02/2023). Then, each paper was analyzed to classify it according to its application field into one of the biotech areas (red, green, white, and grey). Also, another division is added to include those general biotechnological tools that can be applied to more than one area. Most of the contributions found in this area are related to the encapsulation of cells, enzymes, antigens, and antibodies, among others, inside hydrogel micro or nanostructures, like beads, microtubes, etc. The resultant complexes have applications in several fields including tissue engineering, bio-fabrication, organs-on-chips, biotransformation, drug delivery, biosensing, bioremediation, engineering of cells, and disease modeling. Figure 2 shows the contribution of MHC to each one of the biotech areas (red, green, white, grey, and general). As can be seen, the number of contributions of the combination to each area is very variable, the red one being the most favored. In the next sections, some relevant contributions of MHC to each biotech area are presented and described.

MHC for white (industrial) biotechnology

White biotechnology is devoted to the production of valuable bio-products for industries (food, pharmaceutical, agricultural, etc.) using fungi, yeasts, bacteria, and other microorganisms (Kordi et al. 2022). In this branch of biotech, MHC has been used to achieve high-throughput cell culture and enzyme-free cell harvesting (Mohammad et al. 2021), to construct microreactors with enzymes immobilized on hydrogels for continuous flow bio-catalysis (Peschke et al. 2018), for the utilization

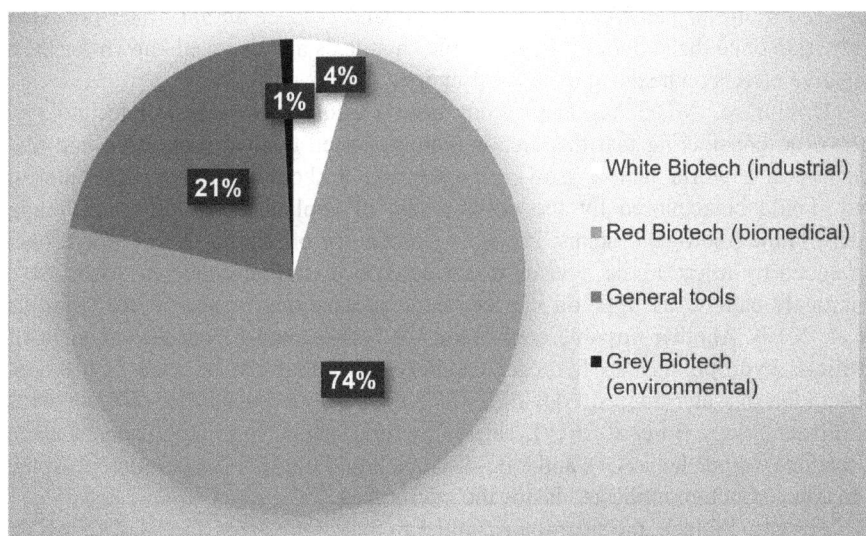

Fig. 2. Circle chart that illustrates the distribution of the MHC devices devoted to each biotechnological area of interest.

of insect oil resources (Wang et al. 2021), and for processing proteomic samples (Luk et al. 2012).

In order to perform biocatalytic flow processes, it is necessary to immobilize enzymes in flow channels, which is more difficult than immobilizing conventional organo (metallic) catalysts (Buchholz et al. 2012). In this context, Peschke et al. (Peschke et al. 2018) proposed a microfluidic reactor filled with *in-situ* polymerized hydrogels loaded with active biocatalysts. They used two homotetrameric enzymes and obtained hydrogels with a mass of 77% enzymes. The use of protein[-containing] hydrogels in biocatalysis is an underdeveloped field and this work is a very interesting contribution to this area. Another microreactor involving a digital microfluidic device and enzyme-loaded hydrogel disks was developed by Luk et al. (Luk et al. 2012) to be used as a miniaturized method for proteome profiling, obtaining better performance than conventional techniques.

In these examples, the hydrogel plays a fundamental role in enzyme immobilization, thanks to its very desirable properties: a near-native environment that preserves enzyme functions, simple immobilization strategies, optimal geometric hydrogel/enzyme congruence, and high enzyme density thanks to the porosity of the hydrogel (Luk et al. 2012).

MHC for green (agricultural) and grey (environmental) biotechnology

Green biotechnology refers to the use of biotechnological procedures to improve every agricultural sector, i.e., crops, cattle, poultry, pesticides, agribusinesses productivity, etc. This subject is devoted to reducing the dependency of agriculture on mechanical and chemical innovations. These kinds of technologies tend to be harmful to the environment (Nasser et al. 2021). Mechanical innovations are usually powered by fossil fuels, directly or indirectly, and produce air and water pollution and waste once their lifetime is over, while chemicals are also well known for their negative effects on health and the environment.

Until now, MHC has almost not been explored for green biotechnology. However, considering that this area is mainly related to food production and that we live in a world with a growing population, it should generate more interest and should be explored for the development of applied technologies, including microfluidic-hydrogel systems. For example, the use of microparticles of hydrogel produced by microfluidic devices to act as crop hydration elements. Also, these microgels can be used as oil carriers with applications in agriculture (Martins et al. 2017). Another possible application for MHC could be the development of artificial hydrophilic PVA hydrogel capillary tubes that can be used as the next generation of water transport devices, which are highly necessary for relevant areas of biotechnology (Li et al. 2019). These structures can perform rapid, spontaneous, directional (diode behavior), and long-distance liquid transport by biomimetic of the peristome structure replicated inside the microtubes.

For grey biotech, the situation is similar to that of the green area, and almost no advantage has been taken from MHC to contribute to environmental issues. Only one approach for bioremediation was proposed by Fujimoto et al. (Fujimoto et al. 2018), who used a triple-coaxial flow microfluidic device to continuously produce hydrogel microtubes for microbial encapsulation and culture. In this case, the hydrogel structures permit oxygen and nutrient diffusion, without microbes escaping through the walls.

Until now, numbers show that green and grey are mostly unexplored biotech areas regarding MHC developments (Fig. 1); new innovations could include plant genetic modification and growth assessment, biosensors for the evaluation of environmental contamination, and water purification and transport systems, among others.

MHC for red (biomedical) biotechnology

In the context of MHC for red biotechnology, which includes tissue engineering, regenerative medicine, cell culture, or simply medical applications, publications have grown exponentially in the past ten years. The great advantages of microfluidic technologies combined with hydrogels have led to the development of many interesting devices or systems.

In the area of red biotechnology, despite the fact that there is a growing number of reports in this area, only three types of applications of microfluidics in combination with hydrogels can be identified: (1) the use of microfluidic chips for the production of hydrogel microcapsules, wherein cells can subsequently be cultivated, for applications in tissue engineering or regenerative medicine; (2) the use of microfluidic devices for the manufacture of hydrogel microfibers or microtubes, which can later be used for the same applications as the hydrogel microcapsules, although the use of microfibers and microtubes is more directed at the vascular part; and (3) the use of microfluidic devices combined with hydrogels for cell culture.

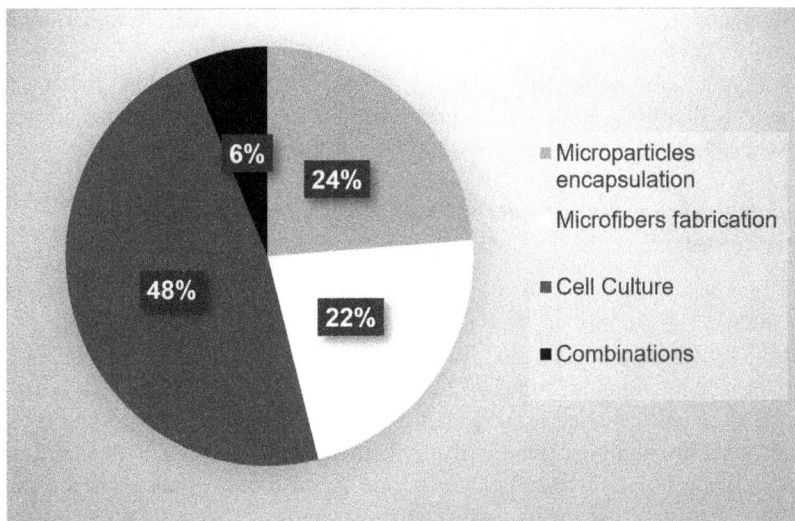

Fig. 3. Circle chart that exposes the distribution of the publications related to red biotechnology classified according to each application.

Only a very small number of applications use this strategy of combining manufacture and cultivation on the same chip.

In the following graph (Fig. 3), all the publications in red biotechnology, classified according to the mentioned applications, can be observed.

Cell culture

As can be seen, cell culture applications are most widely spread, particularly co-culture applications that use MHC are expanding, with applications ranging from microbiology, and ecology to medicine. The knowledge of cell interactions is very important since this allows the progress of synthetic biology. For example, applications in organ engineering for regenerative medicine allow obtaining organs or more complex synthetic structures. On the other hand, the use of hydrogels allows the construction of environments similar to the extracellular matrix (ECM), which increases the options for applications of these technologies. The main characteristics that hydrogels must have to encapsulate cells are: (a) that the polymerization process does not require extreme conditions with sudden changes in temperature or pH; (b) that the morphology of the hydrogel is easily achievable, and (c) that the hydrogel matrix allows adequate cell-cell and cell-matrix interaction (Saeki et al. 2020). Recent advances in new biomaterials, such as these hydrogels, have enabled their use for cell culture, as they can be manufactured with extremely controlled properties. However, not everything has been easy in the development of 3D culture, since although the hydrogel allows a suitable environment for cell development, the diffusion of the necessary nutrients and oxygen to the cells is essential. Microfluidics has become a key enabling technology at this point and its use has been explored in two ways to achieve more successful 3D cultures. One strategy is "top-down" fabrication, where microchannels are fabricated within the cell-laden hydrogel matrix. These channels are analogous to vascular structures that nourish cells. The other strategy involves "bottom-up" manufacturing, where microfluidic chips allow the cell-laden hydrogel to be manufactured in the form of beads, fibers, or strips (Ren et al. 2014). These components then form hollow structures when packed together, and this allows for the exchange of molecules between the fluid and the cells.

Top-down fabrication for microchannels manufactured within a hydrogel matrix

As an example of the first case, the work of Yamahira et al. is noteworthy (Yamahira et al. 2020). The authors presented a method for constructing multiple microscale dynamic concentration gradients around cells encapsulated in a photodegradable hydrogel. The hydrogel is formed within a microfluidic device developed for culture perfusion. Once the device is fabricated, perfusion microchannels are fabricated in the hydrogel by taking advantage of its photodegradation characteristic and using simple and complex micropatterns to generate concentration gradients of the culture medium. These microchannels fabricated in the hydrogel then allow the perfusion of cells in the hydrogel zones where they were encapsulated, called diffusion reservoirs. A simple micropattern consisted of a horizontal rectangular culture cell with perpendicular perfusion channels at different distances, forming diffusion reservoirs of different width. The authors examined the effect of the size of these compartments on cell viability. They found that encapsulated cells remained alive after 3 days of culture in the 300 μm wide compartments, but the proportion of dead cells in the wider compartments was higher, indicating that nutrient supply and waste removal from the perfusion microchannel did not allow longer distances to be reached. They then evaluated different patterns of perfusion channels to obtain zones with nutrients concentration gradients on the microfluidic chip. The concentration gradients are dynamic during perfusion cell culture, allowing the cells encapsulated in the hydrogel to be exposed to a dynamic microenvironment.

Marshall et al. (Marshall et al. 2017) proposed a 3D breast cancer culture system composed of mimic ECM proteins (bovine collagen I and growth factor-reduced Matrigel) backed by an inert backbone and perforated microchannels. As the ECM had low viscosity, a scaffold was required to support the permeability of the precast microchannels and to minimize the shrinkage of the ECM hydrogel. For this, the authors chose a rigid carbon foam (reticulated vitreous carbon), autoclavable and highly porous. The authors demonstrated that with the developed system, breast cancer cells and fibroblasts exhibited increased viability when perfused microchannels were present in the system, compared to cells grown in similar preparations but without microchannels or perfusion.

Another interesting application is the one of Govindasamy et al. (Govindasamy et al. 2021) who presented a platform that allows modeling the implantation of an embryo and its interaction at the embryo-mother interface. For this, they analyzed trophoblast invasion and revealed the first embryonic interactions with the maternal vasculature. The microfluidic device comprises two tubular channels embedded into the hydrogel, which were populated with endothelial cells, generating vascular architectures inside the synthetic environment. Mouse embryos grew in the adhesive hydrogels and the study revealed the dynamics of embryo invasion and the embryo's first contact with the maternal vasculature.

Wang et al. proposed a triple-channel bilayer microfluidic platform incorporating highly biocompatible hydrogel for the simulation of the vascular tumor microenvironment (Wang et al. 2021). They used the platform to build a vascular tumor microenvironment composed of cancer cells, fibroblasts, and epithelial cells with the hydrogel acting as ECM, trying to address the problem of tumor resistance of different loci and their relationship with other cell types, like epithelial cells and fibroblasts, in the microenvironment. They use special self-healing hydrogel, which gels at room temperature simply by combining the same volume of the two components. Furthermore, the biocompatibility and viability of the hydrogel were confirmed by the high viability of the cells in long-term culture on the microfluidic chip. Nutrients and drugs are introduced through the bottom layer of the chip and diffuse into the cancer cells channels (drug-resistant and drug-sensitive tumor mass) on the top of the microfluidic chip separated by a poly-carbonate membrane.

Another example of "top-down" fabrication with an interesting strategy is the microfluidic chip described by Lee et al. (Lee et al. 2013). It consists of a microfluidic chip that simulates liver metabolism. The microfluidic channel contains PEG hydrogel pillars with liver microsomes encapsulated, and geometrically distributed throughout the channel. In this way, the cells are in a suitable environment, and in turn, the nutrients and culture medium are constantly renewed around the pillars. The authors demonstrated a quantitative analysis of the hepatic metabolic response by combining a mathematical modeling approach and experimental analysis. The mathematical models showed good correlations with the results, although with some discrepancies in the high concentration ranges of the substrate. The valid mathematical model developed can be used for chip analysis and improvements. An approach of this type is interesting for other applications for the study of cellular metabolisms on a chip.

Bottom-up manufacturing for cell-laden hydrogel to form microelements

Cell encapsulation is fundamental to virtually all of the biotechnology applications mentioned in this chapter. Furthermore, the study of cell-cell interaction and of cells with their environment is extremely important for advances in tissue engineering and regenerative medicine. However, achieving complex constructs with different cell types and co-cultures has been challenging (Yajima et al. 2018). In recent years, bottom-up manufacturing has become more widespread as it allows interesting advances in the biomanufacturing of these complex architectures. On the other hand, advances in new biomaterials and scaffolds microfabrication technologies have offered better opportunities to recreate 3D organ/tissue models in a controllable manner (Liu et al. 2020). As was mentioned, "bottom-up" manufacturing involves the fabrication of cell-laden hydrogel in the form of beads, fibers, or strips, by using microfluidics technology. Fibers or microtubes, for example, are excellent candidates for the construction of vascular models, which must possess four fundamental attributes of real vascular systems: freestanding, branching, multilayered, and perfusable (Ching et al. 2022). In the case of building functional models of 3D organs *in-vitro*, such as organoids, which are multicellular tissues in a 3D culture that can mimic the corresponding organ *in vivo*, the challenges are even greater. Recent advances have enabled the successful generation of various organoids, including the pancreatic islets (Liu et al. 2020, Wang et al. 2021), brain (Cho et al. 2021), liver (Kobayashi et al. 2013, Yamada et al. 2012), and kidney (Ozcelik et al. 2022), among many others. These developments have shown enormous potential in studies of artificial organs, disease modeling, and drug testing, where hydrogels played a key role. Just as microfibers are useful for vascular models, hydrogel capsules have been recognized as suitable 3D culture scaffolds and cell-laden carriers for organoid models or simply cell culture or co-cultures.

As an example of the great possibilities that microfluidic technology combined with hydrogels can bring, the work of Altay et al. about a 3D model of the intestinal epithelium (Altay et al. 2022) can be mentioned. They created a device to fabricate 3D hydrogels with shapes similar to intestinal villi, based on poly(ethylene glycol) diacrylate (PEGDA), with a controlled mesh size that allows the diffusion of niche biomolecular factors of intestinal stem cells (ISC). They used the photolithography technique to obtain the villi-like hydrogels. Then, using a microfluidic device that includes the designed microstructured hydrogels and a light sheet fluorescence microscope (LSFM), they were able to observe ISC biochemical factor gradients. The microfluidic platform was suitable for the growth, proliferation, and differentiation of organoid-derived intestinal epithelial cells, and can be used to study different tissues as it allows to evaluate *in-vitro* a wide range of essential biochemical factors for maintenance and differentiation of ISC.

The work of Fattahi et al. is another contribution of MCH for cell encapsulation (Fattahi et al. 2021). They evaluated the stirred bioreactor concept for the production in quantity of stem cell spheroids that could be used for disease treatment, that is, for cell therapies based on human pluripotent stem cells (hPSC). They performed microfluidic encapsulation of hPSCs and formed spheroids. The microfluidic device allows the generation of coaxial droplets for the fabrication of 400 μm diameter capsules with a polyethylene glycol hydrogel shell and an aqueous core. This technology can be used to encapsulate different hPSC cell lines.

Gal et al. described a microfluidic device for the formation of cell droplets from a cardiac cell-hydrogel mixture and evaluated cell retention, survival, and maturation within a host tissue (Gal et al. 2020). They used a microfluidic chip to encapsulate the cells in a 3D personalized hydrogel based on an omentum [a fold of the peritoneum] ECM. The omentum can be easily removed from patients by microsurgery and its ECM can be used as a personalized material with the patient's own immunological profile. The fabrication of custom hydrogels is very important in regenerative processes to avoid rejections. The spherical geometry of the microparticles and its surface-to-volume ratio facilitates efficient biomolecular transport, resulting in an equitable distribution of oxygen and nutrients. In this work, human omentum-derived pluripotent stem cells were differentiated into cardiomyocytes and mixed with the liquid ECM hydrogel. The mixture flowed into the microfluidic device to generate the cell droplets that fully matched the immunological and biochemical properties of the patient. The droplets turned into a viscous gel on heating to 37°C and were then injected into a mouse muscle, where cell regeneration was assessed.

Combined top-down, bottom-up, and cell culture MHC systems

The combination of the advantages of each manufacturing technique of combined hydrogel and microfluidic systems is also exploited by a large number of applications. Architectures that allow easy exchange of nutrients and oxygen, adequately perfused cell systems, co-cultures that are close to real organs, and adequate support systems for cells are some of the great possibilities that this combination provides.

One example is the work reported by Berger Fridman et al. where the same microfluidic platform serves for the generation of hydrogel microdroplets with encapsulated cells, as a culture platform, and as a cell perfusion system (Berger Fridman et al. 2021). This platform allows the study of the dynamic microenvironment of breast tumors using hydrogels as tumor microenvironment model (TME) scaffolds. The design of the microfluidic device consists, on one hand, of a matrix layer, with a flow-focusing junction for rapid droplet generation, and another integrated matrix layer with 1000 entrapment sites, each 200 μm in diameter, where the TME scaffolds are trapped and held. On the other hand, the device presents a top layer of 20 μm (width) channels, which extend over the matrix and facilitate efficient perfusion of the TME scaffolds and the matrix channels. The TME scaffolds were generated by mixing the cells with partially crosslinked Alginate (Alg) or alginate-alginate sulfate (Alg/Alg-S), prior to being introduced into the chip via the aqueous phase inlet. Simultaneously, the flow of oil through another inlet allows for the rapid generation of water-in-oil droplets that are trapped in the integrated matrix. The hydrogel-containing droplets were then briefly perfused with $CaCl_2$-supplemented complete medium through a third inlet, to complete cross-linking of the hydrogel and oil removal. After that step, the culture of the TME scaffolds can be performed, which was maintained by constant perfusion throughout the culture at a flow rate that mimics the rate of blood flow in solid tumors *in vivo*.

Another interesting report on the combination of technologies is the work of Yajima et al. (Yajima et al. 2018). In this work, the researchers produce liver micro-organoids on hydrogel microfibers or sheets using microfluidic systems that they developed in previous publications (Kobayashi et al. 2013, Yamada et al. 2012) with which they encapsulated cells at high densities at precisely controlled positions, structurally mimicking linear hepatocyte assemblies *in vivo*. In the present work, the hepatocytes are encapsulated into unitary fibers to create large perfusive liver tissue equivalents. The manufacture of the fibers is carried out by means of a laminar flow microfluidic device specially manufactured for this purpose. These fibers are then pooled and packed in a perfusion chamber, another microfluidic device containing a microchannel where the packed fibers are placed lengthwise. Such a platform allows for dynamic control of the culture environment of the liver cells and, at the same time, allows for the uniform and effective delivery of oxygen and nutrients to the cells between the interfiber voids. The group also carried out the production of vascularized tissues by preparing and joining fibers covered with endothelial cells (EC), forming a structure very similar to the hepatic lobe, which is the unitary structure of the liver. Hydrogel fibers were prepared with sandwich barium alginate (BaAlg) hydrogel, in which HepG2 cells were encapsulated in the nucleus as a template. Using this system, the researchers studied the effects of perfusion flow rate on specific liver cell functions and cell viability.

As a last example, and a more complete demonstration of the enormous advantages that this technology can bring to the area of tissue engineering and regenerative medicine, the work of Grebenyuk et al. is relevant (Grebenyuk et al. 2023). The group developed a platform for organoid and tissue culture that allows, through fully synthetic vascularization, the generation of human tissues of unprecedented complexity. This vascularization is achieved through the generation of synthetic 3D vessel networks at the capillary scale. The cell growth and perfusion device consist of a dense network of regularly spaced tubular hydrogel capillaries, similar to a smooth grid, that allow for diffusion. Cells grow within this smooth mesh between the perfusion tubes which is hydrated via an external pumping system, thus allowing the culture medium to circulate throughout the tissue volume. The grid is printed onto hard plastic containing perfusion holes, forming a hermetic seal. The entire system is then incorporated into a perfusion chip connected to a peristaltic pump that circulates the cell culture medium.

The technology used to fabricate this complex geometric network was two-photon laser scanning light-curing, which allows fabrication in a scale range between 10 μm and 2000 μm. They also had to develop the right hydrogel for the fabrication of the perfusion tubular segments without significant post-polymerization swelling and hydration in order to not distort the geometry and seal of the device. They used a custom-formulated hydrophilic photopolymer based on polyethylene glycol diacrylate (PEGDA) with the addition of a significant amount of pentaerythritol triacrylate (PETA), as a crosslinking agent, and balanced with the addition of an inert "filler" component (Triton-X 100) to retain sufficient porosity to allow rapid diffusion.

The researchers demonstrated the versatility of this platform by perfusing neural and liver tissue cultures over the long term. In the case of neural tissue, they demonstrated that in the perfused system, cell differentiation improved remarkably, since the control organoids without perfusion practically remained in a pluripotent state. The tissue was highly viable and had biomarkers of neural differentiation. In the case of liver tissue, it exhibited improved phenotypic and functional characteristics compared to standard 2D and 3D organoid cultures. Therefore, the designed 3D culture platform provides a physiologically relevant microperfusion that allows greater tissue growth and differentiation, thus being an important advance for tissue engineering and regenerative medicine.

MHC for general biotechnology

In this chapter, general biotechnology refers to the use of microdevices in the field of molecular biology and immunology, for the cultivation of microorganisms such as bacteria, yeast, and foodborne pathogens' detection, for the culture of eukaryotic cells and cellular functions, for producing recombinant proteins and monoclonal antibodies in micro bioreactors, and for the immobilization of enzymes and proteins (Gao et al. 2020). The integration of microfluidics and biosensors provides a powerful tool to replace bulky traditional instruments, with the ability to combine chemical and biological components in a single platform (Gao et al. 2020).

Eukaryotic cells and microorganisms

Droplet-based microfluidics has been used to develop versatile platforms, allowing the efficient mixing of components on the micro-scale, with uniform porosity, elasticity, and distribution of functional groups, generating microenvironments for surface cell cultures and platforms for cascade reactions (Hauck et al. 2018). PNIPAAm (poly-N-isopropylacrylamide) microgels, which are widely used in biomedical applications, are produced through microfluidics and with temperature-sensitive behavior can provide a good insight into cell structure and composition (Hauck et al. 2022). In addition, a microfluidic device was used to validate the cell adhesion and propagation on printed glass slides, which contained two types of substrates (collagen and poly-lysine), with different gradients of adhesion ligands and the results demonstrated the effect of concentration gradients and the cellular activity (Vozzi et al. 2012). Leonaviciene et al. presented a microfluidics-based technique to isolate cells or biological samples, using semi-permeable hydrogel capsules, and work with individual cells simultaneously (Leonaviciene et al. 2020). This system was used for single genome amplification of bacteria and expansion of clones for screening for biodegradable plastic production.

Cell-laden microgels have enormous potential for cell therapy as they provide a 3D environment. The production of monodisperse cell-laden microgels with an ultrafine oil shell in double emulsion has been verified (Choi et al. 2016). In this way, the emulsion can rewet after polymerization of the innermost precursor droplet and facilitate the direct dispersion of microgels in an aqueous phase, resulting in high-yield encapsulation of cells in microgels without compromising cell functionality and viability. Furthermore, GeIMA (photo-cross-linkable biodegradable hydrogel gelatin methacrylate) can be used as three-dimensional tissue-like cell structures *in vitro* inside microfluidic devices, to cultivate RLC-18 rat liver cells (Takeuchi et al. 2020).

Yajima et al. studied two types of microfluidic devices made from alginate and gelatin (Yajima et al. 2014). NHI-3T3 cells were cultivated in the gelatin hydrogel microdevices, and it was confirmed that the cells adhered to the gelatin hydrogel, proliferated, and formed a monolayer maintaining high cell viability.

The process of immobilization of bacteria and biomolecules offers more protection from damage during use, including applications like controlled delivery of biofertilizers, biopesticides, bioreactors, bioremediation, environmental sensors, biomolecule production, and food safety sensing. *Saccharomyces cerevisiae* was desiccated and preserved in nanoliter-scale compartments termed PicoShells (Ng et al. 2022) allowing the diffusion of nutrients. High viability of desiccated and rehydrated *S. cerevisiae* yeast was observed inside PicoShells, with only a 14% decrease in viability compared to non-desiccated yeast over 8.5 weeks.

Escherichia coli and *Bacillus cereus* were encapsulated in hydrogel matrices using a microfluidic device and it was reported that biohybrid fibers maintain 90% cell viability (Daniele et al. 2013). The authors confirmed by LIVE/DEAD BacLight fluorescent (Invitrogen) that both Gram-positive and Gram-negative cells can survive and proliferate during the fiber production process and present metabolic activity in the biohybrid fiber. Lee et al. also demonstrated the efficiency of encapsulating *E. coli* in the biocompatible polyethylene glycol diacrylate (PEGDA) microdroplets (Lee et al. 2010). They demonstrated that uniform size, high porosity, and hydrophilic properties of the microdroplets could be obtained by controlling the continuous flow rate. The results demonstrated that it can be used in biotransformation, biosensing, bioremediation, and artificial cell engineering processes. Other microorganisms, such as *B. subtilis*, *S. aureus*, and *Lactobacillus* widely distributed in nature, water, and food were captured using a microreactor with immobilized monosaccharides, demonstrating the efficiency of the technique (Liu et al. 2014).

Lab-on-a-Chips (LoC) also are used for the cultivation of microorganisms with an environmental application and biomolecule production. Valei et al. reported the development of a microfluidic device containing a micro post array embedded in a microchannel to investigate the development of biofilm streamers of *Pseudomonas fluorescens*, which inhabits and forms biofilms in soil environments (Valiei et al. 2012). The authors found that streamer formation was highly dependent on the fluidic conditions used. Cao et al. studied the bacterial tolerance to different concentrations of the heavy metals, Cu and Zn, by two strains of actinobacteria in a droplet-based microfluidic device in a fast and efficient way to predict conditions in future bioremediation processes (Cao et al. 2013). The ethanol production model was observed by yeast monoculture, in non-core-shell microgel scaffolds (Ou et al. 2023). The spatial organization of cell communities, through the generation of programmable morphologies, demonstrates the increased bioactivities between consortia.

The microfluidic system can also be used for the quantitative determination of food-borne pathogens and toxins (Gao et al. 2020). A method based on droplet microfluidics for the sensitive and rapid detection of *Salmonella* directly from food samples has been described (An et al. 2020). *Salmonella* was encapsulated through microfluidics and detected by fluorescence, with a detection limit of 50 CFU/mL in 5 hours, being a more specific and sensitive method than the conventional one. It can be applied to other microorganisms of interest in food safety. A microfluidic-based impedance biosensor was used for the detection of *Salmonella*, *Legionella,* and *E. coli* O157:H7 in water samples and the results demonstrated that the biosensor was able to detect the pathogens, with the detection sensitivity of 5 to 6,2 times, in 30 to 40 minutes (Muhsin et al. 2022).

Molecular biology and immunology

One of the limitations of molecular detection technologies is an efficient method for target RNA amplification (Kifaro et al. 2022). A novel application of primer-immobilized networks (PIN) hydrogel microparticles was developed for the direct quantitative reversal of transitional PCR without prior purification of RNA from saliva. The viral RNA is captured by the porous PIN particles and converted into complementary DNA for subsequent qPCR (quantitative polymerase chain reaction). The sample processing time was reduced by 50 minutes with an efficiency of 95%, and the IAV M (*gene for influenza A virus*) and 5'UTR (*5' untranslated region*) chicken coronavirus genes were detected demonstrating the reproducibility and suitability of this developed method to detect other RNA virus genes.

Microfluidic devices for immunoassays have great advantages in cost, time, and sensitivity compared to traditional ELISA and immunochromatographic assays (Kasama et al. 2017). The authors performed C-reactive protein (CRP) detection as a model using immune pillar devices, in which antibodies were immobilized on microspheres, and some biomarkers were detected with high sensitivity (a-fetoprotein, PSA-Prostate-specific antigen, CRP, monocyte chemotactic protein 1, angiotensinogen, liver-type fatty acid binding protein) and staphylococcal enterotoxins A, B, C, D, and E. Cheng et al. reported the fabrication of the anisotropic colloidal crystal particles (CPPs) and CPPs were used in the multiplex DNA detection, demonstrating that this technique is highly practical for multiplex coding bioassays (Cheng et al. 2014).

For the development of an alternative monoclonal antibodies approach, an approach called CMAIL (Continuous Microfluidic Assortment of Interactive Ligands) is presented for sorting and screening binding antigens to antibody fragments displayed on bacteriophages (phages) (Hsiao et al. 2016), resulting in the separation of phages of different affinities and allowing the isolation of more than 10^5 CFU (colony forming units) antigen interacting phages, in a process that only took 40 minutes. On the other hand, alginate-based particles carrying encapsulated antibodies, using a triple-flow microdevice to induce hydrogel formation inside droplets before their collection off-chip were evaluated (Mazutis et al. 2015). The antibodies had been encapsulated and the particle biocompatibility within blood samples was confirmed and might be well used for future biomedical applications.

Enzymes and proteins

As in other areas of biotechnology presented in this chapter, microfluidics associated with hydrogel can be used for inmobilization, detection, quantification, and delivery of proteins and enzymes.

Ambrozic and Plazl developed a microfluidic channel device with programmable *in situ* formation of a 3D hybrid hydrogel/alginate by using Fe ions (Ambrožič and Plazl 2021). This system was controlled by electrical signals to adjust the thickness and volume of the hydrogel, and also the porosity of the hydrogel with the Fe ions. Through this novel device, the authors can observe the immobilization of bovine serum albumin (BSA) as a drug-mimicking protein, being able to study the diffusion of these molecules through programmed electrical conditions, and thus highlight that this method has a potential application.

Another example of BSA encapsulation is the work of Jiao et al. (2022). They presented the successful capture and release of bovine serum albumin (BSA) within a microfluidic device, by double crosslinked hydrogels (consisting of poly(N-isopropylacrylamide) (PNiPAAm), with a permanent crosslinker (N,N'-methylenebisacrylamide, BIS) and a reversible redox responsive crosslinker (N,N'-bis(acryloyl) cystamine, BAC), demonstrating the great potential of using Lab On a Chip (LOC) to detect or deliver proteins (Jiao et al. 2022).

ß-galactosidase immobilization was evaluated for increasing the self-life of this enzyme and making them reusable (Suvarli et al. 2022). Polymer nanoparticles were synthesized via aerosol thiol-ene-photopolymerization, and the test confirmed the presence of -SH (sulfhydryl) groups. The results demonstrated that the entrapment within acrylamide (AcAm) hydrogels resulted in particles with no relevant residual activity, whereas the polyethylene glycol diacrylate PEG-DA microparticles preserved a residual activity of 15–25%, compared to the activity of the free unbound enzyme, due to the formation of free radicals from the photoinitiator exposed to UV-LEDs. Hybrid protein-$Cu_3(PO_4)_2$ nanoflowers (NF) were fabricated using a vortex fluidic device (VFD), which has a rapidly rotating angular tube, and observed that the prepared product (laccase-LNF) had a 1.8-fold increase in its catalytic activity (Luo et al. 2020). The quantification of kinase activity (Bcr-Abl) of cell lysates, by a microfluidic platform with hydrogel, is an advantageous alternative in relation to traditional methods, as reported by Lee et al. and the effect of PEG porogen (pore-generating additives) concentration on the porosity of hydrogel pillars formed was recognized for increasing the quantification of activities of large enzymes from cell lysates and their relative performance in quantifying the activity of kinase from cell lysates (Lee et al. 2012).

Simon et al. presented a simple method to integrate hydrogel dots and enzymes in two microfluidic devices and observed the reactions of enzymatic cascades of 5 different enzymes. In the first cascade, the enzymes β-galactosidase, glucose oxidase, and horseradish peroxidase were used and the second cascade consists of the enzymes phospholipase D, choline oxidase, and again horseradish peroxidase. The cascades have a biological importance for applications in biosensing because these allow the detection of glucose and/or lactose and phosphatidylcholine, a relevant molecule for cell signaling pathway in liver diseases (Simon et al. 2019).

Conclusions

This chapter has presented the enormous potential of the use of hydrogels (stimulus-sensitive or not) in combination with microfluidic technology for the development of numerous applications in biotechnology. First of all, the main concepts on which microfluidic technology is based and the device fabrication techniques with their advantages and disadvantages were presented.

The main characteristics of hydrogels, particularly stimulus-sensitive ones, were described highlighting the advantages of their use in biotechnological applications, particularly in combination with microfluidic devices. These new materials have outstanding potential for miniaturized systems, especially for *Lab-on-a-chip* (LoC) technology and its application in biotechnology. While their use in combination with microfluidic devices is becoming increasingly widespread, the design and implementation of active systems is complex and laborious. However, its use as a biomimetic extracellular matrix system has enabled the advancement of a myriad of applications, mainly in the area of tissue engineering and regenerative medicine.

In the context of this chapter, biotechnology has been classified into four main application areas, which are medical (red biotechnology, e.g., tissue engineering applications, therapeutic devices, regenerative medicine, etc.), agricultural (green biotechnology), industrial (white biotechnology), and environmental (grey biotechnology) sectors. The chapter used this classification to analyze the contribution of the combination of microfluidics and hydrogels to each one of these areas. It can be seen that from 2010 to the present, almost three-quarters of the developments reported that appears with the search engine "microfluidics+hydrogel+biotechnology" (in PubMed and Scopus databases) belong to the Red Biotechnology area.

Some of the most relevant contributions of the combination of microfluidic devices with hydrogels in each area were presented. It could be observed that the most used hydrogels for encapsulation in microspheres or fabrication of fibers were MatriGel, Calcium Alginate, methacrylate gelatin (GelMa), poly (ethylene glycol) diacrylate (PEGDA), agarose-based hydrogel, and poly (ethylene glycol). Of the stimulus-sensitive hydrogels, the most commonly used were photodegradable and thermosensitive hydrogels.

In the area of red biotechnology, which includes tissue engineering, regenerative medicine, cell culture, or simply medical applications, the use of MHC has led to the development of many interesting devices or systems. However, it is noteworthy, according to our observations, that most of the reports in the literature can be grouped into only 3 general applications to a larger extent (Cell Encapsulation in Microcapsules, Microfiber Fabrication, and Cell Culture).

In general biotechnology, the use of MHC provides a powerful tool to replace traditional detection and avoid their limitations, such as large consumption, expensive instruments, long cycle, and complicated operation. Microfluidic technology exhibits distinct advantages in detection, including less sample consumption, fast and multiplexed detection, simple operation, multi-functional integration, small size, and portability.

Green, grey, and white biotech areas have benefited less from MHC. In the case of white area, it can be because most developments are in an early or laboratory stage, and also microfluidic and industrial scales are sometimes incompatible. Most works in this section use immobilized enzymes where hydrogels play a key role since they preserve enzyme function, and have a high load capacity thanks to their porosity. Green biotechnology is mainly related to food production in a sustainable way, while grey biotech is devoted to environmental issues. It is quite possible that we will see more applications of MHC in these areas in the near future. New developments could include plant growth assessment, biosensors for the evaluation of environmental contamination, water purification and transport systems, and bioremediation, among others.

The examples described in this chapter illustrate that the use of hydrogels in microfluidics can have diverse applications with great advantages in cost, time, and sensitivity.

References

Ahmed, E. 2015. Hydrogel: Preparation, characterization, and applications: A review. J. Adv. Res. 6: 105–121.

Altay, G., A. Abad-Lázaro, E.J. Gualda, J. Folch, C. Insa, S. Tosi et al. 2022. Modeling biochemical gradients *in vitro* to control cell compartmentalization in a microengineered 3D model of the intestinal epithelium. Adv. Healthc. Mater. 11: 1–15.

Ambrožič, R. and I. Plazl. 2021. Development of an electrically responsive hydrogel for programmable *in situ* immobilization within a microfluidic device. Soft Matter 17: 6751–6764.

Amin, R., S. Knowlton, A. Hart, B. Yenilmez, F. Ghaderinezhad, S. Katebifar et al. 2016. 3D-printed microfluidic devices. Biofabrication 8: 022001–022001.

An, X., P. Zuo and B.C. Ye. 2020. A single cell droplet microfluidic system for quantitative determination of food-borne pathogens. Talanta. 209: 120571–120571.

Babikian, S., G. Li, L. Li Wu, G.P. Li and M. Bachman. 2011. Microfluidic printed circuit boards. 2011 IEEE 61st Electronic Components and Technology Conference (ECTC), Lake Buena Vista, FL, USA. 2011: 1576–1581.

Beck, A., F. Obst, D. Gruner, A. Voigt, P.J. Mehner, S. Gruenzner et al. 2023. Fundamentals of hydrogel-based valves and chemofluidic transistors for lab-on-a-chip technology: a tutorial review. Adv. Mater. Technol. 8: 3, 2200417–2200417.

Becker, H. and L.E. Locascio. 2002. Polymer microfluidic devices. Talanta. 56: 267–287.

Bellan, L.M., M. Pearsall, D.M. Cropek and R. Langer. 2012. A 3D interconnected microchannel network formed in gelatin by sacrificial shellac microfibers. Adv. Mater. 24: 1–5.

Berger Fridman, I., J. Kostas, M. Gregus, S. Ray, M.R. Sullivan, A.R. Ivanov et al. 2021. High-throughput microfluidic 3D biomimetic model enabling quantitative description of the human breast tumor microenvironment. Acta Biomater. 132: 473–488.

Bhardwaj, T., L.N. Ramana and T.K. Sharma. 2022. Current advancements and future road map to develop ASSURED microfluidic biosensors for infectious and non-infectious diseases. Biosensors 12: 357–357.

Buchholz, K., V. Kasche and U. Bornscheuer. 2012. Biocatalysts and Enzyme Technology, Weinheim.

Bulpitt, P. and D. Aeschlimann. 1999. New strategy for chemical modification of hyaluronic acid: Preparation of functionalized derivatives and their use in the formation of novel biocompatible hydrogels. J. Biomed. Mater. Res. 47: 152–169.

Cao, J., D. Kürsten, K. Krause, E. Kothe, K. Martin, M. Roth et al. 2013. Application of micro-segmented flow for two-dimensional characterization of the combinatorial effect of zinc and copper ions on metal-tolerant *Streptomyces* strains. Appl. Microbiol. Biotechnol. 97: 8923–8930.

Chang, Y.J. and Y. Hui. 2019. Progress of microfluidics based on printed circuit board and its applications. Elsevier, Chinese J. Analytical Chem. 47: 965–975.

Cheng, Y., C. Zhu, Z. Xie, H. Gu, T. Tian, Y. Zhao et al. 2014. Anisotropic colloidal crystal particles from microfluidics. J. Colloid Interface Sci. 421: 64–70.

Ching, T., J. Vasudevan, S.Y. Chang, H.Y. Tan, A. Sargur Ranganath, C.T. Lim et al. 2022. Biomimetic vasculatures by 3D-printed porous molds. Small. 18: 39, 2203426–2203426.

Cho, A.N., Y. Jin, Y. An, J. Kim, Y.S. Choi, J.S. Lee et al. 2021. Microfluidic device with brain extracellular matrix promotes structural and functional maturation of human brain organoids. Nat. Commun. 12: 4730–4730.

Choi, C.H., H. Wang, H. Lee, J.H. Kim, L. Zhang, A. Mao et al. 2016. One-step generation of cell-laden microgels using double emulsion drops with a sacrificial ultra-thin oil shell. Lab. Chip. 16: 1549–1555.

Choi, D., E. Jang, Æ.J. Park and W. Koh. 2008. Development of microfluidic devices incorporating non-spherical hydrogel microparticles for protein-based bioassay. Microfluid. Nanofluidics 5: 703–710.

Daniele, M.A., S.H. North, J. Naciri, P.B. Howell, S.H. Foulger, F.S. Ligler et al. 2013. Rapid and continuous hydrodynamically controlled fabrication of biohybrid microfibers. Adv. Funct. Mater. 23: 698–704.

Dabiri, S.M., E. Samiei, S. Shojaei, L. Karperien, B.K. Jush, T. Walsh et al. 2021. Multifunctional thermoresponsive microcarriers for high-throughput cell culture and enzyme-free cell harvesting. Small 17, 2103192: 1–14.

Fattahi, P., A. Rahimian, M.Q. Slama, K. Gwon, A.M. Gonzalez-Suarez, J. Wolf et al. 2021. Core–shell hydrogel microcapsules enable formation of human pluripotent stem cell spheroids and their cultivation in a stirred bioreactor. Sci. Rep. 11: 1–13.

Franke, M., S. Leubner, A. Dubavik, A. George, T. Savchenko, C. Pini et al. 2017. Immobilization of pH-sensitive CdTe Quantum Dots in a Poly(acrylate) hydrogel for microfluidic applications. Nanoscale Res. Lett. 12: 314–314.

Fujimoto, K., K. Higashi, H. Onoe and N. Miki. 2018. Development of a triple-coaxial flow device for fabricating a hydrogel microtube and its application to bioremediation. Micromachines Basel. 9: 76–76.

Gal, I., R. Edri, N. Noor, M. Rotenberg, M. Namestnikov, I. Cabilly et al. 2020. Injectable cardiac cell microdroplets for tissue regeneration. Small. 16; 8, 1904806–1904806.

Gao, H., C. Yan, W. Wu and J. Li. 2020. Application of microfluidic chip technology in food safety sensing. Sens. 20: 6, 1792–1792.

Govindasamy, N., H. Long, H.W. Jeong, R. Raman, B. Özcifci, S. Probst et al. 2021. 3D biomimetic platform reveals the first interactions of the embryo and the maternal blood vessels. Dev. Cell. 56: 3276–3287.

Goy, C., R. Chaile and R.E. Madrid. 2019. Microfluidics and hydrogel: A powerful combination. Elsevier. 145: 104314–104314.

Grebenyuk, S., A.R. Abdel Fattah, M. Kumar, B. Toprakhisar, G. Rustandi, A. Vananroye et al. 2023. Large-scale perfused tissues via synthetic 3D soft microfluidics. Nat. Commun. 14: 1, 193–193.

Hauck, N., N. Seixas, S.P. Centeno, R. Schlüßler, G. Cojoc, P. Müller et al. 2018. Droplet-assisted microfluidic fabrication and characterization of multifunctional polysaccharide microgels formed by multicomponent reactions. Polymers 10: 10, 1055–1055.

Hauck, N., T. Beck, G. Cojoc, R. Schlüßler, S. Ahmed, I. Raguzin et al. 2022. PNIPAAm microgels with defined network architecture as temperature sensors in optical stretchers. Mater. Adv. 3: 6179–6190.

Heintz, K.A., M.E. Bregenzer, J.L. Mantle, K.H. Lee, J.L. West and J.H. Slater. 2016. Fabrication of 3D biomimetic microfluidic networks in hydrogels. Adv. Healthc. Mater. 5: 1–8.

Hsiao, Y.H., C.Y. Huang, C.Y. Hu, Y.Y. Wu, C.H. Wu, C.H. Hsu et al. 2016. Continuous microfluidic assortment of interactive ligands (CMAIL). Scientific Reports 6: 32454–32454.

Ibrahim, M. and M. Richardson. 2017. *In vitro* development of zebrafish vascular networks. Reprod. Toxicol. 70: 102–115.

Ito, T., T. Ota, R. Kono, Y. Miyaoka, H. Ishibashi, M. Komori et al. 2021. Pump-free microfluidic hemofiltration device. Micromachines 12: 8, 992–992.

Jafarkhani, M., Z. Salehi, M.A. Shokrgozar and S. Mashayekhan. 2018. An optimized procedure to develop a three dimensional microfluidic hydrogel with parallel transport networks. Int. J. Numer. Methods Biomed. Eng. 35: 1, e3154.

Jain, V., T.P. Raj, R. Deshmukh and R. Patrikar. 2017. Design, fabrication and characterization of low cost printed circuit board based EWOD device for digital microfluidics applications. Microsyst. Technol. 23: 389–397.

Jiang, N., Y. Montelongo, H. Butt and A.K. Yetisen. 2018. Microfluidic contact lenses. Small. 14: 15, 1704363–1704363.

Jiao, C., F. Obst, M. Geisler, Y. Che, A. Richter, D. Appelhans et al. 2022. Reversible protein capture and release by redox-responsive hydrogel in microfluidics. Polymers. 14: 2, 267–267.

Kasama, T., N. Kaji, M. Tokeshi and Y. Baba. 2017. Fabrication and evaluation of microfluidic immunoassay devices with antibody-immobilized microbeads retained in porous hydrogel micropillars. Methods Mol. Biol. 1547: 49–56.

Katmiwati, E. and T. Nakanishi. 2014. Dye sorption and swelling of poly(vinyl alcohol) hydrogels in Congo red aqueous solution. Macromol. Res. 22: 731–737.

Kifaro, E.G., M.J. Kim, S. Jung, J.Y. Noh, C.S. Song, G. Misinzo et al. 2022. Direct reverse transcription real-time PCR of viral RNA from Saliva samples using hydrogel microparticles. Biochip J. 16: 409–421.

Kim, C.J. 2000. Microfluidics using the surface tension force in microscale. Microfluid. Devices Syst. III. USA. 4177: 18–24.

Kobayashi, A., K. Yamakoshi, Y. Yajima, R. Utoh, M. Yamada and M. Seki. 2013. Preparation of stripe-patterned heterogeneous hydrogel sheets using microfluidic devices for high-density coculture of hepatocytes and fibroblasts. J. Biosci. Bioeng. 116: 761–767.

Koetting, M.C., J.T. Peters, S.D. Steichen and N.A. Peppas. 2015. Stimulus-responsive hydrogels: Theory, modern advances, and applications. Mater. Sci. Eng. R Rep. J. 93: 1–49.

Koh, W. and M. Pishko. 2005. Immobilization of multi-enzyme microreactors inside microfluidic devices. Sens. Actuators B Chem. 106: 335–342.

Koh, A., D. Kang, Y. Xue, S. Lee, R.M. Pielak, J. Kim et al. 2016. A soft wearable microfluidic device for the capture, storage, and colorimetric sensing of sweat. Sci. Transl. Med. 8: 366ra165–366ra165.

Kordi, M., R. Salami, P. Bolouri, N. Delangiz, B.A. Lajayer and E.D. Van Hullebusch. 2022. White biotechnology and the production of bio-products. Springer 2: 413–429.

Kruss, S., L. Erpenbeck, M. Schön and J. Spatz. 2012. Circular, nanostructured and biofunctionalized hydrogel microchannels for dynamic cell adhesion studies. Lab. Chip 12: 3285–3289.

Lee, A.G., D.J. Beebe and S.P. Palecek. 2012. Quantification of kinase activity in cell lysates via photopatterned macroporous poly(ethylene glycol) hydrogel arrays in microfluidic channels. Biomed. Microdevices 14: 247–257.

Lee, J., S.H. Kim, Y.C. Kim, I. Choi and J.H. Sung. 2013. Fabrication and characterization of microfluidic liver-on-a-chip using microsomal enzymes. Enzyme Microb. Technol. 53: 159–164.

Lee, K.G., T.J. Park, S.Y. Soo, K.W. Wang, B.I.I. Kim, J.H. Park et al. 2010. Synthesis and utilization of *E. coli*-encapsulated PEG-based microdroplet using a microfluidic chip for biological application. Biotechnol. Bioeng. 107: 747–751.

Lee, K.Y. and D.J. Mooney. 2001. Hydrogels for tissue engineering. Chem. Rev. 101: 1869–1880.

Leonaviciene, G., K. Leonavicius, R. Meskys and L. Mazutis. 2020. Multi-step processing of single cells using semi-permeable capsules. Lab. Chip 20: 4052–4062.

Li, C., H. Dai, C. Gao, T. Wang, Z. Dong and L. Jiang. 2019. Bioinspired inner microstructured tube controlled capillary rise. Proc. Natl. Acad. Sci. U. S. A. 116: 12704–12709.

Li, J., Y. Wang, E. Dong and H.C. Chen. 2013. USB-driven microfluidic chips on printed circuit boards. Lab. Chip 14: 860–864.

Liu, H., Y. Wang, H. Wang, M. Zhao, T. Tao, X. Zhang et al. 2020. A droplet microfluidic system to fabricate hybrid capsules enabling stem cell organoid engineering. Adv. Sci. 7: 1903739–1903748

Liu, X., Z. Lei, F. Liu, D. Liu and Z. Wang. 2014. Fabricating three-dimensional carbohydrate hydrogel microarray for lectin-mediated bacterium capturing. Biosens. Bioelectron. 58: 92–100.

Logun, M.T., N.S. Bisel, E.A. Tanasse, W. Zhao, B. Gunasekera, L. Mao et al. 2016. Glioma cell invasion is significantly enhanced in composite hydrogel matrices composed of chondroitin 4- and 4,6-sulfated glycosaminoglycans. J. Mater. Chem. B 4: 6052–6064.

Luk, V.N., L.K. Fiddes, V.M. Luk, E. Kumacheva and A.R. Wheeler. 2012. Digital microfluidic hydrogel microreactors for proteomics. Wiley Online Libr. 12: 1310–1318.

Luo, X., A.H.M. Al-Antaki, A. Igder, K.A. Stubbs, P. Su, W. Zhang et al. 2020. Vortex fluidic-mediated fabrication of fast gelated silica hydrogels with embedded laccase nanoflowers for real-time biosensing under flow. ACS Appl. Mater. Interfaces 12: 51999–52007.

Mark, D., S. Haeberle, G. Roth, F. Von Stetten and R. Zengerle. 2009. Microfluidic lab-on-a-chip platforms: requirements, characteristics and applications. Chem. Soc. Rev. 39: 1153–1182.

Marshall, L.E., K.F. Goliwas, L.M. Miller, A.D. Penman, A.R. Frost and J.L. Berry. 2017. Flow–perfusion bioreactor system for engineered breast cancer surrogates to be used in preclinical testing. J. Tissue Eng. Regen. Med. 11: 1242–1250.

Martins, E., D. Poncelet, R.C. Rodrigues and D. Renard. 2017. Oil encapsulation techniques using alginate as encapsulating agent: applications and drawbacks. J. Microencapsul. 34: 754–771.

Mazutis, L., R. Vasiliauskas and D.A. Weitz. 2015. Microfluidic production of alginate hydrogel particles for antibody encapsulation and release. Macromol. Biosci. 15: 1641–1646.

Mohseni, K. 2015. Surface tension, capillarity, and contact angle. Encyclopedia of Microfluidics and Nanofluidics: 1948–1955.

Molinaro, G., J. Leroux, J. Damas and A. Adam. 2002. Biocompatibility of thermosensitive chitosan-based hydrogels: an *in vivo* experimental approach to injectable biomaterials. Biomaterials 23: 2717–2722.

Mross, S., S. Pierrat, T. Zimmermann and M. Kraft. 2015. Microfluidic enzymatic biosensing systems: A review. Biosens. Bioelectron. 70: 376–391.

Mu, X., W. Zheng, L. Xiao, W. Zhang and X. Jiang. 2013. Engineering a 3D vascular network in hydrogel for mimicking a nephron. Lab. Chip 13: 1612–1618.

Muhsin, S.A., M. Al-Amidie, Z. Shen, Z. Mlaji, J. Liu, A. Abdullah et al. 2022. A microfluidic biosensor for rapid simultaneous detection of waterborne pathogens. Biosens. Bioelectron. 203: 113993–114003.

Nasser, H.A., M. Mahmoud, M.M. Tolba, R.A. Radwan, N.M. Gabr, A.A. ElShamy et al. 2021. Pros and cons of using green biotechnology to solve food insecurity and achieve sustainable development goals. Euro-Mediterr. J. Environ. Integr. 6: 1–19.

Ng, S., C. Williamson, M. van Zee, D. Di Carlo and S.R. Santa Maria. 2022. Enabling clonal analyses of yeast in outer space by encapsulation and desiccation in hollow microparticles. Life 12: 1168–1185.

Nie, J., Q. Gao, Y. Wang, J. Zeng, H. Zhao, Y. Sun et al. 2018. Vessel-on-a-chip with hydrogel-based microfluidics. Small 14: 1802368–1802392.

Nielsen, J.B., R.L. Hanson, H.M. Almughamsi, C. Pang, T.R. Fish and A.T. Woolley. 2020. Microfluidics: innovations in materials and their fabrication and functionalization. Analytical Chemistry 92: 150–168.

Noviana, E., T. Ozer, C.S. Carrell, J.S. Link, C. McMahon, I. Jang et al. 2021. Microfluidic paper-based analytical devices: from design to applications. Chem. Rev. 121: 11835–11885.

Novotný, J. and F. Foret. 2017. Fluid manipulation on the micro-scale: Basics of fluid behavior in microfluidics. J. Sep. Sci. 40: 383–394.

Ortseifen, V., M. Viefhues, L. Wobbe and A. Grünberger. 2020. Microfluidics for biotechnology: bridging gaps to foster microfluidic applications. Front. Bioeng. Biotechnol. 8: 589074–589086.

Ou, Y., S. Cao, Y. Zhang, H. Zhu, C. Guo, W. Yan et al. 2023. Bioprinting microporous functional living materials from protein-based core-shell microgels. Nat. Commun. 14: 322–336.

Ozcelik, A., B.I. Abas, O. Erdogan, E. Cevik and O. Cevik. 2022. On-chip organoid formation to study CXCR4/CXCL-12 chemokine microenvironment responses for renal cancer drug testing. Biosensors 12: 1177–1188.

Peschke, T., P. Bitterwolf, S. Gallus, Y. Hu, C. Oelschlaeger, N. Willenbacher et al. 2018. Self-assembling all-enzyme hydrogels for flow biocatalysis. Angew. Chem. 130: 17274–17278.

Piao, Y., D. Ju, M. Reza, M. Park and T. Seok. 2015. Enzyme incorporated microfluidic device for *in-situ* glucose detection in water-in-air microdroplets. Biosens. Bioelectron. 65: 220–225.

Qiu, Y. and K. Park. 2001. Environment-sensitive hydrogels for drug delivery. Adv. Drug Deliv. 53: 321–339.

Regmi, S., C. Poudel, R. Adhikari and K.Q. Luo. 2022. Applications of microfluidics and organ-on-a-chip in cancer research. Biosensors 12: 459–498.

Ren, K., J. Zhou and H.W. Hongkai. 2013. Materials for microfluidic chip fabrication. ACS Publ. 46: 2396–2406.

Ren, K., Y. Chen and H. Wu. 2014. New materials for microfluidics in biology. Curr. Opin. Biotechnol. 25: 78–85.

Rogers, C.I., K. Qaderi, A.T. Woolley and G.P. Nordin. 2015. 3D printed microfluidic devices with integrated valves. Biomicrofluidics 9: 016501–016501.

Sackmann, E.K., B.P. Casavant, S.F. Moussavi-Harami, D.J. Beebe and J.M. Lang. 2014. Cell-based microfluidic assays in translational medicine. Engineering in Translational Medicine 927–956.

Saeki, K., H. Hiramatsu, A. Hori, Y. Hirai, M. Yamada, R. Utoh et al. 2020. Sacrificial alginate-assisted microfluidic engineering of cell-supportive protein microfibers for hydrogel-based cell encapsulation. ACS Omega 5: 21641–21650.

Scott, S. and A. Zulflqur. 2021. Fabrication methods for microfluidic devices: An overview. Micromachines 12: 319–357.

Sechriest, V.F., Y.J. Miao, C. Niyibizi, A.M. Westerhausen-Larson, H.W. Matthew, C.H. Evans et al. 2000. GAG-augmented polysaccharide hydrogel: A novel biocompatible and biodegradable material to support chondrogenesis. J. Biomed. Mater. Res. 49: 534–541.

Shin, S. and J. Hyun. 2017. Matrix-assisted three-dimensional printing of cellulose nanofibers for paper microfluidics. ACS Appl. Mater. Interfaces 9: 26438–26446.

Simon, D., F. Obst, S. Haefner, T. Heroldt, M. Peiter, F. Simon et al. 2019. Hydrogel/enzyme dots as adaptable tool for non-compartmentalized multi-enzymatic reactions in microfluidic devices. React Chem. Eng. 4: 67–77.

Squires, T.M. and S.R. Quake. 2005. Microfluidics: Fluid physics at the nanoliter scale. Rev. Mod. Phys. 77: 977–1026.

Štumberger, G. and B. Vihar. 2018. Freeform perfusable microfluidics embedded in hydrogel matrices. Materials 11: 2529–2529.

Suvarli, N., L. Wenger, C. Serra, I. Perner-Nochta, J. Hubbuch and M. Wörner. 2022. Immobilization of β-galactosidase by encapsulation of enzyme-conjugated polymer nanoparticles inside hydrogel microparticles. Front Bioeng. Biotechnol. 9: 818053–818067.

Takeuchi, M., T. Kozuka, E. Kim, A. Ichikawa, Y. Hasegawa, Q. Huang et al. 2020. On-chip fabrication of cell-attached microstructures using photo-cross-linkable biodegradable hydrogel. J. Funct. Biomater. 11: 18–32.

Tiwari, S., S. Bhat and K.M. Krishna. 2020. Design and fabrication of low-cost microfluidic channel for biomedical application. Sci. Rep. 10: 9215–9229.

Toepke, M.W., N.A. Impellitteri, J.M. Theisen and W.L. Murphy. 2013. Characterization of thiol-ene crosslinked PEG hydrogels. Macromol. Mater. Eng. 298: 699–703.

Valiei, A., A. Kumar, P.P. Mukherjee, Y. Liu and T. Thundat. 2012. A web of streamers: Biofilm formation in a porous microfluidic device. Lab Chip. 12: 5133–5137.

Vozzi, G., T. Lenzi, F. Montemurro, C. Pardini, F. Vaglini and A. Ahluwalia. 2012. A novel method to produce immobilised biomolecular concentration gradients to study cell activities: Design and modelling. Mol. Biotechnol. 50: 99–107.

Vutha, A.K., R. Patenaude, A. Cole, R. Kumar, J.N. Kheir and Polizzotti, B.D. 2022. A microfluidic device for real-time on-demand intravenous oxygen delivery. Proc. Natl. Acad. Sci. USA. 119: e2115276119–125.

Waheed, S., J. Cabot, N. Macdonald and T.L. Chip. 2016. 3D printed microfluidic devices: enablers and barriers. Lab Chip. 16: 1993–2013.

Wang, J., C. Wu, C.H. Yan, H. Chen and S. You. 2021. Nutritional targeting modification of silkworm pupae oil catalyzed by a smart hydrogel immobilized lipase. Food Funct. 14: 6240–6253.

Whitesides, G. 2006. The origins and the future of microfluidics. Nature 442: 368–373.

Yajima, Y., M. Yamada, E. Yamada, M. Iwase and M. Seki. 2014. Facile fabrication processes for hydrogel-based microfluidic devices made of natural biopolymers. Biomicrofluidics 8: 024115–024125.

Yajima, Y., C.N. Lee, M. Yamada, R. Utoh and M. Seki. 2018. Development of a perfusable 3D liver cell cultivation system via bundling-up assembly of cell-laden microfibers. J. Biosci. Bioeng. 126: 111–118.

Yamada, M., R. Utoh, K. Ohashi, K. Tatsumi, M. Yamato, T. Okano et al. 2012. Controlled formation of heterotypic hepatic micro-organoids in anisotropic hydrogel microfibers for long-term preservation of liver-specific functions. Biomaterials 33: 8304–8315.

Yamahira, S., T. Satoh, F. Yanagawa, M. Tamura, T. Takagi, E. Nakatani et al. 2020. Stepwise construction of dynamic microscale concentration gradients around hydrogel-encapsulated cells in a microfluidic perfusion culture device. R Soc Open Sci. 7: 200027–200039.

Zhang, X., L. Li and C. Luo. 2016. Gel integration for microfluidic applications. Lab Chip. 16: 1757–76.

Zhao, S., Y. Chen, B.P. Partlow, A.S. Golding, P. Tseng, J. Coburn et al. 2016. Bio-functionalized silk hydrogel microfluidic systems. Biomaterials 93: 60–70.

Chapter 4

Electrochemical Biosensors Based Upon Stimuli-Sensitive Gels

Kamil Marcisz, Klaudia Kaniewska and *Marcin Karbarz**

Introduction

The unique properties of smart hydrogels combined with conducting surfaces opened new possibilities in the construction of sensors and biosensors. Electrodes modified with thin layers of gel exhibit properties that are usually unattainable for bare electrodes. Typically, a thin gel layer provides a highly permeable matrix for analytes, and can also serve as an immobilizing/impermeable matrix for larger molecules. Hydrogel layer is able to protect the electrode surface from unwanted environmental influences and insulate the environment from the usually metallic, rough, and hard electrode surfaces. Such properties are strongly desired for the construction of biosensors. In addition, the ability of smart gel materials to undergo a reversible volume phase transition related to significant changes in their volume is of great interest from the point of view of controlling of electrochemical signal by an external factor.

Biosensors

Biosensors are analytical devices to detect and measure the concentration of specific biological or chemical substances. Their construction consists of a biological sensing element (a living organism or biological molecules, specially enzymes or antibodies),

Faculty of Chemistry, Biological and Chemical Research Center, University of Warsaw, 1 Pasteura Str., PL-02-093 Warsaw, Poland.
Emails: kmarcisz@chem.uw.edu.pl, kkaniewska@chem.uw.edu.pl
* Corresponding author: karbarz@chem.uw.edu.pl

a transducer (which transforms information about concentration of a target substance/analyte into an analytical signal), and an electronic system. From the point of view of the principle of transducer, sensors can be divided into the following groups: electrochemical, optical, electronic, piezoelectric, gravimetric, pyroelectric and magnetic (Chadha et al. 2022). Among them, electrochemical biosensors have proven to be useful in a wide range of applications and have become the most commonly used class of biosensors (Chaubey and Malhotra 2002). In this kind of biosensors, the biological sensing element selectively reacts with the target substance/analyte and an electrical signal, which is proportional to the concentration of the analyte, appears. There are four main approaches that are used to detect the electrical signal that occurs during a biorecognitive event; therefore, electrochemical biosensors can be classified as: amperometric, potentiometric, impedance and conductometric. Electrochemical biosensors are usually based on reactions catalyzed by redox enzymes. During these reactions, electrons are produced or consumed. The main part of a biosensor usually contains three electrodes: a working electrode, a reference electrode and a counter electrode. The reference electrode acts as a reference in measuring and controlling the potential of the working electrode, without passing current. The only role of the counter electrode is to pass all the current needed to balance the current observed at the working electrode. The working electrode is modified with a redox enzyme, and the target analyte participates in the reaction occurring on its surface. Electrons can be transferred through the double layer (producing a current) or can contribute to the double layer potential (producing a voltage). The current is proportional to the analyte concentration measured at a given potential, or the potential can be measured at zero current, giving a logarithmic response. In the second case, a two-electrode system is used, which includes a working electrode and a reference electrode.

Hydrogels

Definition and properties

Hydrogels are useful materials in construction of electrochemical biosensors, serving as an immobilization matrix for the biosensing elements. Polymeric hydrogels belong to class of soft materials that consist of a hydrophilic network filled with an aqueous solvent. These networks can be formed by various natural and synthetic polymers cross-linked through covalent or/and noncovalent interactions. Hydrogels typically have a very high fluid content, often exceeding 95%. Due to their high solvent content and solid consistency, hydrogels combine the properties of solids and liquids. The presence of a polymer network effectively "immobilizes" the solvent, causing it to lose its fluid nature. On a macroscopic scale, the three-dimensional gel network is responsible for maintaining the hydrogel's shape, storing mechanical energy, and participating in deformation processes. Meanwhile, on a microscopic scale, diffusion processes involving small molecules occur within the hydrogel. Furthermore, polymeric hydrogels have the remarkable ability to easily trap larger entities such as enzymes, antibodies, nucleic acids, tissues, microorganisms, and inorganic nanoparticles within their polymeric networks (Ahmed 2015).

The unique properties of polymeric hydrogels, such as absorption of large amount of water, three-dimensional network that gives specific mechanical properties, thermal and chemical resistance, flexibility, non-toxicity, often biocompatibility, biodegradability and sorption of heavy metal ions and organic compounds, make them versatile materials with applications in various fields (Dhanjai et al. 2019, Karbarz et al. 2017, Mahinroosta et al. 2018).

Sensitivity to stimuli

In addition to the aforementioned advantageous characteristics, many polymeric gels exhibit another intriguing property: they undergo a phenomenon known as volume phase transition (VPT). This process entails the gel's conversion from the swollen to shrunken phase, or vice versa. The extent of the reversible change in the gel's volume during the phase transition can be up to a thousand times, resulting in large changes in its properties. The transition of the gel from one phase to another may be caused by a slight change in environmental conditions, such as temperature, pH, ionic strength, the presence of specific ions/molecules or an electric/magnetic field. These changes originate from a shift in the balance between repulsive intermolecular forces that make the polymer network expand, and attractive forces that make it shrink (Caló and Khutoryanskiy 2015, Ullah et al. 2015). Repulsive forces often involve electrostatic interactions between similarly charged groups, while osmotic pressure also contributes significantly to the expansion of polymer networks. The attractive interactions may be of the van der Waals type or may involve hydrophobic interactions between the polymer chains, hydrogen bonding, electrostatic interactions between functional groups of opposite charge, and the interactions of specific, added species with two or more sites on the polymeric chains. This remarkable ability to adapt to changing external conditions/stimuli sensitivity categorizes polymer gels as "smart" materials (Liu et al. 2022, 2020, Oh et al. 2013).

Thermo- and pH-sensitive gels are among the most extensively studied and utilized types of smart gels. A number of thermosensitive polymers have been already reported. Most of them are built from polymers that possesses a lower critical solution temperature (LCST) and exhibit a soluble-to-insoluble phase transition when temperature becomes higher than the LCST. The group of polymers characterized by LCST include poly(N-isopropylacrylamide) (Oh et al. 2013), poly[oligo(ethylene glycol)methacrylate] (Tian et al. 2016), poly(N-2-(diethylaminoethyl acrylamide)) (Garnier et al. 2003), poly(N,N-dimethylaminoethyl methacrylate) (Hu et al. 2006), and N-vinylcaprolactam (Bian et al. 2015). Another type of thermosensitive polymers is characterized by upper critical solution temperature (UCST) and undergoes the opposite, insoluble-to-soluble reversible phase transition when temperature becomes higher than the UCST. The cross-linked polymers with a LCST-type thermal behavior, swollen in a given solvent, undergo a reversible shrinking process (volume phase transition) stimulated by an increase in temperature. In the case of cross-linked UCST polymers, an increase in temperature leads to a transition from the shrunken- to the swollen state. When the groups able to change their charge in response to a change in pH (weak acid, base, or their salts) are introduced to the polymeric network, gels become sensitive to a change in pH (Hua et al. 2016, Karbarz et al. 2009, Kato et al. 2016).

Also, it's worth noting that for many applications, the rate of VPT process is a critical parameter. The rate of the gel swelling/shrinking transition is diffusion-controlled and strongly depends on size of the gel. The characteristic time, τ, of gel-volume changes is determined by the gel size, R, and the collective diffusion coefficient, D, of the polymer network: $\tau = R^2/D$ (Tanaka and Fillmore 1979). In typical polymeric gels, D is in the order of 10^{-7}–10^{-6} cm^2 s^{-1} and depends on polymer concentration, cross-link density, etc. (Hirose and Shibayama 1998). Calculations reveal that for the nanosize gels (smaller than 100 nm), microsize gels (0.1–100 µm) and macrogels (bigger than 100 µm), the characteristic time may be: up to 0.1 ms, from 0.1 ms to a few hours (~ 3 h), and more than ~ 3 h, respectively. From the point of view of the utility of the VPT phenomenon in electrochemical biosensors, this time should not exceed rather several tens of minutes. Consequently, working with very small gel particles has gained significant attention in recent years as it addresses the challenge of achieving a sufficiently rapid VPT process. Additionally, another approach to overcome the limitations associated with traditionally sized gels involves using a thin hydrogel layer attached to the surface of a substrate, such as an electrode (Bandehali et al. 2021, Marcisz et al. 2020, Thorne et al. 2011). Moreover, altering the structure of hydrogels and the hydrophilic/hydrophobic properties of the polymeric network can have a significant impact on the rate of gel swelling/shrinking transitions. Introducing a heterogeneous microporous or mesoporous structure, for example, allows for rapid water release or absorption during phase transitions (Zhang et al. 2008). The incorporation of hydrophilic chains into the polymer network can enhance the rate of hydrogel deswelling (Kaneko et al. 1998), and copolymerization with more hydrophilic monomers has been shown to expedite the response time (Zhang et al. 2002). Additionally, increasing the hydrophobicity of a microgel can lead to a faster transition to its deswollen state (Ahiabu and Serpe 2017).

Gel as part of electrochemical biosensor

Combining hydrogels sensitive to external stimuli with conducting surfaces opens up new possibilities in electrochemistry (Kaniewska 2017, 2018a). From the perspective of applying hydrogels for the modification of electrode surfaces, their transport properties play a crucial role. The transport of molecules or ions within the gel layer strongly depends on several factors, including the percentage of the polymer, cross-linker concentration, polymeric chain mobility, the presence of charged groups in the gel structure, and swelling ratio (Kaniewska et al. 2016, 2018b, Karbarz et al. 2013). Two extreme cases can be observed: diffusional transport of small probes, which closely resembles transport in a regular solution, and immobilization within an impermeable matrix for larger entities. In the latter scenario, organic, inorganic, and biological micro- and nanoparticles can be readily entrapped within the polymeric networks, and the electrode surface can be shielded from the unwanted influence of certain components, such as proteins, present in the solution.

In addition, modification of electrode surface with environmentally sensitive/smart gels allows control of electrochemical signal by external stimulus

(Fandrich et al. 2017). These smart interfaces, which can switch the ON-OFF signal, have found applications in the development of sensors and biosensors. In this field, many fundamental (Katz 2016, 2018a, b, Kaniewska et al. 2017) and application research papers (Katz et al. 2013) have been published. The VPT exhibited by thin gel layers attached to electrode surfaces can induce alterations in analytical signals through two distinct mechanisms. The shrinking process can cause decrease/abolish or increase of an electrochemical signal. Furthermore, the swelling and shrinking processes can be harnessed for the introduction and immobilization of enzymes and nanoparticles onto electrode surfaces.

Various techniques have been employed to modify electrode surfaces with hydrogel films. One such method is the drop-on technique, where a droplet of a pre-gel solution is placed on the surface and allowed to undergo solvent evaporation. The pre-gel solution typically comprises monomers, an initiator, and optionally an accelerator (Yao and Hu 2011). The thickness of the resulting films can be controlled by adjusting the volume of the pre-solution applied to the surface. However, a common drawback of this method is the relatively low stability of the formed layers. To enhance stability, the droplet deposition can be preceded by surface functionalization, which facilitates the formation of bonds with the hydrogel layer (Karbarz et al. 2005). An alternative approach involves the use of electrochemically-induced free radical polymerization. This method eliminates the need for prior surface functionalization and results in well-adhered, thin, smooth, and stable hydrogel layers (Marcisz et al. 2021a, Reuber et al. 2006). Spin coating and layer-by-layer techniques are also utilized in this context (Gentile et al. 2017, Matsukuma et al. 2006). In a newer trend, electrode functionalization is achieved by covering their surfaces with a monolayer of microgels. Two primary approaches are employed: the droplet method and a self-assembling process (Carvalho et al. 2021, Kaniewska et al. 2023, Marcisz et al. 2021b).

Electrochemical biosensors based on smart thin gel layer

Hydrogels as intelligent materials are excellent matrix for immobilization of biologically active compounds (Culver et al. 2017). The huge water content provides gentle and good environment for compounds like enzymes, proteins, etc. (Roquero et al. 2022). The smart interface electrode/hydrogel film with entrapped compounds allows to develop the novel biosensors (Tavakoli and Tang 2017).

The immobilization process is usually beneficial for enzyme operation (DiCosimo et al. 2013). The entrapment of enzymes in three-dimensional network allows for maintaining stability, whereas the ability to undergo volume phase transition can be applied for altering the enzyme activity. The change in volume upon phase transition is related to change in a pore size, and this parameter regulates the transport of species along the polymer network. Hydrogel deposited on electrode surface maintains the ability to swell/shrink in response to specific stimulus (Katz 2018a, b). In swollen state, the compounds that are smaller than the pore size can diffuse inside the gel structure, substrate molecules can reach active center of enzyme, and products can be transported through polymer network. During collapse, the pores diameter decrease, and transport of substrate is restricted (Gawlitza et al. 2012).

This behavior enables control of the enzyme action, and depending on hydrogel network composition, chosen trigger can control enzyme activity. In addition, the hydrogel matrix increases life of hydrogel-enzyme system due to protective function of network. The gel protects against denaturation/deactivation of such systems. The description of such smart electrode systems modified with environmentally sensitive gel layer was given by Klis et al. (Klis et al. 2009). They anchored thermosensitive poly(*N*-isopropylacrylamide) gel layer to the indium−tin oxide (ITO) electrode surface. The redox enzyme laccase from *Cerrena basidiomycetes* was physically entrapped in the polymer network of the gel layer attached to the electrode surfaces. Playing with temperature between temperature below (20°C) and above (35°C), volume phase transition in the modified electrode was switched ON and OFF. At 20°C, the gel layer is in swollen state and substrate, oxygen and redox probe can diffuse through the polymer network to enzyme active centers and catalytic reaction can occur. Elevation of temperature above VPT led to collapse of hydrogel film, and significantly decreases the efficiency of the enzymatic reaction. The modified electrode can be used for the oxygen concentration monitoring in solutions. Additionally, temperature control of the enzymatic efficiency of the gel layer is possible because the efficiency is restored after lowering back temperature. There were no undesirable changes in the enzyme structure during the temperature cycle. The electrode was stable, and over time no leakage of enzyme was observed. If the gel layer exhibits the upper critical solution temperature (UCST) behavior, the enzyme activity will be limited at low temperature, and increase after exceeding the USCT. Kappauf et al. described operation of hydrogel-enzyme system where UCST-type hydrogel poly(N-acryloyl glycinamide), (pNAGA), was employed as a smart matrix for immobilization of *Bacillus megaterium transaminase* (Kappauf et al. 2021). This system possessed high activity (97%) and confirmed that the enzyme added to the hydrogel pre-gel solution prior to free radical polymerization did not deactivate.

The glucose oxidase was used in biosensor construction with electrodeposition method on the glassy carbon (GC) electrode surface. The composite nanofilm obtained from chitosan, covalently linked with ferrocene and Au nanoparticles, was used as a matrix for the enzyme immobilization. The electrochemical properties were examined with electrochemical impedance spectroscopy and cyclic voltammetry. The ferrocene groups linked with chitosan chains were entrapped into film layer and were used as an electroactive center. However, chitosan polymer chains exhibit poor conductivity, and to improve these properties, gold nanoparticles were employed. The presence of the nanoparticles strongly enhanced electron transfer during the enzymatic reaction. It also has a strong effect on biosensory properties. It was found that the sensor response decreased from about 15s to 5s, and the sensitivity increased by at least 2 times compared to the results without Au nanoparticles. The biosensing nanocomposite film was characterized with a fast response to the glucose presence and very good reproducibility and stability in time. The detection limit was on the satisfactory level (circa 5.6 μM) (Qiu et al. 2009).

In another example, the electroactive hydrogel multilayer on the gold electrode surface was used to immobilize GOx enzyme (Li et al. 2012). To obtain the biosensor, spin coating layer-by-layer method was employed. Poly(2-methacryloyloxyethyl phosphorylcholine-*co-p*-vinylphenylboronic acid-*co*-vinylferrocene) (PMVF) and poly(vinyl alcohol) (PVA) chains bound with GOx were used to prepare the hydrogel layer. Vinylferrrocene groups play the role of polymeric electron transfer mediator and GOx acts as the biological sensing part. The covalent bond between PVA and GOx ensured that the enzyme molecules were entrapped in the hydrogel layer. The electrochemical properties were examined with cyclic voltammetry and chronoamperometric techniques. It was found that the ferrocene oxidation current increased following glucose addition. The increase in numbers of biosensor multilayers on the electrode surface led to increase in registered current response, which was indicative of the electron transfer process between each layer. The coupling preparation method with electroactive and enzyme modified polymers seems to be useful in other enzyme based electrochemical biosensors construction.

The conductive polymer, polypyrrole, and GOx were immobilized in poly(2-hydroxyethyl methacrylate), p(HEMA), hydrogel layer on the electrode surface. The conductive polymer was responsible for electron transfer from GOx to electrode surface, and strongly increase biosensor activity. The glucose sensitive biotransducer was tested for ascorbic acid interferences. The oxidized polypyrrol presence in hydrogel membrane led to blocking the negatively charged ascorbic acid diffusion. As a consequence, ascorbic acid molecules, in physiological important concentration range of scorbic acid, affected the current response only by 3–5%. After the biosensor storage for even 17 days, the decrease in sensing activity was not observed (Kotanen et al. 2013).

In another example the chitosan hydrogel layer on the GC electrode surface was used as a matrix for immobilization of platinum nanoparticles and GOx. The platinum nanoparticles were electrodeposited in the chitosan hydrogel film and in the next step, the glucose oxidase was introduced to the formed membrane. Platinum nanoparticles increase the electron transfer yield and it exhibits the appropriate biocompatibility for glucose oxidase immobilization. The platinum nanoparticles and chitosan gel layer coupling strongly enhanced the electrocatalytic H_2O_2 sensing properties, which is a product of GOx-glucose reaction. The obtained composite biosensor had a quick response time and a detection limit of circa 2 μM (Chen et al. 2010).

Another example of a biosensor for detecting hydrogen peroxide used horseradish peroxidase (HRP) as the detection element. The HRP enzyme was entrapped in the polyhydroxyl cellulose (PHC) hydrogel matrix on the ITO electrode surface. To anchor the gel-enzyme layer on the ITO electrode, a surface polymerization mixture was spread on it. The PHC was the appropriate microenvironment for HRP immobilization and allowed direct electron transfer between HRP and electrode surface upon catalyzed H_2O_2 reduction. This third-generation H_2O_2 biosensor was characterized with good stability and reproducibility (Feng et al. 2009).

The Au electrode modified with gelatin B hydrogel layer along with cytochrome c peroxidase (CCP) and horse heart cytochrome c (HHC) was used as an electrochemical hydrogen peroxide biosensor. To attach gel layer on the Au electrode surface, firstly it was treated with 6-mercaptohexanol and next gelatin B was added with drop drying method. Finally, the CCP and HHC were introduced into the formed layer. The CCP was responsible for biologically active center and HHC for mediating process. The HHC molecules were stabilized into the hydrogel layer through the electrostatic interaction with gelatin B, HHC was positively charged and gelatin B negatively. Hydrogen peroxide was electrocatalytically determined with modified Au electrode. The biosensor exhibited very fast response time and a linear response range from 0 to 0.3 mM (Wael et al. 2012).

In yet another case (Zhou et al. 2014) the indium tin oxide electrodes (ITO) modified with molecular hydrogel with embedded enzyme Cytochrome c (Cyt c) was used as third generation H_2O_2 biosensor. The ITO electrode surface was modified with gel layer with drop on method. For this, small portion of polymerizing solution of Fmoc-l-lysine, Fmoc-l-phenylalnine and sodium carbonate was placed on the electrode surface. Next, enzyme molecules were introduced to the polymer network. The hydrogel structure was responsible for stabilization of the Cyt c moieties near the electrode surface and enhanced the direct electron transfer process between the enzyme and the electrode. The biosensor exhibited high selectivity against interfering substances, wide linear range and low detection limit. The constructed biosensor had a long-term stability and could be successfully used for detection of hydrogen peroxide released from live cells.

The hydrogel-electrochemical system can be designed to obtain desired response. Polymer networks give the possibility to introduce functional group, like electroactive units acting as charge transfer mediator (Tatsuma et al. 1994). They can be covalently bound to polymer chains which have high stability. The combination of smart redox active hydrogel layer with biocompounds can be implemented in several ways. An example can be a modified electrode surface with linear thermoresponsive brush based on poly(N-isopropylacrylamide) copolymerized with vinylferrocene (pNIPA-Fc) (Nagel et al. 2007). This redox polymer enabled electrical contact between the cofactor pyrrolinoquinoline quinone (PQQ) of soluble glucose dehydrogenase (sGDH) and the electrode and can be used for sensitive detection of this enzyme as a prospective protein label. The porous structure of polymer significantly increased active surface area and allowed for effective regeneration of the sGDH cofactor by redox process. Due to presence of pNIPA, the temperature can switch the modified electrode ON-OFF. The increase in temperature above LCST shrank the (pNIPA-Fc) layer and limited electron transfer between enzyme, redox groups and electrode surface. At 20°C, detection limit of the hydrogel modified electrode was 0.5 nM of PQQ–sGDH.

Schuhmann and co-workers designed the thermoresponsive amperometric glucose biosensor with local heating at the interface electrode/hydrogel layer without heating analyzed solution (Pinyou et al. 2016). For this purpose, the heatable screen printed Au electrode was functionalized with the statistical copolymer [polymer I: poly(ω-ethoxytriethylenglycol methacrylate-*co*-3-(N,N-dimethyl-N-2-methacryloyloxyethyl ammonio) propanesulfonate-*co*-ω-butoxydiethylenglycol methacrylate-*co*-2-(4-benzoyl-phenoxy)ethyl methacrylate)] with a LCST temperature at 28°C and a pH-responsive redox-polymer [polymer II: poly(glycidyl methacrylate-*co*-allyl methacrylate-*co*-poly(ethylene glycol)methacrylate-*co*-

butyl acrylate-*co*-2-(dimethylamino)ethyl methacrylate)-[Os(bpy)$_2$(4-(((2-(2-(2-aminoethoxy)ethoxy)ethyl)amino)methyl)-*N*,*N*-dimethylpicolinamide)]$^{2+}$] and pyrroloquinoline quinone-soluble glucose dehydrogenase (PQQ−sGDH) acting as biological recognition element. This layer was obtained via electrochemically induced co-deposition. Role of Polymer I was immobilization of enzyme and introduction of sensitivity to temperature changes. Polymer II served as mediators for shuttling electrons between the active centers of enzyme and the Au surface, due to presence of Os-complexes. Due to the presence of the temperature-sensitive hydrogel layer, the biocatalytic activity of the modified electrode can be switched ON-OFF by locally heating the electrode surface.

Development of such electrochemical smart sensors allows design of multi-responsive hydrogel layers modified with bioactive compounds to obtain multi-controllable biosensors. This multi-sensitivity is achieved by selecting the appropriate gel composition. The work presented by Liang et al. illustrates this idea in the example of simple two component network (Liang et al. 2011). Authors electrochemically deposited pNIPAM-GOx gel layer on pyrolytic graphite (PG) electrode. As pNIPA gel exhibit sensitivity to more than one stimuli, the modified electrode responds to temperature, ionic concentration and methanol presence. The enzymatic reaction of glucose oxidase mediated by ferrocene carboxylic acid was modulated by listed factors. The temperature sensitivity was typical for pNIPA layer, reversible thermo switching property between the ON and OFF states. The altering ionic strength by changing the Na$_2$SO$_4$ concentrations between 0 and 0.30 M leads to swelling (low concentration) or shrinking (high concentration) of hydrogel film. The effect of methanol fraction in water/methanol mixture is more complex; the pNIPA hydrogel layer was in a shrunken state (where the electrochemical oxidation of glucose catalyzed by immobilized enzyme is not possible) for methanol concentration between 20 and 35%, whereas below and above this fraction, the gel layers swelled and the bioelectrocatalytic reaction took place. Authors presented the triply switchable electrochemical biosensors based on intelligent hydrogel interface. Intelligent multiresponsive system are evolving and becoming more complex, and this applies to both the recognition elements and the polymer network (Parlak et al. 2015, Yao et al. 2014, Zhang et al. 2012). Liu and co-workers presented interesting example of preparation of an intelligent interface (Zhang et al. 2014a). The semi interpenetrating polymer network (semi-IPN) film based on PDEA-HA-GOx (poly(N,N'-diethylacrylamide) with hyaluronic acid (HA)) was prepared on the electrode surface by enzyme-initiated radical polymerization method. This system demonstrated reversible thermo-, pH-, and ionic- (SO$_4$$^{2-}$) sensitivity, and can be used as a switchable glucose sensor in the presence of mediator in solution. Another possibility to achieve multisensitivity is to deposit copolymer layer, for example Hu and co-workers presented triply switchable bioelectrocatalysis based on poly(*N,N*-diethylacrylamide-co-4-vinylpyridine) copolymer hydrogel film with entrapped glucose oxidase (Liang et al. 2012).

Examples of intelligent interfaces have established a foundation for preparing biosensors with stimuli-responsive materials controlled by logic gates operation (Katz et al. 2012, Liu et al. 2012). Qu et al. have shown that electrodes with the

ON-OFF switch of hydrogel can be used in construction of a smart bioelectronic device (Qu et al. 2021). Authors created switchable biosensors by precise positioning of biocatalytic sites in hierarchical polymer brushes attached to electrode surface. The enzyme glucose oxidase was placed within diblock-copolymer brush nanostructures. pH-responsive poly(dimethylaminoethyl methacrylate) (pDMAEMA) brushes were obtained on the gold electrodes via surface-initiated atom transfer radical polymerization. The enzyme was absorbed on pDMAEMA blocks, and then the growth of temperature sensitive pNIPA blocks was initiated. This approach allows the layer to be properly designed to exhibit reversible switch ON-OFF behavior upon pH and temperature and to achieve sufficient enzyme activity (Fig. 1). The biosensor, gold electrode modified with pDMAEMA/GOx–*b*–pNIPA, was used to detect glucose. As electrochemical technique, differential pulse voltammetry (DPV) and FcCOOH as a mediator were selected. In ON state (pH = 5.0, T = 20°C), the pDMAEMA/GOx–*b*–pNIPAM modified Au electrode exhibited the linear response in range of 0.05–12.8 mM of glucose, sensitivity 5.87 μA mM^{-1} cm^{-2} and limit of detection equaled 23.9 μM. The OFF state was at pH = 8 and/or 45 degree C. Authors presented that architecture of co-polymer brush has significant impact on the sensitivity and responsiveness of such smart biosensors.

Yu et al. modified electrode surface with poly(*N*-isopropylacrylamide-*co*-N,N′-dimethylaminoethylmethacrylate) copolymer layer with the enzyme horseradish peroxidase (HRP); p(NIPA-*co*-DMEM)-HRP (Yu et al. 2016). The shrinking and swelling processes were controlled by the external stimuli, pH, CO_2, temperature and SO_4^{2-} concentration; therefore, the electrochemical response was controlled by 5-input/3-output logic gate. The electrochemical signal originated from reduction of hydrogen peroxide catalyzed by HRP (embedded in film, enzymatic reaction was mediated by $Fe(CN)_6^{3-}$ in solution). The system was expanded by addition of another enzyme, GOx. In that way, the cascade enzymatic reaction was employed to produce electrochemical signal, and made p(NIPA-*co*-DMEM)-HRP-GOx film sensitive to glucose concentration, simultaneously remaining sensitive to the other triggers. Presented system can be used as sophisticated glucose biosensor. The intelligent hydrogel materials combined with enzymes are good platform for electrochemical multianalyte biosensing.

At this point, let us discuss another class of sensors - imprinting biosensor. These sensors consist of molecularly imprinted polymers (MIPs), which selectively recognize molecules by size and shape, and can be deposited on electrode surface. The polymer composition can be designed to possess affinity sites to targeted molecules, which increase selectivity of molecular recognition; monomers can be selected to create chemical or physical interaction with the functional groups of the analyte (Cieplak and Kutner 2016). In general, hydrogel layer is polymerized with the template molecule, then this molecule is removed from the polymer network, and created cavity selectively recognizes target molecule. The usage of stimuli responsive polymers to build recognition element introduces additional function, the swelling and shrinking processes of the hydrogel allow for reversible adsorption/desorption of analyte. In this case, we should rather use the term artificial biosensor due to presence of artificial recognition units. With close to natural

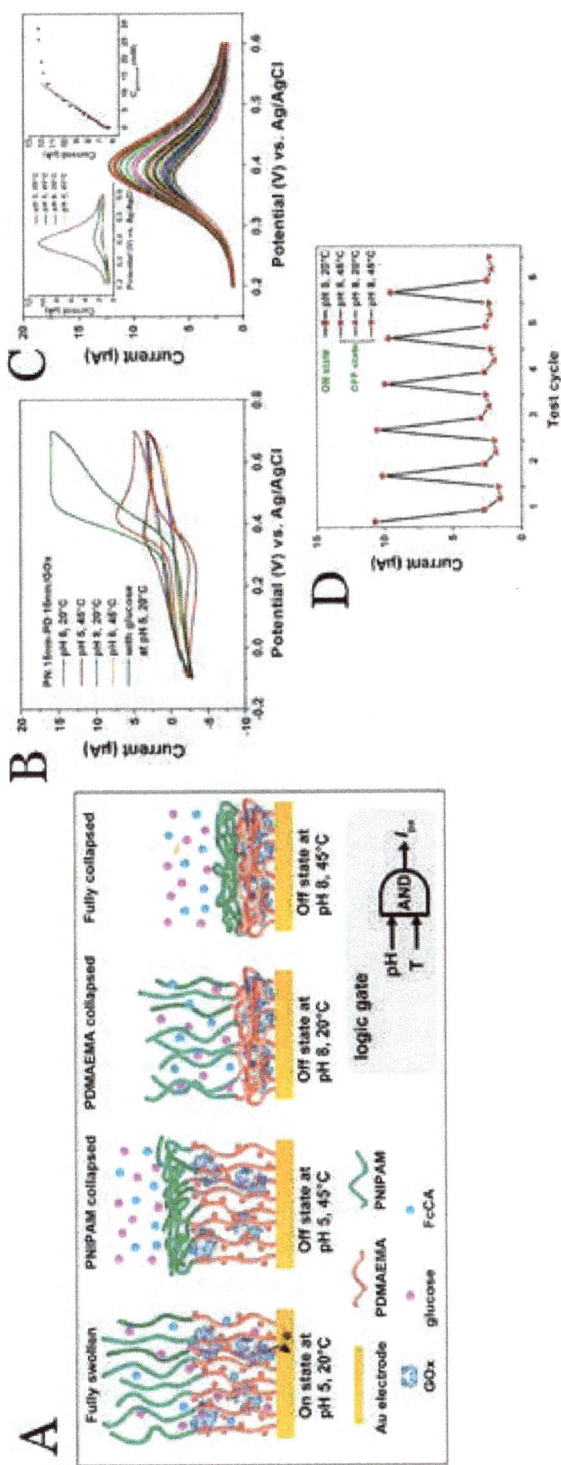

Fig. 1. (A) The architecture of electrode modified with PDMAEMA/GOx–*b*–PNIPAM and its operation as a logic gate. (B) Electrochemical response in different conditions for Au- PDMAEMA/GOx–*b*–PNIPAM electrode. (C) Differential pulse voltammograms (DPVs) registered for PDMAEMA/GOx–*b*–PNIPAM modified Au electrode with increasing addition of glucose. Right inset: the calibration curves for glucose. Left inset: DPVs for different conditions pH and temperature in the presence of 10 mM glucose (condition: PBS buffer with 1 mM FcCA at pH 5 and 20°C). (D) Oxidation peak current (DPV) of FcCOOH towards 10 mM glucose in consecutive cycles. Cycling was carried out at pH = 5 and 20°C, pH = 5 and 45°C, pH = 8 and 20°C, and in the presence of 10 mM glucose at pH 5 and 20°C. Scan rate: 50 mV s⁻¹.
Reprinted with permission from Ref. (Qu et al. 2021). Copyright 2018 Elsevier.

selectivity, they are characterized by much greater long-term stability, resistance to harsh external conditions, and lower costs. Modification of electrode surface with the MIP allow for easy to perform electrochemical detection of peptides, proteins, bacteria, drugs, etc. (Lee et al. 2022, 2023, Yasmeen et al. 2021, Jyoti et al. 2021). Wang and co-workers presented Bovine serum albumin electrochemical biosensor based on smart MIPs made of the thermosensitive pNIPA gel and cross-linked polyacrylamide (Wei et al. 2018). The advantage of this system is ability to self-clean from template molecules by cycling modified electrode in an appropriate potential range at 37 C in a PBS buffer. The operation of modified electrode with temperature, and presence of analyte BSA is presented in Fig. 2A with the idea of electrochemical detection. Authors employed both DPV and EIS for detection. In presence of targeted molecule, the cavities are occupied and the transport of redox probe to electrode surface is hindered; due to high selectivity, only the BSA can be re-bound to the polymer network, and fill the pores in hydrogel structure restricting diffusion of electroactive species. The calibration curves for the sensor, see Fig. 2B,

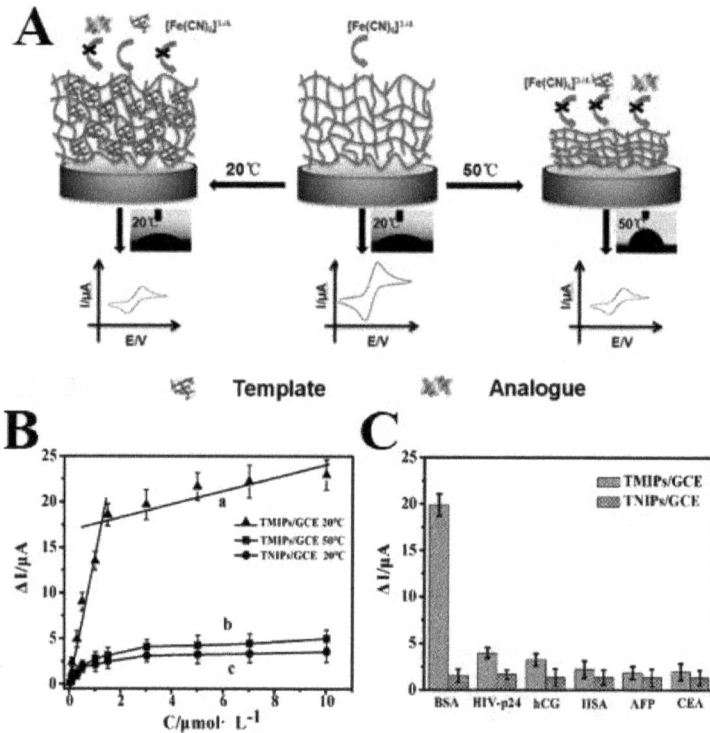

Fig. 2. (A) Scheme of electrode modified with TMIP layer and its performance as a BSA sensor in different conditions. (B) DPV peak current variation (ΔI) versus BSA concentration for TMIP modified electrodes at 20 and 50°C, and for TNIP modified electrode (thermosensitive non-molecular imprinted polymer layer) at 20°C. (C) DPV peak current variation in the presence of various interferents at 20°C for TMIP modified electrode. Interfering entities: HIV-p24 human immunodeficiency virus, HCG - human chorionic gonadotropin, HAS - human serum albumin, AFP - α-fetoprotein, CEA - carcino-embryonic antigen. Reprinted with permission from Ref. (Wei et al. 2018). Copyright 2018 Elsevier.

has two slopes related to existing two recognition sites: one specific and one with non-specific binding; nevertheless, the electrode is modified with same layer of TMIPs and measurement is made at 50°C. When hydrogel is in a contracted state, it has no response to analyte, just like electrode modified with the gel layer without molecular imprinting with analyte BSA (TNIPs). Figure 2C shows that TIMPs electrochemical biosensor is characterized by high selectivity even in presence of molecules with similar structure. Authors claim that TIMPs electrochemical biosensor can be used to detect BSA in a real sample (daily milk), and possess good stability and recovery.

Electrochemical biosensors based on smart microgel layer

Microgels are suspensions of colloidal-scale particles that contain a cross-linked polymer network. They possess many interesting features that make them very useful in sensors and biosensors construction (Caputo et al. 2019, Li et al. 2019, Ma et al. 2020, Mutharani et al. 2019, Shah et al. 2019, Wang et al. 2018, Zhang et al. 2019, Zhu et al. 2019). Similar to normal size gels, micro-sized polymeric particles exhibit ability to undergo a reversible phase transition. Because of the small sizes, this phenomenon occurred very quickly in response to specific stimuli. The electrode surfaces' modification with micro-sized hydrogel particles could lead to significant increase in possible applications of both materials. Microgels could be simply placed on the electrode surfaces with several methods, like "drop on", "paint on", dip coating or chemisorption process (Mackiewicz et al. 2018, Mutharani et al. 2020, Sigolaeva et al. 2017, Zhang et al. 2019).

The modification of p(NIPA) microgel polymer network with appropriate pH-sensitive groups, e.g., *N,N*-dimethylaminopropyl methacrylamide (DMAPMA), led to the introduction of the ionized groups to particle structures. The presence of the positively charged groups was used to immobilize negatively charged enzyme in the microgel polymer network. The attaching p(NIPA-co-DMAPMA) microgel particles on the graphite surface, by employing hydrophilic-hydrophobic interactions, led to first step in biosensor formation. The appropriate microgel particles' adsorption conditions were studied. It was found that at temperatures above the volume phase transition and at pH 9, with deprotonated DMAPMA groups, the packing on the electrode surface was most effective. To the anchored microgel layer, the enzyme – tyrosinase – was introduced and immobilized through the electrostatic interactions between the positively charged groups in the microgel polymer network and negatively charged enzyme molecules. To increase the entrapment/immobilization efficiency, a sponge like adsorption was used. For this purpose, the volume phase transition from the shrunken to swollen state was used. The system so obtained was examined as an electrochemical biosensor sensitive to phenol molecules. Firstly, the enzyme catalyzed the phenol oxidation to catechol, and subsequent catechol reduction on the electrode surface was monitored as an amperometric response (Fig. 3A). The activity of the sensors was strongly related to the temperature at which the adsorption of the microgel particles was carried out. The best results of biosensor response were obtained when the microgel particles were attached to the electrode surface at temperature above the volume phase transition (Fig. 3B) (Sigolaeva et al. 2014).

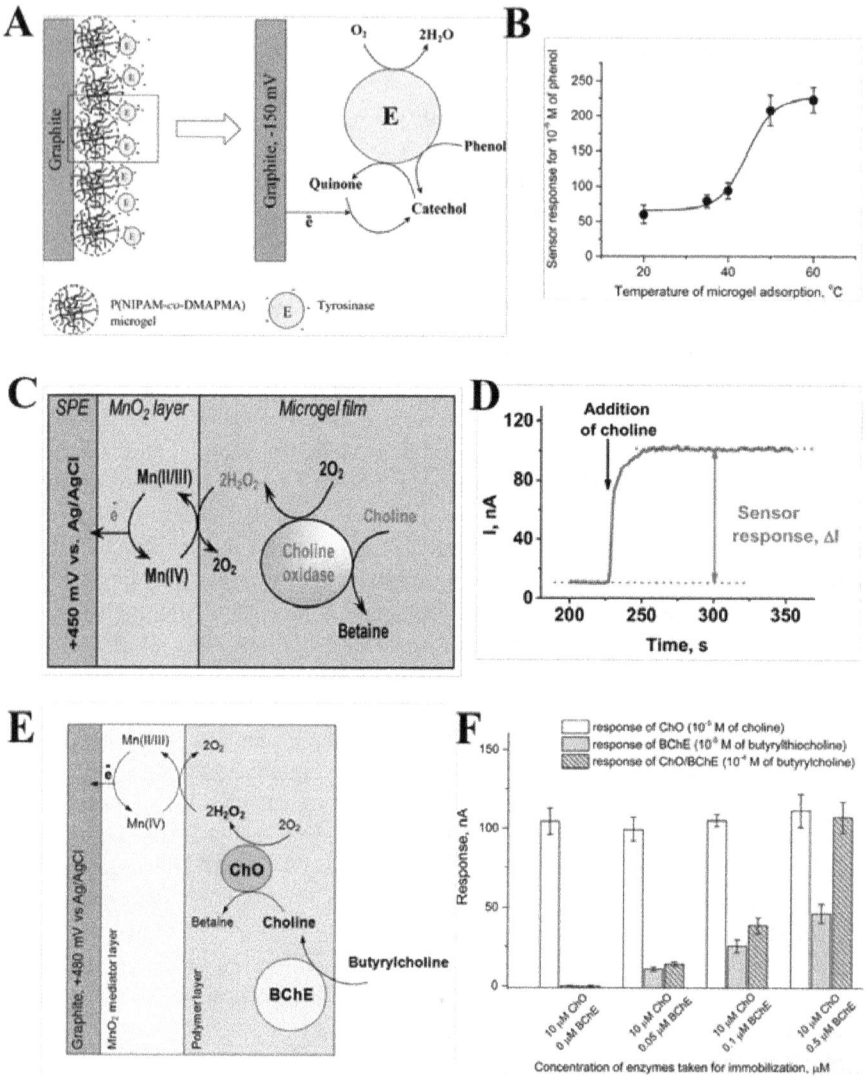

Fig. 3. (A) Scheme of phenol biosensor based on microgel with tyrosinase enzyme and detection mechanism. (B) Biosensor response signal as a function of microgel absorption temperature on the electrode surface [93]. Reprinted with permission from (Sigolaeva et al. 2014). Copyright (2014) American Chemical Society. (C) Scheme of choline biosensor on the SPE electrode surface modified with MnO₂ mediator layer. (D) Biosensor's electrochemical response to the choline presence. Reprinted with permission from Ref. (Sigolaeva et al. 2018). Copyright 2018 MDPI. (E) Scheme of bi-enzymatic microgel based biosensor's electrochemical response for butyrylcholine detection. (F) Electrochemical responses of biosensors with different enzyme concentrations immobilized in the microgel polymer network. Reprinted with permission from Ref. (Sigolaeva et al. 2015). Copyright 2018 American Chemical Society.

The pNIPA microgel with added N-3-(aminopropyl)methacrylamide (APMA) as a copolymer was used for modification of Screen-Printed Electrode (SPE) to obtain an enzymatic biosensor sensitive to choline presence (Sigolaeva et al. 2018). SPEs are small electrochemical measurement devices, usually made of plastic and equipped with a system of three electrodes (working, reference and counter). SPEs can be characterized with low cost of preparation and possibility to use with portable electrochemical devices allows to perform quick analysis. In the first step, the SPE surface was covered with a MnO_2 mediator layer. For better microgel particles packing on the electrode surface, attaching process was carried out at temperature above the microgel volume phase transition temperature. To immobilize the enzyme (choline oxidase) into the p(NIPAM-co-APMA) microgel layer, the presence of amino groups in the polymer network was used to covalently bind enzyme with microgel through glutaraldehyde cross-linking. This process significantly improved the biosensor stability. The obtained modified electrode was examined as an electrochemical biosensor sensitive to choline (Fig. 3C). The sensor response was monitored with amperometric technique (Fig. 3D).

In another work, p(NIPA-co-DMAPMA) microgel layer on the SPE surface, modified with MnO_2, was used as a matrix for glucose oxidase (GOx) immobilization. The microgel particles were anchored on SPE/MnO_2 electrode surface with dip-coating method at temperature above the volume phase transition temperature. To introduce GOx moieties to the microgel layer, absorption method via electrostatic interactions was employed. The resultant system worked as an electrochemical biosensor sensitive to glucose presence (Sigolaeva et al. 2022).

A thermoresponsive p(NIPA-co-DMAPMA) microgel with positively charged amine groups was also used to obtain a bi-enzymatic biosensor. Microgel particles were anchored on the graphite surface modified with MnO_2 mediator layer. In the next step, enzymes choline oxidase (ChO) and butyrylcholinesterase (BChE) were introduced to the microgel layer through the electrostatic interactions. The biosensor response was related to cascade enzymatic reaction. Firstly, butyrylcholine was hydrolyzed to choline with BChE, and secondly, choline was used by ChO to form betaine with simultaneously generation of hydrogen peroxide. The H_2O_2 molecules were then oxidized with the MnO_2 mediator layer. This process was observed as an electrochemical response (Fig. 3E) (Sigolaeva et al. 2015).

The pNIPA microgel was also used as a matrix in glucose amperometric sensor construction. The GOx enzyme and carbon microbeads were entrapped in the microgel polymer network. The carbon microbeads addition was used both as a micro-bioreactor and as a working electrode. The glucose oxidase presence caused the system to work as the biosensor sensitive to glucose molecules. The concentration of H_2O_2, a byproduct of the GOx enzymatic reaction that is correlated with glucose concentration, was monitored using electrochemical techniques. The microgel based biosensor was characterized by very fast response (circa 5 s), over a wide linear glucose detection range, low detection limit, low sample consumption and low fabrication cost (Mugo et al. 2019).

Another application of thermosensitive pNIPA microgel in design of glucose biosensor was described by Marcisz et al. (2018). Authors prepared microgel with

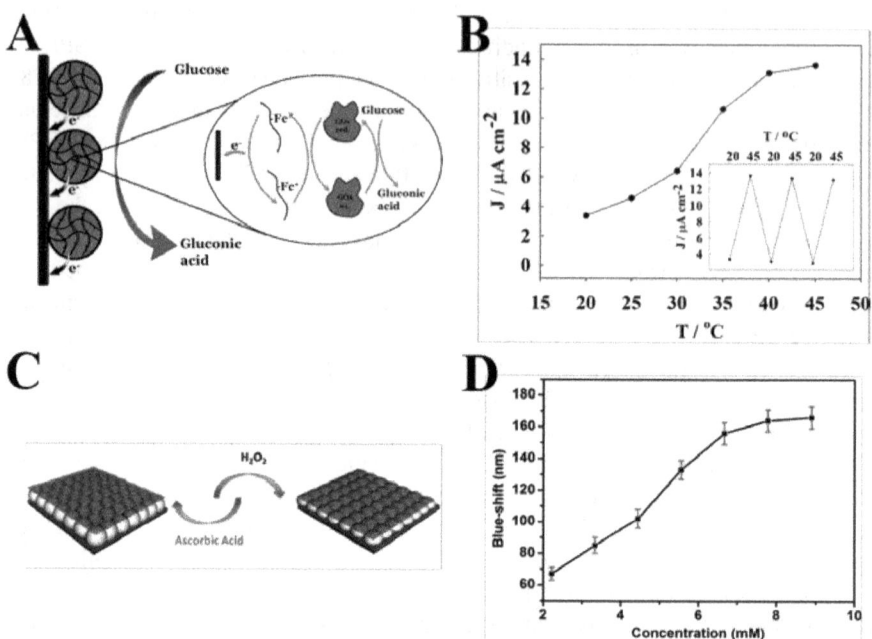

Fig. 4. (A) Scheme of thermoresponsive pNIPA microgel biosensor layer on the electrode surface and glucose sensing mechanism. (B) pNIPA and GOx microgel based biosensor with temperature dependent response. Inset: biosensor switchable ON-OFF temperature properties. Reprinted with permission from Ref. (Marcisz et al. 2018). Copyright 2018 Wiley. (C) Scheme of microgel particles between Au overlayers swelling and shrinking in response to oxidation and reduction of microgel ferrocene groups. (D) Blue-shift microgel sensor response as a function of glucose concentration. Reprinted with permission from (Zhang et al. 2016a). Copyright (2016) American Chemical Society.

covalently bound electroactive ferrocene species and also glucose oxidase enzyme. The resultant microparticles were anchored on the glassy carbon electrode surface. The GOx and ferrocene presence in the microgel layer on the conductive surface led to the system working as an electrochemical glucose sensor. The enzyme moieties played a role of active center and ferrocene electroactive groups were responsible for mediating process (Fig. 4A). It was found that the glucose biosensor works better at temperature higher than microgel volume phase transition temperature. In a shrunken state, registered response was significantly higher than obtained in swollen state (lower temperature) (Fig. 4B). As a result, the thermo-reversible properties were studied for the swollen and shrunken states. The reversible increase and decrease in registered current depended on the microgel swelling ratio. It was a typical behavior characteristic of ON-OFF systems. The electrochemical biosensor response was characterized with very good linearity in the glucose physiological concentration range and what is also important is that the satisfying results were obtained around normal temperature of human body (Marcisz et al. 2018).

Ferrocene modified pNIPA microgel was also used in optical sensors (etalon) construction. Etalons are sandwich-like sensors where microgel layer is entrapped between two Au layers. Light beams impinging on the top layer could enter into its center and resonate between the two Au layers; in consequence, light beams with specific wavelength were reflected/transmitted. The change in microgel volume related with swelling ratio change caused a blue shift in the reflectance spectra, which could be used for sensing application. Ferrocene modified pNIPA microgel based etalon could work as sensor sensitive to hydrogen peroxide presence. The hydrogen peroxide presence led to microgel ferrocene oxidation. The ferrocene oxidation state change led to microgel shrinking, which caused the decrease in distance between Au overlayers (Fig. 4C). This phenomenon was observed as a shift in the reflectance spectrum. It was found that the increase in H_2O_2 concentration caused appropriate increase in the blue shift. This property was used in monitoring the enzymatic reaction of glucose oxidase with glucose (Fig. 4D). The sensor had ability to work in the physiologically glucose concentration range (Zhang et al. 2016a). Appropriate microgels, used for etalon preparation and modification, result in new application of those optical sensors for several analytes. Etalons sensitive to triglycerides, Tabun, Cu^{2+}, CO_2, proteins and DNA have been described in the literature (Hu and Serpe 2013, Islam and Serpe 2013, 2014, Zhang et al. 2014, 2015a, b, 2016b).

Wei et al. (2019) presented very interesting application of surface bound pNIPA microgels co-polymerized with glycosyloxyethyl methacrylate (p(NIPAm-*co*-GEMA)). Next, microgel polymer network was modified with protein, concanavalin A (ConA), which could bind glucose molecules. This property was used to study the glucose detection. The p(NIPAm-*co*-GEMA) with ConA groups microgel swelled after glucose addition. This was related with glucose binding by ConA groups, interaction between ConA and GEMA in the polymer network was weakened and as a consequence, ConA bound with glucose diffused from the polymer network. The swelling processes of microgels attached to Au surface in response to glucose concentration could be monitored with the surface plasmon resonance (SPR) technique. The mechanism and detection method of p(NIPAm-*co*-GEMA) with ConA based sensor is shown in Fig. 5A and B (Wei et al. 2019).

Two kinds of microgels based on NIPA were used for construction of an electrochemical immunosensor. The first one was co-polymerized with 3-(methacrylamide) propyltrimethylammonium chloride (MPTC) and glycidyl methacrylate (GMA) and the second one with acrylic acid (AA) and zwitterionic ionic liquid 1-propyl-3-vinylimidazole sulfonate (PVIS). Then, the microgels with GMA were conjugated with anti-STR, which specifically recognizes streptomycin (STR). The two microgels were anchored to the surface of a glassy carbon electrode modified with graphene oxide with self-assembly process. Then, modified electrode was used as an antifouling immunosensor for streptomycin (STR) detection. The sensor preparation scheme and electrochemical response are shown in Fig. 5C. In this system, microgels modified with PVIS were used for resisting nonspecific absorption. The antigen specific bonding with anti-STR led to decrease in registered current density, which was observed as a decrease in electrochemical signal. The presented microgel electrochemical immunosensor was characterized with good sensitivity to biomolecules' detection with circa 2 pg mL^{-1} limit (He et al. 2019).

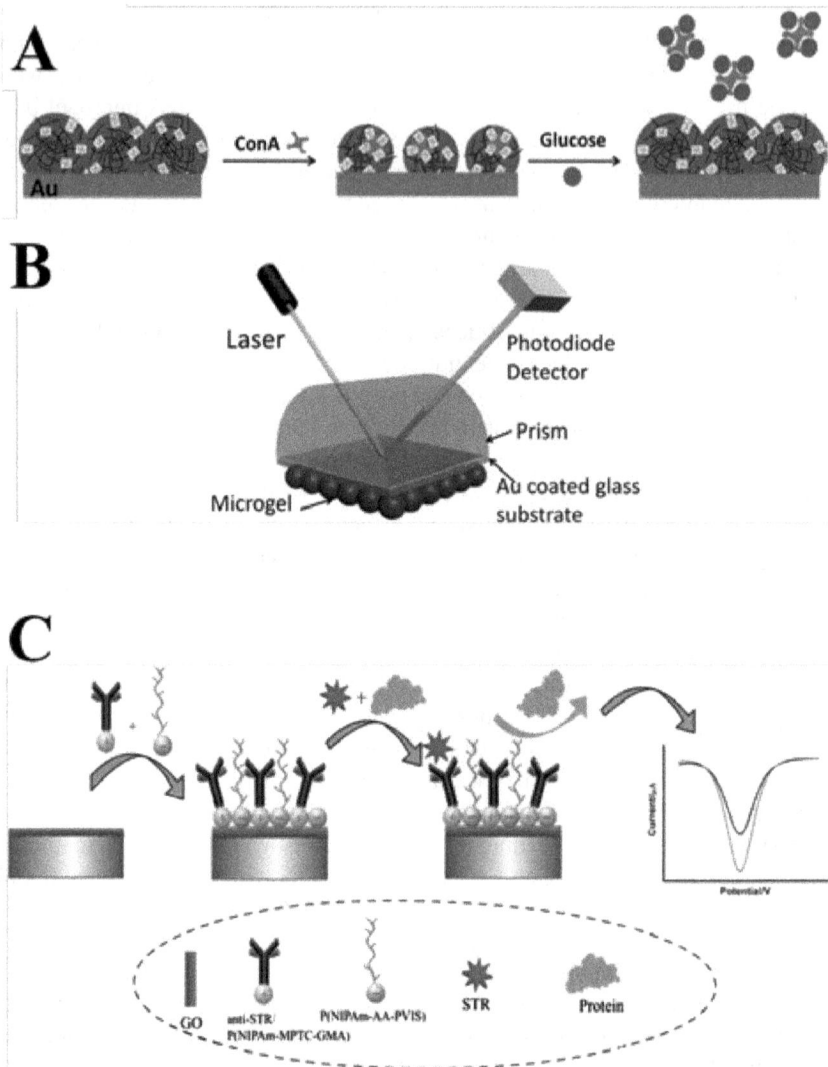

Fig. 5. (A) Scheme of p(NIPAm-co-GEMA) microgel on the Au surface glucose sensor work mechanism. (B) Scheme of the SPR detection method of p(NIPAm-co-GEMA) microgel film thickness. Reprinted with permission from Ref. (Wei et al. 2019). Copyright 2019 American Chemical Society. (C) Scheme of self-assembled microgel layer formation on the electrode surface towards electrochemical immunosensor formation. Reprinted with permission from Ref. (He et al. 2019). Copyright 2019 American Chemical Society.

The other biological materials have also been used for construction of novel label-free biosensor based on p(NIPA) microgel such as with immobilized *Escherichia coli* bacteria, sensitive to antibiotic presence in diluted solutions. The sensor was able to monitor the bacterial response to ampicillin (antibiotic) presence. The metabolic processes and release of ionic products was registered with the rapid impedance technique and sensor resistance changes. This bacteria-based sensor for antibiotics' monitoring opens new possibilities in electroanalysis (Brosel-Oliu et al. 2019).

Thermosensitive microgels containing pNIPA have been often used in the area of biosensor. However, papers on sensors obtained from microgels synthesized from other polymers also can be found. One example is presented by Yu et al. [2017] where authors synthesized alginate composite with graphene (CA–graphene) microspheres and additionally immobilized myoglobin (Mb). In the next step, Mb–CA–graphene microspheres were anchored on the GC electrode surface. The graphene sheets were responsible for increase in the myoglobin electrocatalytical and electrochemical properties. The sensor could improve the electroreduction of various substrates, e.g., hydrogen peroxide, sodium nitrite and trichloroacetic acid. It is worth mentioning that the Mb–CA–graphene microspheres based sensor had a wide linear range and low detection limits (Yu et al. 2017).

He et al. [2018] used a nano-sized core-shell composite obtained with hollow carbon spheres coated with needle-like polyaniline moieties, with immobilized acetylcholine esterase (AChE), to prepare an electrochemical biosensor, on the GC electrode surface, sensitive to malathion. The malathion presence caused the inhibition of acetylthiocholine chloride electrochemical catalytical oxidation, which was used for electrochemical detection. The biosensor had low detection limit for malathion concentration (0.16 ng mL^{-1}) (He et al. 2018).

The alginate hydrogel microspheres with additionally introduced CdZnTeS quantum dots and urate oxidase enzyme were used as a uric acid in human urine biosensor. The alginate microparticles enhanced quantum dots stability and they also limited the protein interference. Fluorescence color cards were used for detection. It was found that the detection method is universal and could be used for sensing of different small molecules. This method seems to be very attractive for the first screening of diabetic patients. The combination of sensor with other fluorescence colors could led to simultaneous detection of multiple targets, which could be very useful for health monitoring (Lu et al. 2021).

Dou et al. [2020] synthesized microgel based on 3-(acrylamido) phenylboronic acid (3-APB), glycidyl methacrylate (GMA), acrylamide (AM) and N,N'-methylenebisacrylamide (BIS), which was attached at quartz-crystal-microbalance (QCM) electrode surface. The above system was used as a sensor for the saliva glucose detection. The presence of boric acid groups in the microgel polymer network was responsible for glucose sensing properties; on the other hand, the gel overlayer and amino acid parts made a barrier for protein interferences. The glucose detection, based on cyclic lactone formation through the boric acid and glucose, was observed as a decrease in registered quartz crystal frequency shift. The examined results showed that glucose sensor could work at physiological conditions and was characterized with very good linear response to glucose concentrations. It is very important that the above glucose microgel-QCM sensors were resistant to the saliva proteins' interference presence (Duo et al. 2020).

Retama et al. [2004] described an electrochemical biosensor based on polypyrrole-polystyrensulfonate and GOx immobilized in polyacrylamide microsphers with circa 3.5–7 μm sizes. Next, microparticles were attached to the platinum electrode. The polpyrrole present in the polymer network was responsible for mediating the electron transfer from enzymatic reaction to the electrode surface. The combination of GOx and polypyrrole allowed the successful glucose electrochemical detection under anaerobic conditions. The above biosensor was found to have low interferences from many metabolites, e.g., ascorbic and uric acids, hence making it useful for glucose detection in human blood samples (Retama et al. 2004).

Summary and outlook

The unique properties of polymeric hydrogels, i.e., high permeability for analytes, the ability to immobilize larger individuals like enzymes and nanoparticles, and environmental sensitivity make them very useful in construction of various types of sensors and biosensors. Combining hydrogels sensitive to external stimuli with conducting surfaces opens new possibilities in electrochemistry. Considerable efforts and progress have been made in the field of modifying electrode surfaces with smart hydrogels, resulting in the recent publications of a significant number of papers on switchable sensors/biosensors, switchable electrochemical systems and signal-responsive interfaces. Smart interfaces that can modulate electroanalytical signals are very promising in terms of obtaining sensors/biosensors that work optimally under strictly defined conditions, e.g., under physiological conditions. Such functionalities of the biocompatible hydrogel layers, which are able to protect the electrode surface from unwanted environmental influences and insulate the environment from the usually metallic, rigid and hard electrode surfaces, are desirable in implantable bioelectronic devices. Among future innovations related to these devices, the development of electrochemical biosensors that can monitor important body parameters *in situ* is expected.

Acknowledgments

This work was supported by the National Science Centre of Poland through grant number 2021/43/D/ST5/01082.

References

Ahiabu, A. and M.J. Serpe. 2017. Rapidly responding pH- and temperature-responsive poly (N-Isopropylacrylamide)-based microgels and assemblies. ACS Omega 31: 1769–1777.

Ahmed, E.M. 2015. Hydrogel: Preparation, characterization, and applications: A review. Journal of Advanced Research 6: 105–121.

Bandehali, S., F. Parvizian, S.M. Hosseini, T. Matsuura, E. Drioli, J. Shen et al. 2021. Planning of smart gating membranes for water treatment. Chemosphere 283: 131207.

Bian, S., J. Zheng, X. Tang, D. Yi, Y. Wang and W. Yang. 2015. One-pot synthesis of redox-labile polymer capsules via emulsion droplet-mediated precipitation polymerization. Chem. Mater. 27: 1262–1268.

Brosel-Oliu, S., O. Mergel, N. Uria, N. Abramova, P.V. Rijn and A. Bratov. 2019. 3D impedimetric sensors as a tool for monitoring bacterial response to antibiotics. Lab Chip 19: 1436.

Caló, E. and V.V. Khutoryanskiy. 2015. Biomedical applications of hydrogels: A review of patents and commercial products. Eur. Polym. J. 65: 252–267.

Caputo, T.M., E. Battista, P.A. Netti and F. Causa. 2019. Supramolecular microgels with molecular beacons at the interface for ultrasensitive, amplification-free, and SNP-Selective miRNA fluorescence detection. ACS Appl. Mater Interfaces 11: 17147–17156.

Chadha, U., P. Bhardwaj, R. Agarwal, P. Rawat, R. Agarwal, I. Gupta et al. 2022. Recent progress and growth in biosensors technology: A critical review. J. Ind. Eng. 109: 21–51.

Chaubey, A. and B. Malhotra. 2002. Mediated biosensors. Biosens. Bioelectron. 17: 441–456.

Chen, H., R. Yuan, Y. Chai, J. Wang and W. Li. 2010. Glucose biosensor based on electrodeposited platinum nanoparticles and three-dimensional porous chitosan membranes. Biotechnol. Lett. 32: 1401–1404.

Cieplak, M. and W. Kutner. 2016. Artificial biosensors: how can molecular imprinting mimic biorecognition? Trends Biotechnol. 34: 922–941.

Culver, H.R., J.R. Clegg and N.A. Peppas. 2017. Analyte-responsive hydrogels: intelligent materials for biosensing and drug delivery. Accounts Chem. Res. 50: 170–178.

Dhanjai, A. Sinha, P.K. Kalambate, S.M. Mugo, P. Kamau, J. Chen et al. 2019. Polymer hydrogel interfaces in electrochemical sensing strategies: A review. Trends Anal. Chem. 188: 488–501.

DiCosimo, R., J. McAuliffe, A.J. Poulose and G. Bohlmann. 2013. Industrial use of immobilized enzymes. Chem. Soc. Rev. 42: 6437–6474.

Dou, Q., S. Wang, Z. Zhang, Y. Wang, Z. Zhao, H. Guo et al. 2020. A highly sensitive quartz crystal microbalance sensor modified with antifouling microgels for saliva glucose monitoring. Nanoscale 12: 19317.

Fandrich, A., J. Buller, H. Memczak, W. Stöcklein, K. Hinrichs, E. Wischerhoff et al. 2017. Responsive polymer-electrode interface—study of its thermo- and pH-sensitivity and the influence of peptide coupling. Electrochim Acta 229: 325–333.

Feng, L., L. Wang, Z. Hu, Y. Tian, Y. Xian and L. Jin. 2009. Encapsulation of horseradish peroxidase into hydrogel, and its bioelectrochemistry. Microchim. Acta 164: 49–54.

Garnier, R., L. Levy-Amon and L. Malingrey. 2003. Occupational contact dermatitis from N-(2-(diethylamino)-ethyl)acrylamide. Contact Dermatitis 48: 343–344.

Gawlitza, K., C. Wu, R. Georgieva, M. Ansorge-Schumacher and R. von Klitzing. 2012. Temperature controlled activity of lipase B from Candida antarctica after immobilization within p-NIPAM microgel particles. Z. Phys. Chem. 226: 749.

Gentile, P., C. Ghione, A.M. Ferreira, A. Crawford and P.V. Hatton. 2017. Alginate-based hydrogels functionalised at the nanoscale using layer-by-layer assembly for potential cartilage repair. Biomater. Sci. 5: 1922–1931.

He, L., B. Cui, J. Liu, Y. Song, M. Wang, D. Peng et al. 2018. Novel electrochemical biosensor based on core-shell nanostructured composite of hollow carbon spheres and polyaniline for sensitively detecting malathion. Sens. Actuators B 258: 813.

He, X., H. Han, L. Liu, W. Shi, X. Lu, J. Dong et al. 2019. Self-assembled microgels for sensitive and low-fouling detection of streptomycin in complex media. ACS Appl. Mater. Interfaces 11: 13676.

Hirose, H. and M. Shibayama. 1998. Kinetics of volume phase transition in poly(N-isopropylacrylamide-co-acrylic acid) gels. Macromolecules 31: 5336–5342.

Hu, H., Z. Cao, X. Fan and H. Yi. 2006. Synthesis and characterization of the thermo- and pH-sensitive dendritic polymers based on polyamidoamine dendrimer and poly(N,N-dimethylaminoethyl methacrylate) and study on their controlled drug release behaviour. Acta Polym. Sin. 4: 574–580.

Hu, L. and M.J Serpe. 2013. Controlling the response of color tunable poly (N-isopropylacrylamide) microgel-based etalons with hysteresis. Chem. Commun. 49: 2649–2651.

Hua, J., Z. Li, W. Xia, N. Yang, J. Gong, J. Zhang et al. 2016. Preparation and properties of EDC/NHS mediated crosslinking poly (gamma-glutamic acid)/epsilon-polylysine hydrogels. Mater. Sci. Eng. A 61: 879–892.

Islam, M.R. and M.J. Serpe. 2013. Label-free detection of low protein concentration in solution using a novel colorimetric assay. Biosens. Bioelectron. 49: 133–138.

Islam, M.R. and M.J. Serpe. 2014. A novel label-free colorimetric assay for DNA concentration in solution. Anal. Chim. Acta 843: 83–88.

Jyoti, G.C., T. Zolek, D. Maciejewska, A. Kutner, F. Merlier, K. Haupt et al. 2021 Molecularly imprinted polymer nanoparticles-based electrochemical chemosensors for selective determination of cilostazol and its pharmacologically active primary metabolite in human plasma. Biosens. Bioelectron. 193: 113542.

Kaneko, Y., S. Nakamura, K. Sakai, T. Aoyagi, A. Kikuchi, Y. Sakurai et al. 1998. Rapid deswelling response of poly(N-isopropylacrylamide) hydrogels by the formation of water release channels using poly(ethylene oxide) graft chains. Macromolecules 31: 6099–6105.

Kaniewska, K., M. Karbarz, K. Ziach, A. Siennicka, Z. Stojek and W. Hyk. 2016. Electrochemical examination of the structure of thin hydrogel layers anchored to regular and microelectrode surfaces. J. Phys. Chem. B 120: 9540–9547.

Kaniewska, K., M. Karbarz, Z. Stojek and W. Hyk. 2017. Mass transport affected by electrostatic barrier in ionized gel layers attached to microelectrode surface. Electrochem. Comm. 81: 24–28.

Kaniewska, K., K. Kyriacou, M. Donten, Z. Stojek and M. Karbarz. 2018a. Micro- and nanoelectrode array behavior at regularly sized electrode modified with a thin film of thermoresponsive polymeric gel. Electrochim. Acta 90: 595–604.

Kaniewska, K., W. Hyk, Z. Stojek and M. Karbarz. 2018b. Diffusional and migrational transport of ionic species affected by electrostatic interactions with an oppositely charged hydrogel layer attached to an electrode surface. Electrochem. Comm. 88: 97–100.

Kaniewska, K., K. Marcisz and M. Karbarz. 2023. Transport of ionic species affected by interactions with a pH-sensitive monolayer of microgel particles attached to electrode surface. J. Electroanal. Chem. 931: 117183.

Kappauf, K., N. Majstorovic, S. Agarwal, D. Rother and C. Claassen. 2021. Modulation of transaminase activity by encapsulation in temperature-sensitive poly(N-acryloyl glycinamide) hydrogels. ChemBioChem 22: 3452–3461.

Karbarz, M., M. Gniadek and Z. Stojek. 2005. Electroanalytical properties of ITO electrodes modified with environment-sensitive poly(N-isopropylacrylamide) gel and prussian blue. Electroanalysis 17: 1396–1368.

Karbarz, M., W. Hyk and Z. Stojek. 2009. Swelling ratio driven changes of probe concentration in pH- and ionic strength-sensitive poly(acrylic acid) hydrogels. Electrochem. Commun. 11: 1217–1220.

Karbarz, M., A. Lukaszek and Z. Stojek. 2013. Modified with environmentally sensitive poly(N-Isopropylacrylamide) gel via electrochemically induced free-radical polymerization. Electroanalysis 25: 875–880.

Karbarz, M., M. Mackiewicz, K. Kaniewska, K. Marcisz and Z Stojek. 2017. Recent developments in design and functionalization of micro- and nanostructural environmentally-sensitive hydrogels based on N-isopropylacrylamide. Appl. Mater. Today 9: 516–532.

Kato, M., Y. Tsuboi., A. Kikuchi and T.-A. Asoh. 2016. Hydrogel adhesion with wrinkle formation by spatial control of polymer networks. J. Phys. Chem. B 120: 5042–5046.

Katz, E., V. Bocharova and M. Privman. 2012. Electronic interfaces switchable by logically processed multiple biochemical and physiological signals. Journal of Materials Chemistry 22: 8171–8178.

Katz, E., S. Minko, J. Halamek, K. MacVittie and K. Yancey. 2013. Electrode interfaces switchable by physical and chemical signals for biosensing, biofuel, and biocomputing applications. Anal. Bioanal. Chem. 405: 3659–3672.

Katz, E. 2016. Modified electrodes and electrochemical systems switchable by temperature changes. Electroanalysis 28: 1916–1929.

Katz, E. 2018a. Modified electrodes and electrochemical systems switchable by light signals. Electroanalysis 30: 759–797.

Katz, E. 2018b. Modified electrodes and electrochemical systems switchable by temperature changes. pp. 71–99. *In*: Katz, E. (ed.). Signal-Switchable Electrochemical Systems: Materials, Methods, and Applications. Wiley-VCH Verlag GmbH & Co, Weinheim, Germany.

Klis, M., M. Karbarz, Z. Stojek, J. Rogalski and R. Bilewicz. 2009. Thermoresponsive poly(N-isopropylacrylamide) gel for immobilization of laccase on indium tin oxide electrodes. J. Phys. Chem. B 113: 6062–6067.

Kotanen, C.N., C. Tlili and A. Guiseppi-Elie. 2013. Amperometric glucose biosensor based on electroconductive hydrogels. Talanta 103: 228–235.

Lee, M.-H., C.-C. Lin, P.S. Sharma, J.L. Thomas, C.-Y. Lin, Z. Iskierko et al. 2022. Peptide selection of MMP-1 for electrochemical sensing with epitope-imprinted poly(TPARA-co-EDOT)s. Biosensors 12: 1018.

Lee, M.-H., C.-C. Lin, W. Kutner, J.L. Thomas, C.-Y. Lin, Z. Iskierko et al. 2023. Peptide-imprinted conductive polymer on continuous monolayer molybdenum disulfide transferred electrodes for electrochemical sensing of Matrix Metalloproteinase-1 in lung cancer culture medium. Biosens. Bioelectron. 13: 100258.

Li, F., D. Lyu, S. Liu and W. Guo. 2019. DNA hydrogels and microgels for biosensing and biomedical applications. Adv. Mater. 1806538.

Li, Z., T. Konno, M. Takai and K. Ishihara. 2012. Fabrication of polymeric electron-transfer mediator/enzyme hydrogel multilayer on an Au electrode in a layer-by-layer process. Biosens. Bioelectron. 34: 191–196.

Liang, Y., S. Song, H. Yao and N. Hu. 2011. Triply switchable bioelectrocatalysis based on poly(N-isopropylacrylamide) hydrogel films with immobilized glucose oxidase. Electrochim. Acta 56: 5166–5173.

Liang, Y., H. Liu and N. Hu. 2012. Triply switchable bioelectrocatalysis based on poly(N,N-diethylacrylamide-co-4-vinylpyridine) copolymer hydrogel films with immobilized glucose oxidase. Electrochim. Acta 60: 456–463.

Liu, D., H. Liu and N. Hu. 2012. pH-, sugar-, and temperature-sensitive electrochemical switch amplified by enzymatic reaction and controlled by logic gates based on semi-interpenetrating polymer networks. J. Phys. Chem. B 116: 1700–1708.

Liu, H., T. Prachyathipsakul, T.M. Koyasseril-Yehiya, S.P. Le and S. Thayumanavan. 2022. Molecular bases for temperature sensitivity in supramolecular assemblies and their applications as thermoresponsive soft materials. Mater. Horiz. 9: 164–193.

Liu, Z., Y. Wang, Y. Ren, G. Jin, C. Zhang, W. Chen et al. 2020. Poly(ionic liquid) hydrogel-based anti-freezing ionic skin for a soft robotic gripper. Mater. Horiz. 7: 919–927.

Lu, F., Y. Yang, Y. Liu, F. Wang, X. Jia and Z. He. 2021. A fluorescence color card for point-of-care testing (POCT) and its application in simultaneous detection. Analyst 146: 949.

Ma, Y., Y. Kong, J. Xu, Y. Deng, M. Lu, R. Yu et al. 2020. Carboxyl hydrogel particle film as a local pH buffer for voltammetric determination of luteolin and baicalein. Talanta 208: 120373–120381.

Mackiewicz, M., K. Marcisz, M. Strawski, J. Romanski, Z. Stojek and M. Karbarz. 2018. Modification of gold electrode with a monolayer of self-assembled microgels. Electrochim Acta 268: 531–538.

Mahinroosta, M., Z.J. Farsangi, A. Allahverdi and Z. Shakoori. 2018. Hydrogels as intelligent materials: A brief review of synthesis, properties and applications. Mater. Today. Chem. 8: 42–55.

Marcisz, K., K. Kaniewska, M. Mackiewicz, A. Nowinska, J. Romanski, Z. Stojek et al. 2018. Electroactive, mediating and thermosensitive microgel useful for covalent entrapment of enzymes and formation of sensing layer in biosensors. Electroanalysis 30: 2853.

Marcisz, K., K. Kaniewska and M. Karbarz. 2020. Smart functionalized thin gel layers for electrochemical sensors, biosensors and devices. Curr. Opin. Electrochem. 23: 57–64.

Marcisz, K., K. Kaniewska, Z. Stojek and M. Karbarz. 2021a. Electroresponsiveness of a positively charged thin hydrogel layer on an electrode surface. Electrochem. Commun. 125: 106981.

Marcisz, K., J. Romanski and M. Karbarz. 2021b. Electroresponsive microgel able to form a monolayer on gold through self-assembly. Polymer 229: 123992.

Matsukuma, D., K. Yamamoto and T. Aoyag. 2006. Stimuli-responsive properties of N-isopropylacrylamide-based ultrathin hydrogel films prepared by photo-cross-linking. Langmuir 22: 5911–5915.

Mugo, S.M., D. Berg and G. Bharath. 2019. Integrated microcentrifuge carbon entrapped glucose oxidase poly (N-Isopropylacrylamide) (pNIPAm) microgels for glucose amperometric detection. Anal. Lett. 52: 825.

Mutharani, B., P. Ranganathan and S.M. Chen. 2019. Highly sensitive and selective electrochemical detection of antipsychotic drug chlorpromazine in biological samples based on Poly-N-isopropylacrylamide microgel. J. Taiwan Inst. Chem. Eng. 96: 599–609.

Mutharani, B., P. Ranganathan, S.M Chen and D. Vishnu. 2020. Stimuli-enabled reversible switched aclonifen electrochemical sensor based on smart PNIPAM/PANI-Cu hybrid conducting microgel. Sensors and Actuators: B. Chemical 304: 127232.

Nagel, B., A. Warsinke and M. Katterle. 2007. Enzyme activity control by responsive redoxpolymers. Langmuir. 23: 6807–6811.

Oh, S.Y., H.J. Kim and Y.C. Bae. 2013. Molecular thermodynamic analysis for phase transitions of linear and cross-linked poly(N-isopropylacrylamide) in water/2-propanol mixtures. Polymer 54: 6776–6784.

Parlak, O., M. Ashaduzzaman, S.B. Kollipara, A. Tiwari and A.P.F. Turner. 2015. Switchable bioelectrocatalysis controlled by dual stimuli-responsive polymeric interface. ACS Appl. Mater. Interfaces 7: 23837–23847.

Pinyou, P., A. Ruff, S. Pöller, S. Barwe, M. Nebel, N.G. Alburquerque et al. 2016. Thermoresponsive amperometric glucose biosensor. Biointerphases 11: 011001.

Qiu, J.D., R. Wang, R.P. Liang and X.H. Xia. 2009. Electrochemically deposited nanocomposite film of CS-Fc/Au NPs/GOx for glucose biosensor application. Biosens. Bioelectron. 24: 2920–2925.

Qu, F., X. Ma and J.E. Gautrot. 2021. Precise positioning of enzymes within hierarchical polymer nanostructures for switchable bioelectrocatalysis. Biosens. Bioelectron. 179: 113045.

Reuber, J., H. Reinhardt and D. Johannsmann. 2006. Formation of surface-attached responsive gel layers via electrochemically induced free-radical polymerization. Langmuir 22: 3362–3367.

Roquero, D.M. and E. Katz. 2022. "Smart" alginate hydrogels in biosensing, bioactuation and biocomputing: State-of-the-art and perspectives. Sens. Actuators Rep. 4: 100095.

Rubio Retama, J., E. Lopez Cabarcos, D. Mecerreyes and B. Lopez-Ruiz. 2004. Design of an amperometric biosensor using polypyrrole-microgel composites containing glucose oxidase. Biosens. Bioelectron. 20: 1111.

Shah, M., L.A. Shah, M.S. Khan, M.Q. Nasar and S. Rasheed. 2019. Synthesis, fabrication and characterization of polymer microgel/photochromic dye-based sandwiched sensors. Iran Polym. J. 28: 515–525.

Sigolaeva, L.V., S.Y. Gladyr, A.P.H. Gelissen, O. Mergel, D.V. Pergushov, I.N. Kurochkin et al. 2014. Dual-stimuli-sensitive microgels as a tool for stimulated spongelike adsorption of biomaterials for biosensor applications. Biomacromolecules 15: 3735–3745.

Sigolaeva, L.V., O. Mergel, E.G. Evtushenko, S.Yu. Gladyr, A.P.H. Gelissen, D.V. Pergushov et al. 2015. Engineering systems with spatially separated enzymes via dual-stimuli-sensitive properties of microgels. Langmuir 31: 13029–13039.

Sigolaeva, L.V., S.Y. Gladyr, O. Mergel, A.P.H. Gelissen, M. Noyong, U. Simon et al. 2017. Easy-preparable butyrylcholinesterase/microgel construct for facilitated organophosphate biosensing. Anal. Chem. 89: 6091–6098.

Sigolaeva, L.V., D. Pergushov, M. Oelmann, S. Schwarz, M. Brugnoni, I. Kurochkin et al. 2018. Surface functionalization by stimuli-sensitive microgels for effective enzyme uptake and rational design of biosensor setups. Polymers 10: 791.

Sigolaeva, L.V., D.V. Pergushov, S.Yu. Gladyr, I.N. Kurochkin and W. Richtering. 2022. Microgels in tandem with enzymes: tuning adsorption of a pH- and thermoresponsive microgel for improved design of enzymatic biosensors. Adv. Mater. Interfaces 9: 2200310.

Tanaka, T. and D.J. Fillmore. 1979. Kinetics of swelling of gels. J. Chem. Phys. 70: 1214–1218.

Tatsuma, T., K.-I. Saito and N. Oyama. 1994. Enzyme electrodes mediated by a thermoshrinking redox polymer. Anal. Chem. 66: 1002–1006.

Tavakoli, J. and Y. Tang. 2017. Hydrogel based sensors for biomedical applications: an updated review. Polymers 9: 364.

Thorne, J.B., G.J. Vine and M.J. Snowden. 2011. Microgel applications and commercial considerations. Colloid Polym. Sci. 289: 625–646.

Tian, Y., S. Bian and W. Yang. 2016. A redox-labile poly(oligo(ethylene glycol) methacrylate) based nanogel with tunable thermosensitivity for drug delivery. Polym. Chem. 7: 1913–1921.

Ullah, F., M.B.H. Othman, F. Javed, Z. Ahmad and H.M. Akil. 2015. Classification, processing and application of hydrogels: A review. Mater. Sci. Eng. C 57: 414–433.

Wael, K.D., Q. Bashir, S.V. Vlierberghe, P. Dubruel, H. Heering and A. Adriaens. 2012. Electrochemical determination of hydrogen peroxide with cytochrome c peroxidase and horse heart cytochrome c entrapped in a gelatin hydrogel. Bioelectrochemistry 83: 15–18.

Wang, Y., Z. Liu, H.Y. Peng, F. He, L. Zhang, Y. Faraj et al. 2018. A simple device based on smart hollow microgels for facile detection of trace lead(II) ions. ChemPhysChem 19: 2025–2036.

Wei, M., X. Li and M.J. Serpe. 2019. Stimuli-responsive microgel-based surface plasmon resonance transducer for glucose detection using a competitive assay with Concanavalin A. ACS Appl. Polym. Mater. 1: 519.

Wei, Y., Q. Zeng, Q. Hu, M. Wang, J. Tao and L. Wang. 2018. Self-cleaned electrochemical protein imprinting biosensor basing on a thermo-responsive memory hydrogel. Biosens. Bioelectron. 99: 136–141.

Welsch, N., A. Wittemann and M. Ballauff. 2009. Enhanced activity of enzymes immobilized in thermoresponsive core–shell microgels. J. Phys. Chem. B 113: 16039–16045.

Yao, H. and N. Hu. 2011. Triply responsive films in bioelectrocatalysis with a binary architecture: combined layer-by-layer assembly and hydrogel polymerization. J. Phys. Chem. B 115: 6691–6699.

Yao, H., L. Lin, P. Wang and H. Liu. 2014. Thermo- and sulfate-controllable bioelectrocatalysis of glucose based on horseradish peroxidase and glucose oxidase embedded in poly(N,N-diethylacrylamide) hydrogel films. Appl. Biochem. Biotechnol. 173: 2005–2018.

Yasmeen, N., M. Etienne, P.S. Sharma, S. El-Kirat-Chatel, M.B. Helú and W. Kutner. 2021. Molecularly imprinted polymer as a synthetic receptor mimic for capacitive impedimetric selective recognition of *Escherichia coli* k-12. Anal. Chim. Acta 1188: 339177.

Yu, C., H. Sun and S. Hou. 2017. Direct electrochemistry and electrocatalysis of myoglobin immobilized in calcium alginate–graphene microsphere films. Anal. Methods 9: 4873.

Yu, X., W. Lian, J. Zhang and H. Liu. 2016. Multi-input and -output logic circuits based on bioelectrocatalysis with horseradish peroxidase and glucose oxidase immobilized in multiresponsive copolymer films on electrodes. Biosens. Bioelectron. 80: 631–639.

Zhang, K.N., Y. Liang, D. Liu and H.Y. Liu. 2012. An on–off biosensor based on multistimuli-responsive polymer films with a binary architecture and bioelectrocatalysis. Sensor. Actuator. B Chem. 173: 367–376.

Zhang, K., W. Lian, S. Liu and H. Liu. 2014a. Multi-switchable bioelectrocatalysis based on semi-interpenetrating polymer network films prepared by enzyme-induced polymerization. J. Electrochem. Soc. 161: H493–H500.

Zhang, Q.M., D. Berg, S.M. Mugo and M.J. Serpe. 2015a. Lipase-modified pH-responsive microgel-based optical device for triglyceride sensing. Chem. Commun. 51: 9726–9728.

Zhang, Q.M., A. Ahiabu, Y. Gao and M.J. Serpe. 2015b. CO_2-Switchable Poly(N-isopropylacrylamide) microgel-based etalons. J. Mater. Chem. C 3: 495–498.

Zhang, Q.M., D. Berg, J. Duan, S.M. Mugo and M.J. Serpe. 2016a. Optical devices constructed from ferrocene-modified microgels for H_2O_2 sensing. ACS Appl. Mater. Interfaces 8: 27264–27269.

Zhang, Q.M., W. Wang, Y.-Q. Su, E.J. Hensen and M.J. Serpe. 2016b. Biological imaging and sensing with multiresponsive microgels. Chem. Mater. 28: 259–265.

Zhang, Q., D. Li, X. Cao, H. Gu and W. Deng. 2019. Self-assembled microgels arrays for electrostatic concentration and surface-enhanced raman spectroscopy detection of charged pesticides in seawater. Anal. Chem. 91: 11192–11199.

Zhang, Q.M., W. Xu and M.J. Serpe. 2014b. Optical devices constructed from multiresponsive microgels. Angew. Chem. Int. Ed. 53: 4827–4831.

Zhang, X.-Z., Y.-Y. Yang, F.-J. Wang and T.-S. Chung. 2002. Thermosensitive poly(N-isopropylacrylamide-co-acrylic acid) hydrogels with expanded network structures and improved oscillating swelling–deswelling properties. Langmuir 18: 2013–2018.

Zhang, X.-Z., X.-D. Xu, S.-X. Cheng and R.-X. Zhuo. 2008. Strategies to improve the response rate of thermosensitive PNIPAAm hydrogels. Soft Matter 4: 385–391.

Zhang, Y., W.S.P. Carvalho, C. Fang and M.J. Serpe. 2019. Volatile organic compound vapor detection with responsive microgel-based etalons. Sensors and Actuators B: Chemical 290: 520–526.

Zhou, J., C.A. Liao, L.M. Zhang, Q.G. Wang and Y. Tian. 2014. Molecular hydrogel-stabilized enzyme with facilitated electron transfer for determination of H_2O_2 released from live cells. Anal. Chem. 86: 4395.

Zhu, Z., J. Xue, B. Wen, W. Ji, B. Du and J. Nie. 2019. Ultrasensitive and selective detection of MnO_4^-in aqueous solution with fluorescent microgels. Sensors & Actuators: B. Chemical 291: 441–450.

Chapter 5

Aptamer-based Stimuli-Responsive Surfaces

Agnivo Gosai,[1,]* *Pranav Shrotriya*[2] and *Xiao Ma*[3,]*

Introduction

Aptamers are engineered DNA or RNA oligonucleotides which present binding affinity and specificity to their targets, such as proteins, carbohydrates, small molecules, toxins, metal ions, and even live cells, by folding into unique tertiary structures (Ellington and Szostak 1990, Tuerk and Gold 1990). Aptamers have been extensively used in fundamental biological studies, clinical diagnostics, serve as therapeutic and imaging agents, and can potentially be employed to maintain food safety and monitor the environmental changes. Compared to antibodies which share quite a few common functionalities with aptamers upon specific target binding, recognition, and sensing, aptamers are typically more structurally stable, possess a longer shelf life, and can be produced through a relatively simple and inexpensive process in short time.

The term "Aptamer" was first defined by Andy Ellington and Jack Szostak in 1990, in their *Nature* paper for *in-vitro* selection of RNA ligands to specifically bind their target organic dyes (Ellington and Szostak 1990). The "Apta" stems from the Latin terms "aptus", which means to fit, while "mer" comes from the Greek "meros", which means part or region. This definition reveals the essence of aptamers binding towards their targets, i.e., aptamers fit their targets for the specific binding. More essentially, rather than the primary sequence of the nucleic acid, it is the tertiary structure of aptamers and three-dimensional, shape-dependent interactions between aptamers and their targets that determine the binding affinity

[1] Department of Science and Technology, Corning Research and Development Corporation, Painted Post, NY 14870, United States.
[2] Department of Mechanical Engineering, Iowa State University, Ames, IA 50011, United States.
[3] Department of Biomedical Engineering, New York University, Brooklyn, NY 11201, United States.
* Corresponding authors: agnivo2007@gmail.com, xm8@nyu.edu

and specificity. Aptamers are selected from a large oligonucleotide library through an *in-vitro* enrichment process, known as SELEX, i.e., sequential evolution of ligands by exponential enrichment, which was developed by Craig Tuerk and Larry Gold and published in their *Science* paper in 1990 for selection of RNA ligands to specifically bind their target bacteriophage T4 DNA polymerase (Tuerk and Gold 1990). This process starts with a blind oligonucleotide library including either a pool of ssRNA or ssDNA, then an affinity column functionalized with target molecules is employed to bind a portion of oligonucleotides with high affinity towards their targets, while the remaining non-binding oligonucleotides are washed out and discarded. Then the bound oligonucleotides are eluted by the elution buffer, and polymerase chain reaction (PCR) is conducted to amplify or enrich the binders pool. Subsequently, the enriched binders pool is carried to the affinity column again, and the same procedures as above are run repeatedly, typically 6–12 rounds, to finally screen out the aptamers with high affinity and specificity towards their targets. In 1992, Louis Bock and John Toole et al. developed the first generation of DNA aptamers to target human coagulation enzyme thrombin (Bock et al. 1992). In 1999, Matthias Homann and Ulrich Göringer selected a RNA aptamer to target the live cells for the first time in African trypanosomes (Homann and Goringer 1999), which is later known as cell-SELEX. In 2004, the first generation of aptamer drug, Macugen, also known as Pegaptanib was approved by FDA to treat neovascular age-related macular degeneration (Gragoudas et al. 2004).

Stimuli-responsive surfaces, also known as smart surfaces, are a category of functional materials that exhibit dynamic changes in their surface properties including molecular conformations and surface energy in response to specific external stimuli, such as metal ions, small molecules, enzymes, pH value, temperature, electrical fields, and light exposure (Mendes 2008). For example, polymers functionalized on substrates presenting acidic or basic end groups like carboxy, sulfonyl, and amino groups, are sensitive to pH changes which result in the potential ionization of those end groups. Therefore, in response to pH changes, the stimuli-responsive surfaces composed of pH-sensitive polymers may undergo changes in conformation, hydrophilicity, solubility, and volume. As another example, DNA molecules are negatively charged because of the presence of phosphate groups in their molecular structure. By applying electrical fields with different polarities, it is possible to manipulate the conformational states of DNA molecules functionalized stimuli-responsive surfaces by means of the electrostatically attractive or repulsive forces exerting on the DNA molecules. The desired switchable properties of stimuli-responsive surfaces promote and contribute to a large number of applications in different engineering fields, such as environmental sensing, data storage, and microfluidic device design. Among them, the applications of stimuli-responsive surfaces in the biological, physiological and medical fields have attracted enormous attention and interests, ranging from modulation of biomolecular activity, protein separation or purification, regulation of cellular processes, tissue engineering, and regenerative medicine (Mendes 2008, Gomes et al. 2018). For example, in nearly neutral solution (pH 7.2–7.4), thrombin binding aptamers (TBA) present their natural folding motifs and can specifically capture the target thrombin molecules. While in acidic solution (pH 3.0–3.2), TBAs denature and lose their binding specificity towards thrombin molecules, which results in thrombin release accordingly. Therefore, TBA coated stimuli-responsive surfaces can be used to separate or purify thrombin based on the above capture and release actions modulated by pH changes. As another example, thermo-responsive polymer, such as poly-N-isopropylacrylamide (PNIPAM), exhibits hydrophobic properties at body temperature (37°C), while presenting hydrophilic properties at room temperature (20°C). Bovine serum albumin (BSA) is functionalized on the target cell surface via a series of conjugation to biotin, streptavidin, and biotinylated anti-epithelial cell adhesion molecule (EpCAM) antibody towards the EpCAM located on the cell membrane. The BSA can then bind PNIPAM coated substrate at body temperature via preferable hydrophobic interactions, so that cell attachment to the surface is achieved. In contrast, at room temperature, the PNIPAM shows an extended confirmation and its interaction with BSA is disrupted, leading to the cell release. In this way, by applying a moderate temperature change, the cell adhesion behavior can be modulated towards the thermo-responsive polymer functionalized stimuli-responsive surfaces.

By incorporating aptamers into stimuli-responsive surfaces, and using them as the essential sensing unit, aptamer-based stimuli-responsive surfaces have been developed. In combination of the binding affinity (which indicates the strength of the binding interaction between two molecules) and specificity (which indicates a molecule binds to its target molecules specifically, without binding to other molecules; it should be pointed out here that high affinity does not necessarily guarantee high specificity) presented by the aptamers, and switchable and versatile performance of stimuli-responsive surfaces, aptamer-based stimuli-responsive surfaces manifest optimal performance and multiple advantages. In this book chapter, we start with more fundamental backgrounds and historical review upon the invention and development of stimuli-responsive surfaces, and the precursors of aptamer-based smart surfaces, i.e., electrically controlled biointerfaces composed of DNA and its target protein. Subsequently and logically, we review and discuss the aptamer-based stimuli-responsive surfaces in terms of molecular sensing and cellular targeting. Finally, we briefly summarize the potential challenges and future directions of the investigations on aptamer-based stimuli-responsive surfaces.

Smart bio-interfaces modulated and controlled by external stimulus

A surface that responds to stimuli and can attach or release specific biomolecules when exposed to external triggers could dynamically turn external stimuli into biochemical signals and effectively transmit these signals within biological systems. These surfaces could be utilized for targeted delivery of signaling agents, identifying disease locations, fixing or altering genetic information, and adjusting biological system activity (Mendes 2008, Wong et al. 2010, Kuroki et al. 2012). Smart surfaces based on electrostatic stimuli have numerous benefits for transmitting information in device-biological interfaces.

Nanoscale biological devices, also known as nano-biodevices, are tiny structures or systems that are designed and engineered at the nanoscale level (typically ranging from 1 to 100 nanometers) for applications in the field of biology, including biotechnology, biomedicine, and bioengineering. These devices can be fabricated using nanotechnology, which involves manipulating and controlling matter at the nanometer scale, allowing for precise control over the structure, properties, and functionality of materials and devices. Nanoscale devices controlled by electrical mechanisms are easily scalable and offer full reversibility, as well as the capability for massive parallel activation and measurement (Wong et al. 2010). This means, the devices can be operated simultaneously and in a large-scale manner with continuous data acquisition. Additionally, the well-established techniques and infrastructure for semiconductor processing can be used for nanoscale manufacturing and synthesis. However, electrical mechanisms with limited specificity are not typically the mode of communication in biological systems. In comparison, biomolecules possess remarkable nanoscale abilities, including the ability to detect changes in shape, chemical modification and specific binding to ligands, as well as carrying out mechanical activation, membrane signaling, chemical transportation and catalysis (Wong et al. 2010, Wong and Melosh 2010). Therefore, smart surfaces that allow for electrical field-based modulation of biomolecular responses can take advantage of these capabilities to control biological behavior or improve the performance of artificial devices.

Self-assembled monolayers (SAMs) of polymers that change shape in response to stimuli such as changes in temperature, ionic strength, light, or electricity have been extensively studied for adjusting and changing surface properties. For instance, thermoresponsive polymers like poly-N-isopropylacrylamide (PNIPAm) change their configuration when the temperature changes, affecting protein absorption and cell adhesion (Akiyama et al. 2004, Burkert et al. 2010). The change in ionic strength or pH value has also been explored as a way to control ligand binding and cellular adhesion on surfaces coated with polyelectrolyte polymers (such as poly-2-dimethylamino-ethyl methacrylate (PDMAEMA)) through the effect of surrounding ions on shielding (Ballauff and Borisov 2006, Chiang et al. 2011). However, it's worth mentioning that thermoresponsive surfaces and polyelectrolyte brushes have limitations in terms of specific ligand recognition and protein absorption due to nonspecific hydrophobic and columbic interactions (Hianik et al. 2005, 2007). These interactions manifest between nucleic acid base pairs and protein side chains and thus receptors can interact with biomolecules or chemical species other than the target molecules. There have been several studies examining the optical mechanism governing responsive surfaces (Auernheimer et al. 2005, Nakanishi et al. 2007, Liu et al. 2009). Photoresponsive surfaces with photocleavable 2-nitrobenzyl groups enable specific protein adsorption and cell adhesion through the cleavage of functional groups, leading to the formation of specific cell patterns. However, applying optical stimuli leads to localized electron transfer and functional group cleavage (Willner and Katz 2005), which restricts the extent of conformational change and reversibility of control. Hence, electrical actuation offers greater control over system response. Proper choice of electrical field strength conserves the biomolecule structure and permits regulation of system response which results in tunable or smart interfaces.

Electrically controlled bio-interfaces composed of DNA and proteins

Compared to previously mentioned mechanisms of modulation, application of electrostatic fields is a feasible and efficient way to generate substantial conformational transition in polar or charged molecules (Lahann et al. 2003, Ma and Shrotriya 2012, 2015). Formation of ionic double layer near the electrode surface leads to the generation of strong electrical fields within several nanometers, and an exponential decrease of electrical field strength beyond that range. This scenario implies that relatively low external voltages may induce considerable conformational transition on the polar or charged biomolecules near the electrode surface without any detrimental effects on the system, which is very useful for modulating ligand sensing and binding. Rant et al. demonstrated that the conformation of short DNA oligomers immobilized on gold substrate with low grafting densities can be reversibly switched by the application of electrical fields (Rant et al. 2004, 2007), as shown in Fig. 1.

Wong et al. discovered that electrostatic fields can effectively influence self-assembly and hybridization of DNA (Wong and Melosh 2009). Positive voltage of 300 mV applied to single-strand DNA functionalized gold electrodes shows a three-fold enhancement in complementary DNA hybridization, while under negative voltage of –300 mV the hybridization level decreases by an order of magnitude. In another paper, Wong et al. described a method for electrically activating actin

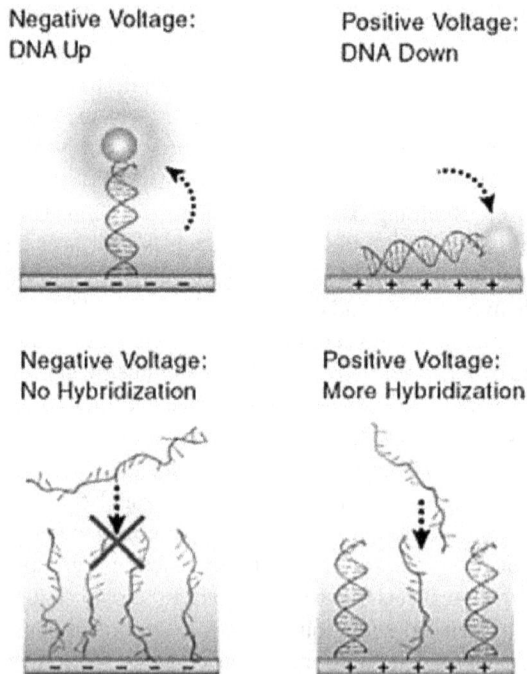

Fig. 1. DNA actuation through electric field and effect on hybridization (Wong et al. 2010). Reproduced with permission from Elsevier.

protein polymerization and alignment (Wong et al. 2008). Actin is a protein that plays a crucial role in cell structure and movement. The authors developed a method that uses an electric field to align actin fibers and stimulate their polymerization, resulting in the formation of actin gels. They found that the strength of the electric field and the concentration of actin protein influenced the rate of gel formation and the mechanical properties of the resulting gels. The authors suggest that this method could potentially be used to manipulate the organization and function of actin in cells and tissues. Mendes et al. utilized negative electrical potential to attract positively charged biotinylated oligolysine peptides (biotin-Lys-Lys-Lys-Lys-Cys) onto gold electrode surface, and dramatically altered the binding activity between the biotin functionalized on the end of peptides and its counterpart neutravidin (Yeung et al. 2010). Huang et al. adopted electrical field to generate conformational changes in thrombin binding aptamer (TBA) molecules immobilized on Au nanowire for ultrasensitive detection of fluorescent-labeled thrombin (Huang and Chen 2008), and quantified the binding affinity between thrombin and its aptamer to be ~ 30 nM in the titration assay, which is within the previously reported range (25–200 nM) (Bock et al. 1992), and comparable to high binding affinity between antigen and antibody which is usually less than 100 nM.

The paper "Switchable DNA interfaces for the highly sensitive detection of label-free DNA targets" by Rant et al. describes the use of a switchable DNA interface for the sensitive and specific detection of DNA targets (Rant et al. 2007). The authors developed a method using a thin layer of DNA on a metal surface that can be switched between a conductive and non-conductive state. When the DNA layer is in the non-conductive state, the presence of a specific DNA target can be detected by a change in the electrical resistance of the system. The authors showed that this method can detect DNA targets at concentrations as low as picomolar levels, making it a highly sensitive method for DNA detection. The authors also demonstrated that the switchable DNA interface can be used for the detection of multiple DNA targets in a single sample, making it a useful tool for DNA analysis. The conformation of ds-DNA and ss-DNA on the surface is influenced by the ionic strength of the solution they are in. In solutions with high ionic strength, electric interactions are mostly suppressed, and both ds-DNA and ss-DNA adopt different conformations. While single strands are compact, double strands stand upright, likely due to interactions with the underlying surface. These conformations are robust, not significantly affected by variations in surface potential or temperature. In an intermediate range of salinity, when DNA are grafted on electrode surface, the application of positive or negative potentials can change the DNA conformation. The change is more effective for double-stranded DNA because it behaves like a rigid rod (see Fig. 1), while the change for single-stranded DNA is more like a folding/unfolding process. At low ionic strength, the properties of ss-DNA change as electrostatic repulsions between charged sites along its backbone increase, making the single strands stiffer and easier to manipulate with electric fields (Kaiser and Rant 2010). The conformation of single- and double-stranded DNA molecules is affected differently by temperature changes, except at high salt concentrations where both are insensitive. At intermediate salinity, conformation changes are due to melting of base pairs, i.e., dsDNA transitions towards ssDNA, while in low-salinity solutions, ssDNA stretches out with increasing temperatures due to entropy or disruptions in base stacking. Analysis of the potential of conformation transition shows that increasingly negative electrode potentials must be applied in decreasing ionic strength solutions to effectively repel DNA from the surface. This is due to the screened image charge attraction that must be overcome by the external negative bias. In the case of DNA strands grafted on electrodes, the negatively charged DNA molecules will induce a positive image charge on the surface of the electrode. Hence, more negative electrode potential is required for repulsion of DNA from the surface.

In another publication, Langer et al. formulated a model that explains the steady and dynamic behavior of short, double-stranded DNA molecules attached to an electrically charged surface (Langer et al. 2014). The model examines how thermal energy and electrostatic interactions interact, and how the added hydrodynamic drag from a protein at the end of the DNA affects the molecular motion. The size of the protein can be determined from the slowed motion. The study found that the protein charge has a minor impact on the molecular dynamics but affects the orientation of the DNA in a steady state. Results from time-resolved fluorescence measurements were compared to the model's predictions and showed a good match. The model can be used to determine the size of different proteins with an accuracy of 0.3 nm. The limitations of the model become apparent when dealing with longer DNA molecules and very large proteins. The described analytical principle also enables detection of changes in protein conformation by measuring changes in hydrodynamic friction. This leads to a new type of chip-based biosensors that can detect biomolecules and determine their biophysical parameters in a single measurement. The development of electrically powered DNA origami nanomachines requires effective means of actuation by external electric fields. They are created by folding a long single-stranded DNA molecule, known as a scaffold, into a desired shape using short complementary "staple" strands that bind to specific regions of the scaffold, guiding its folding. By carefully designing the sequence of the scaffold and staple strands, researchers can create a wide range of complex and functional nanoscale structures with precise control over their shape, size, and surface properties. DNA origami nanomachines can be functionalized with various components, such as DNA aptamers, enzymes, or nanoparticles, to perform specific tasks. For example, they can be designed to respond to specific molecular cues, perform logic operations, or act as molecular shuttles or tweezers for manipulating other molecules. DNA origami nanomachines hold great promise for a wide range of applications, including drug delivery, biosensing, molecular computing, and nanoscale fabrication. Kroener et al. demonstrate how origami nanolevers on an electrode can be manipulated by alternating voltages with a fast response time of less than 100 μs, making dynamic control of the induced motion possible (Kroener et al. 2017). DNA origami structures can be attached to surfaces without extensive purification steps using a technique called voltage-assisted capture, and they can be interfaced with standard electronics to be used as moving parts in nanoscale devices with both electrical and mechanical properties.

Aptamer-based stimuli-responsive surfaces applied in molecular sensing and targeting

a) Electrical modulation

The Shrotriya research group were among the first to demonstrate the use of electric field to release targets bound on aptamers immobilized to conductive substrates (Gosai et al. 2016, Ma et al. 2017). In their first publication, they used continuum mechanics based electrostatic models and molecular dynamics to investigate the mechanism of binding/unbinding of thrombin from its aptamer, as shown in Fig. 2.

The continuum model was based on DNA electrostatics to compute the electric field produced by the TBA grafted on the gold electrode surface. The MD model determined the interaction between the TBA and thrombin due to the computed electric fields. It was demonstrated that with certain positive electrical fields, the thrombin can be dissociated from TBA, whereas it would remain bound when negative electrical fields were used. The Poisson-Boltzmann theory, a simple but effective model for understanding the electrostatic interaction between particles, was used to show that at positive electrode potentials, the DNA layer creates a repulsive area for the thrombin molecule, hindering the formation of the complex and thus augments the dissociation. Calculations of the binding free energy for the complex in different electric fields from umbrella sampling simulations indicate that the free energy of binding ($\Delta G_{binding}$) is lower for positive electric fields than for neutral/negative fields. The electrical stimulus affects the non-bonded interaction energies and H-bonding between thrombin and TBA. A positive electric field weakens the non-bonded interaction energy and H-bonding between the TBA/aptamer complex, leading to a reduced $\Delta G_{binding}$. The molecular dynamics simulations of the TBA/thrombin complex under positive electric fields showed that the complex breaks apart spontaneously and that the binding between protein-ligand complexes can be controlled through electrical stimulation. Subsequently, a full description of the experiments was published that provided more details on fabrication microstructured

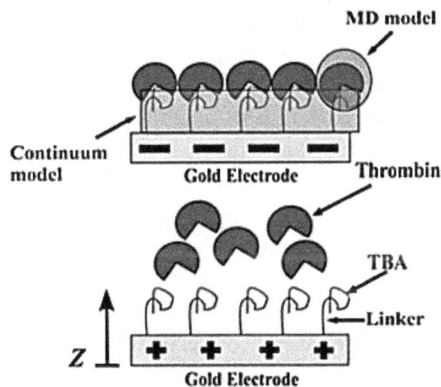

Fig. 2. A schematic of the electrical actuation of thrombin from its bound state with the thrombin binding aptamer (TBA) (Gosai et al. 2016). The domains for the continuum and molecular dynamics (MD) models are marked in the scheme. Modified and reproduced with permission from Springer Nature.

surfaces utilized in creating smart aptamer interfaces (Ma et al. 2017). It was proven that the structure of the protein is not damaged for certain field magnitudes like + 0.5 to 1.0 V/nm. Later studies by the same authors quantified the force of binding between TBA and thrombin using specialized atomic force microscopy (Ma et al. 2019, 2020). It was also proven that the unbinding happens through the breaking of the hydrogen bonds between TBA and thrombin due to the lowering of the energy landscape by the positive electric field.

Researchers have further explained how this method could be utilized for advanced wearable sensors, which is a hot research topic. Tu et al. hypothesized that the novel technique can be applied to overcome the deficiencies in previously known methods (Tu et al. 2020). Specifically, using earlier methods that relied on electrochemical desorption, the receptors would be dislodged, when the protein bound to them needed to be released. This is not a desirable situation when a reuse is required. However, Tu et al. stated that electrical actuation method is more advantageous and viable as it is reversible, and the receptors (aptamers) are not removed (Tu et al. 2020).

Use of electric voltage for releasing targets has also been recently exploited by another research group. The study tested a paper-based platform called "Aptapaper" for analyzing small molecules in complex matrices using two known aptamers (Martínez-Jarquín et al. 2022), the quinine binding aptamer (QBA) and serotonin binding aptamer (SBA), as shown in Fig. 3. The aptapaper was incubated under conditions to enable proper aptamer folding and then used to concentrate target analytes from complex matrices. The aptapaper was rinsed, dried, and the target analyte was analyzed using paper spray ionization and high-resolution mass spectrometry, either immediately or up to 4 days later. The article describes the design of a paper spray platform that is functionalized with aptamers for the analysis of compounds in complex matrices. The platform uses glass microfiber paper, making it inexpensive and accessible for inexperienced individuals to collect samples from remote locations. The platform was tested using QBA-quinine and SBA-serotonin model systems and could be applied to other aptamers for analyzing compounds in low concentrations. The minimum concentrations detectable were 81 pg/mL and 1.8 ng/mL for quinine and serotonin, respectively. The glass microfiber paper, after being bound, rinsed, and dried, can be read at a later time, making it suitable for remote sampling. The target analyte was separated from the QBA aptamer, which has a mechanism for folding based on the target (Reinstein et al. 2013), by reducing the pH with a spray solution and applying a high voltage to the tip of the paper. This method of using electrical currents to cause changes in the folding of aptamers has been reported previously, by Gosai et al. (Gosai et al. 2016), in electrochemical aptasensors and is a common method used to regulate binding and unbinding in other aptamers. To optimize the conditions for each aptamer, it needs to be tested separately by adding the required salts in a preconditioning step. The rinsing protocol after the binding step ensures that salts from preconditioning or the binding matrix will not interfere with MS detection.

Biologically modified nanopore sensors are a type of nanoscale sensing technology that utilize nanopores, which are tiny openings on a nanometer scale,

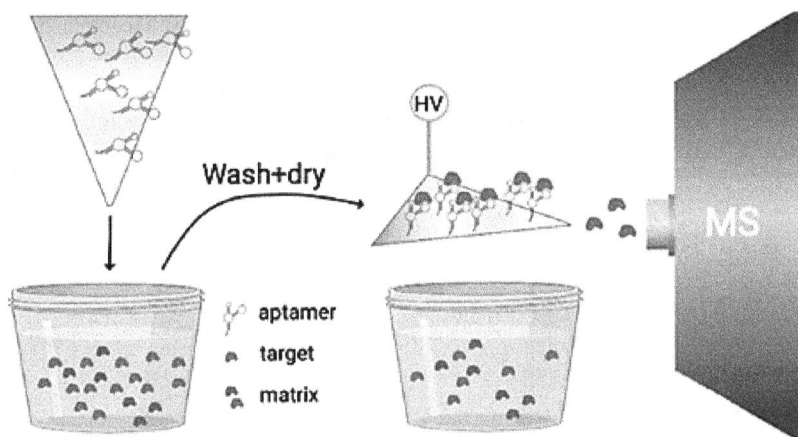

Fig. 3. High voltage (HV) was used to release target from aptamer for analysis by mass spectrometer (MS) (Martínez-Jarquín et al. 2022). Modified and reproduced with permission from American Chemical Society (ACS).

to detect and analyze molecules or particles. These nanopores can be modified or functionalized with biological molecules, such as proteins, enzymes, DNA, or other biomolecules, to impart specific sensing capabilities or selectivity to the sensor. Biologically modified nanopore sensors offer a promising platform for a wide range of applications in biotechnology like DNA and protein sensing, diagnostics, drug discovery, environmental monitoring, and fundamental research, due to their high sensitivity, selectivity, and versatility in detecting and analyzing various biomolecules or particles at the nanoscale. Pal et al. have recently published a study demonstrating the utilization of aptamer-decorated DNA origami structures in the functionalization of silicon nitride nanopores (Pal et al. 2022). Aptamers, with their high sensitivity and specificity, are advantageous as chemical agents for analyte detection. However, chemically modifying a silicon nitride nanopore with biomolecules is challenging and requires intricate genetic engineering. The study presented a method of attaching a G-quadruplex or G-q hosting DNA aptamer to a solid-state nanopore, which undergoes a conformational change influenced by potassium, leading to modulation of ionic current. This method allowed for the detection of G-q folds. In previous studies (Bošković et al. 2019, Si et al. 2019), the detection of G-q structures using solid-state nanopores by translocating DNA carriers with folded quadruplex structures took 0.5–1 ms. In the current study, the entrapment of G-q-hosting aptamers on the cis side of the pore resulted in increased observation events by multiple orders of magnitude, allowing for the detailed examination of thrombin-TBA interactions. The short length of the TBA overhangs, only 15 nucleotides, led to less crowding and more accurate detection of G-q kinetics. The optimum G-q formation of TBA takes place upon binding to thrombin and hence the binding events are recorded. The binding between thrombin and TBA is a stochastic process with various modes of interaction. In the nanopore, when thrombin bound to TBA, it caused blockage of ionic current across the nanopore. The electric field around the nanopore region kept the structures in a dynamic state of binding and unbinding, with the protein-aptamer complex lasting several milliseconds. In fact, this has been demonstrated previously by Gosai et al. (Gosai et al. 2016). The use of DNA origami with nitride pores replicated biologically modified nanopore sensors, while also offering the benefits of adjustable pore size, easy incorporation of the aptamer ligand, and a robust and stable nanopore design (Shim et al. 2009). The selectivity of the aptamer allowed for differentiation of protein structures, for, e.g., thrombin and streptavidin, that differ in size by only ± 1.5 nm. As the TBA only binds to thrombin, the G-q kinetic response is maximized only when thrombin is present in the system. Additionally, the hybridization of DNA origami with oligonucleotides is a straightforward process, eliminating the need for designing separate sensors for various analytes. This modular approach to single-molecule sensing using nanopores can be easily expanded to include any aptamer, thereby enabling detection of a broad spectrum of molecules.

b) Non-electrical modulation

Researchers have created switchable aptamer-diacyllipid conjugates on a molecular level to form micelle flares for measuring ATP in a lab setting and for molecular imaging within living cells (Wu et al. 2013), as shown in Fig. 4. These switchable aptamer micelle flares maintain their binding capability and have potential for delivering cargo without the use of toxic materials and for imaging the expression and distribution of biomolecules inside cells. The hydrophobic diacyllipid tails are combined with hydrophilic switchable aptamer heads through a solid phase phosphoramidite chemistry process using an automatic DNA synthesizer. This results in a uniform spherical nanostructure known as the switchable aptamer micelle flare (SAMF). The similarity in intermolecular forces between the diacyllipid in DNA micelle flares and dynamic phospholipid bilayers in cell membranes allows SAMFs to interact easily with cell membranes and enter cells. SAMFs offer many benefits over previous methods, such as easy probe modification, self-delivery capability, high signal-to-background ratio, strong target specificity, and excellent biocompatibility. These properties make SAMFs an efficient platform for delivering and imaging living cells without the use of potentially toxic nanomaterials. Furthermore, these switchable aptamer micelle flares can track changes in ATP levels of the cells in real time. Future plans may include using different types of nucleic acid probe micelle flares for accurately determining the number of biomolecules in living cells.

In another study, researchers designed water-soluble and stimulus-responsive luminescent hydrogels with aptamer-functionalized covalent organic framework (COF) probe to capture and protect latent fingerprint information (Hai et al. 2019). COFs have been proposed as alternative candidates for "smart" materials due to their ordered π-columnar structures. The researchers designed and synthesized a water-dispersible and smart responsive luminescent carboxymethyl cellulose-COF hydrogel encapsulated 5-(dimethylamino)-N,N-bis(pyridin-2-ylmethyl) napthalene-1-sulfonamide, named CMC-COF-LZU1⊃DPYNS. The combination of carboxymethyl cellulose, i.e., CMC and COF-LZU1 increased the biocompatibility of the nanocomposites and the inclusion of dimethylamino-napthalene-sulfonamide, i.e., DPYNS provided smart fluorescence response capabilities. The nanocomposites

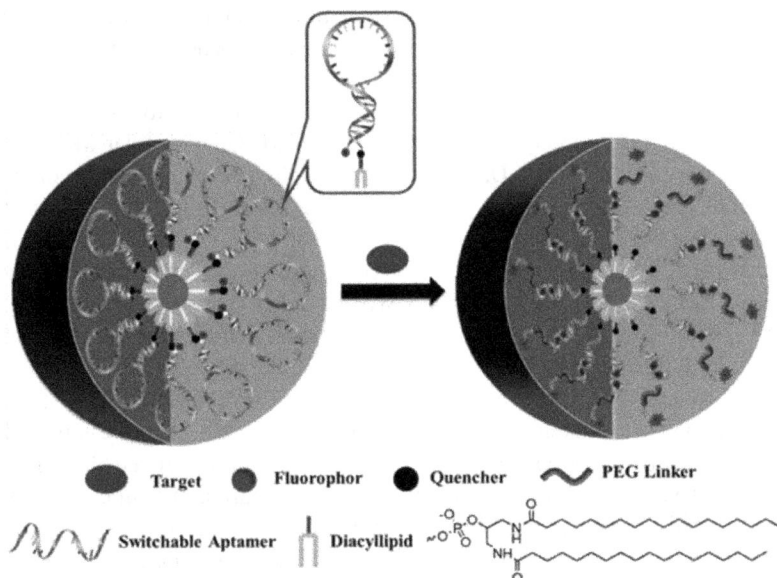

Fig. 4. The functioning of switchable aptamer micelle flares is based on the self-assembly of aptamer switch probes and diacyllipid conjugates into a micelle flare nanostructure (Wu et al. 2013). When the target is not present, the aptamer switch probe remains in a loop-stem structure, causing the fluorescence to be suppressed due to the close proximity between the fluorophore and quencher. However, when the target binds, the shape of the switchable aptamer changes, causing the fluorescence signal to be restored. Reproduced with permission from American Chemical Society (ACS).

were found to have good dispersibility in water and displayed changes in fluorescence when exposed to Cu^{2+}/H_2O in both liquid and solid forms. These luminescent CMC-COF-LZU1⊃DPYNS hydrogels can be used as an invisible security probe for latent fingerprint information storage and retrieval, which can only be revealed by adding H_2O under UV light. By using optical imaging techniques, they were able to achieve high-resolution fingerprint patterns at three different levels. This approach has the potential to pave the way for the development of smart luminescent COF materials in security protection and COFs modified with cellulose could have applications in clinical settings.

Aptamer-based stimuli-responsive surfaces applied in cell targeting

Separation of specific type of cells from heterogeneous types of cells is of practical importance in fundamental biological or physiological studies, clinical screening, diagnostics, and treatment. Quite a few physical properties, such as density, dimension or size, and electrical charges can be utilized to isolate certain types of cells. However, it is evident that those physical separation methods are non-specific since different type of cells may exhibit similar physical properties that are difficult to distinguish. To this end, chemical separation methods based on affinity binding between surface-immobilized ligands, such as antibodies or aptamers, and target cells can be utilized to implement specific cell screening. The benefits of using aptamers instead of antibodies include better stability, longer shelf life, higher reproducibility, easier modification and immobilization, and enhanced access to tissues owing to their smaller size (an antibody is typically ~ 150–170 kDa in molecular weight, while an aptamer is typically 12–30 kDa in molecular weight, i.e., an aptamer is commonly much smaller in size compared to an antibody). Meanwhile, stimuli-responsive surfaces enable the reversible capture and release of cells, i.e., effective modulation of the binding behavior between target cells and surface-immobilized ligands. Therefore, aptamer-based stimuli-responsive surfaces can serve as an advanced affinity isolation method to capture, release, separate, and purify specific type of cells with high throughput, outstanding efficiency and lower cost.

Lin et al. implemented the capture of specific CCRF-CEM cells, the human acute lymphoblastic leukemia (ALL) cell line, by adopting surface-immobilized aptamers in a microchamber (Zhu et al. 2012), while they realized the release of the captured CCRF-CEM cells in response to the application of temperature stimuli in an integrated device composed of microfluidic channel, heaters and temperature sensors, as shown in Fig. 5. More specifically, the DNA aptamers sgc8c presented regularly folded conformation to specifically bind to their target CCRF-CEM cells at room temperature, while the aptamers sgc8c did not recognize the Toledo cells, the human diffuse large-cell lymphoma cell line which serves as a negative control to represent the nonspecific binding. After being washed by Dulbecco's phosphate-buffered saline (D-PBS), the non-target Toledo cells were removed from the binding structure composed of aptamers sgc8c and target CCRF-CEM cells. Subsequently, a moderate temperature elevation from room temperature to 48°C was applied by gold heater units located in the passivation layer of the microfluidic device. The aptamers sgc8c underwent conformational transition from folded state to unfolded state and lost the binding specificity to target CCRF-CEM cells, thus releasing them into the microfluidic channel which were then washed away by D-PBS. The following experiments also demonstrated the cell viability after a cycle of specific capture and temperature-mediated release of target CCRF-CEM cells by harnessing the above aptamer-based temperature-stimuli response surfaces.

Qu et al. reported an aptamer-based stimuli-responsive surface design which can realize specific cancer cell release based on the supramolecular interaction between β-cyclodextrin (β-CD) and ferrocene (Fc) conjugated to DNA aptamer AS1411 (Feng et al. 2015), as shown in Fig. 6. To enable the capture of HeLa cancer cells, β-CD molecules were first functionalized on graphene substrate, then aptamers AS1411 conjugated by Fc were exposed to the surface. Due to the host-guest interaction between β-CD and Fc, the aptamer AS1411 – Fc – β-CD construct was formed on the graphene substrate. Aptamer AS1411 naturally contains stable G-quadruplex structures, which presents high binding affinity to nucleolin, the receptor on HeLa cell surface, so HeLa cells can be specifically bound by aptamer AS1411, and thus stabilized onto the aptamer AS1411 – Fc – β-CD modified graphene substrate. By applying an electrical potential of 1 V, the Fc moieties conjugated on aptamer AS1411 were oxidized, and their binding affinity to β-CD was evidently decreased,

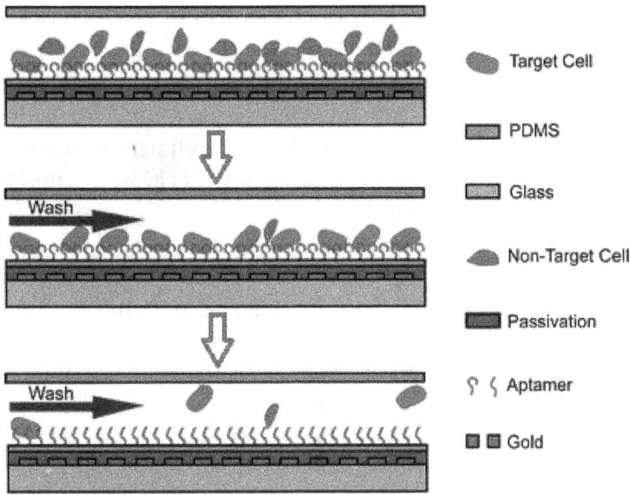

Fig. 5. Scheme of target cell capture and temperature-mediated release (Zhu et al. 2012). First, specific capture of target CCRF-CEM cells by aptamer sgc8c at room temperature, while non-target Toledo cells cannot be captured by the aptamer. Second, the non-target Toledo cells were washed away by D-PBS. Third, release of target CCRF-CEM cells was implemented by applying a moderate temperature stimulus. Modified and reproduced with permission from Royal Society of Chemistry (RSC).

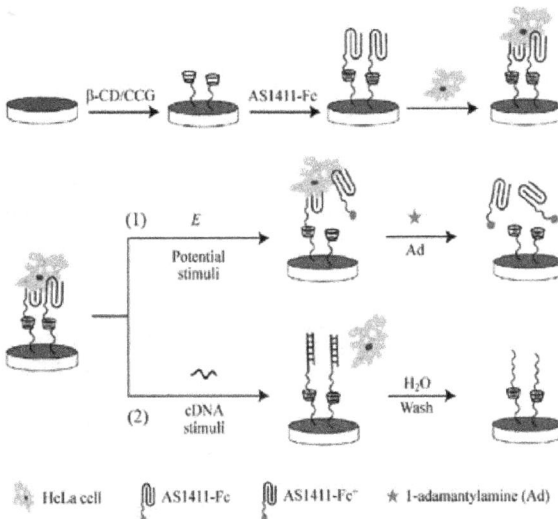

Fig. 6. Schematic illustration of aptamer AS1411 functionalized graphene platform (Feng et al. 2015). Top: the sequential process leading to the HeLa cell - aptamer AS1411 – Fc – β-CD modified graphene substrate, i.e., cancer cell capture. Bottom: application of electrochemical stimuli to implement cancer cell release. (1) By applying electrical potential, the Fc moieties are oxidized with decreased binding affinity to β-CD, so the HeLa cell was released from the surface. Competitive binding of Ad to β-CD can further amplify the cancer cell release. (2) By inducing cDNA, the duplex structure was formed by the aptamer AS1411 and its cDNA, while the G-quadruplex is disrupted which results in decreased binding affinity of aptamer AS1411 to HeLa cell, so the HeLa cell was released from the surface. Reproduced with permission from Springer Nature.

thus initiating the HeLa cell release. By further incorporating 1-adamantylamine (Ad) molecules onto the surface, the competitive binding of Ad to β-CD can amplify the HeLa cell release from the surface. On the other aspect, by inducing the chemical stimuli, i.e., the complementary DNA (cDNA) of aptamer AS1411, to the HeLa cells - aptamer AS1411 – Fc – β-CD modified graphene substrate, aptamer AS1411 and its cDNA formed duplex structure. Then the aptamer AS1411 loses its the G-quadruplex structure, and thus cannot specifically bind to nucleolin on HeLa cell surface anymore, which leads to the release of HeLa cell from the surface accordingly. Through the application of above label-free electrochemical stimuli, the specific capture and release of cancer cells can be implemented via the aptamer functionalized graphene platform.

Wang et al. employed DNA aptamer S2.2 to conjugate the breast cancer cells MCF-7 and azobenzene (Azo) with high affinity and specificity (Bian et al. 2016), as shown in Fig. 7. The MCF-7 cells – Aptamer S2.2 – Azo constructs can then be functionalized onto cyclodextrin (CD)-modified silicon substrate to form the biointerface. Subsequently, UV light irradiation can be utilized as the stimuli to trigger the cancer cell release from the biointerface, i.e., the UV light induced the conformational transition of Azo from trans- to cis-conformation state, which was not recognized by CD owing to the unmatched host-guest interaction between CD and Azo, which thus led to the release of the originally captured MCF-7 cells in the biointerface. This light-triggered specific cancer cell release using the aptamer-based stimuli-responsive surfaces present high efficiency and significant potential applications for the separation, screening, and analysis of circulating tumor cells. Meanwhile, the photo-switching mechanism could potentially avoid the excessive input of thermal energy or chemicals, e.g., from temperature and pH stimuli.

The selective recognition of aptamers on specific types of cells are highly desirable for cell screening, clinic diagnostics and targeted therapeutics. However,

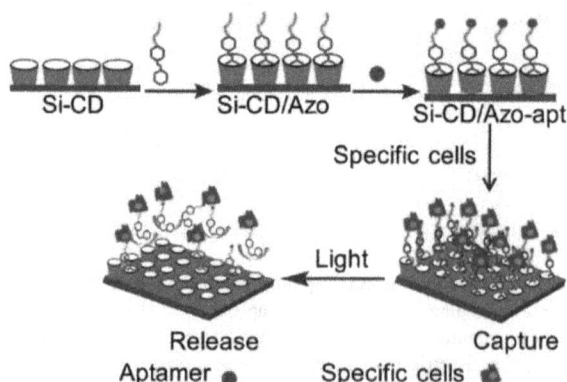

Fig. 7. Aptamer based photo-stimuli-responsive biointerface including cyclodextrin-azobenzene host-guest interaction, DNA aptamer S2.2 and MCF-7 breast cancer cell (Bian et al. 2016). The UV light irradiation can trigger the specific MCF-7 cell release from the biointerface. Reproduced with permission from American Chemical Society (ACS).

both target cells and nontarget cells may present the identical or similar types of cell surface receptors to a kind of aptamer, which can significantly compromise the selective recognition of aptamers on the target cell. To cope with this issue, the microenvironment of a cell should be taken into consideration. For example, tumor cells commonly present acidic microenvironment owing to the high glycolytic rate, while regular cells do not. Then if the aptamer can specifically recognize both the tumor cell surface receptor and exclusively acidic microenvironment near the cell, such kind of aptamer-based pH-stimuli-responsive surfaces can be developed to implement the switchable binding behavior between target and nontarget cells.

Based on such a strategy, Tan et al. designed and constructed structure-switchable aptamer (SW-Apt) isgc8-5 presenting reconfigurable binding affinity by exploiting their pH value in the microenvironment (Li et al. 2018), as shown in Fig. 8. Specifically, this single-stranded SW-Apt contains (1) a recognition region which can bind with the membrane receptor of the target cell in their circular, confined conformation, while losing the binding affinity towards the membrane receptor in their relaxed, open conformation; (2) a linker that conjugates the recognition region and modulator region; (3) a modulator region, i.e., i-motif, the quadruplex structure rich in cytosine to contribute to the binding affinity of the aptamer to its target cells. At acidic pH value (pH 6.5) in the microenvironment of target cells, the quadruplex structure in i-motif is well stabilized, so that the linker

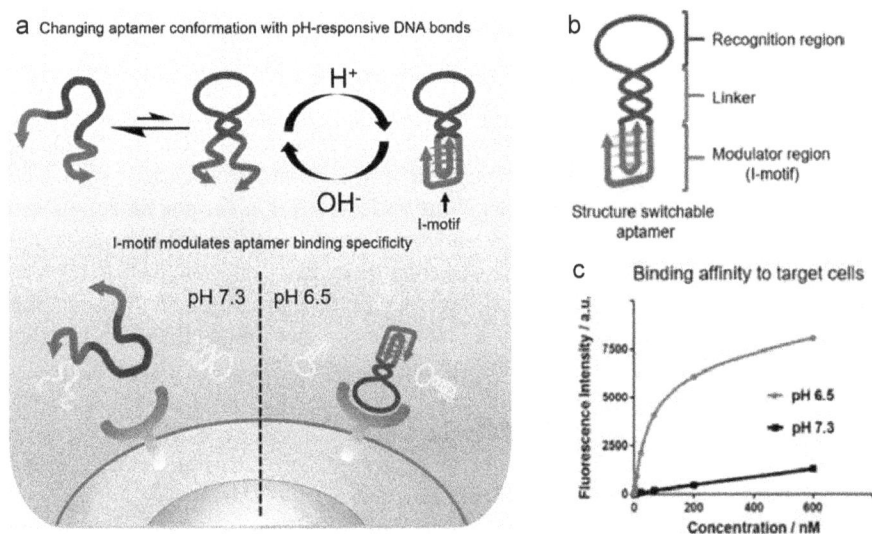

Fig. 8. Aptamer based pH-stimuli-responsive surfaces including structure-switchable aptamer isgc8-5 for selective binding to the target cell surface receptors by exploiting specific pH value in the microenvironment (Li et al. 2018). (a) Acidic pH value can trigger the specific binding confirmation of the aptamer while physiological pH value lead to unfolding conformation of the aptamer. (b) The structure-switchable aptamer includes three characteristics components: recognition region, linker and modulator region (I-motif). (c) At pH 6.5, structure-switchable aptamer presents much higher binding affinity towards its target cell compared to that at ph 7.3. Modified and reproduced with permission from American Chemical Society (ACS).

and recognition region also stay in the confined state; thus, the SW-Apt presented high binding ability towards its target cells. While at physiological pH value (pH 7.3) in the microenvironment of nontarget cells, the quadruplex structure in i-motif is disrupted and the modulator region presents more relaxed, denatured state, so that the linker and recognition region are affected and consequently present relaxed, open state, and thus no obvious binding between the SW-Apt and cells was revealed. The conformational switch of SW-Apt attributed to the reconfigurable quadruplex i-motif structure was demonstrated by Förster resonance energy transfer (FRET) and circular dichroism (CD) spectroscopy. Such a simple and efficient design of the aptamer-based pH-stimuli-responsive surfaces via chemical modulation can be potentially employed in tumor cell targeting and screening.

In summary, the common feature of the aptamer-based stimuli-responsive surfaces to specifically capture and release target cells is the conformational transition of either the aptamer structure itself, or its conjugated structure in the biointerface in response to various types of stimuli, e.g., temperature change, electrical potential application, light exposure, chemical agent exposure, or pH value change. More generalized reports about the cellular targeting and regulation by using aptamer-based stimuli-responsive surfaces can be found in various research and review articles (Mendes 2008, Nandivada et al. 2010, Peng and Bhushan 2012, Rant 2012, Krol and Chmielarz 2014, Conde et al. 2015, Bordoni et al. 2016, Gao et al. 2018, Gomes et al. 2018, Esmaeilzadeh and Groth 2019, Cheng et al. 2021, Friedl et al. 2021, Zhang et al. 2021, Qu et al. 2022, McCorry et al. 2023).

Future directions

Contemporarily, many studies have been reported upon the molecular sensing and cellular capture and release. In the near future, more investigations would be necessary for expanding and deepening our understanding about the actuation and modulation of various cellular and physiological processes using the aptamer-based stimuli-responsive surfaces, such as cell migration, cell adhesion, cell cycle and division, transcription and translation, cell signaling, and cell metabolism, so that we may potentially and successfully employ the aptamer-based stimuli-responsive surfaces in physiological studies, clinical diagnostics, therapeutic treatment, tissue engineering, and regenerative medicine. For example, intercellular adhesion molecules (ICAM) are supposed to keep the integrity of a sheet of cells and promote collective cell migration. Based on SELEX process, it is plausible to design and develop the ICAM specific aptamers, but how this type of aptamers and associated stimuli-responsive surfaces affect or modulate the intercellular adhesion and collective cell migration needs further investigation and clarification. As another example, various cell signaling pathways start at the binding between the specific ligands with their target cell surface receptors, e.g., either the ion channel linked receptors, or G protein–coupled receptors, or enzyme-linked receptors. Using SELEX process, we may routinely design and generate the specific aptamers as the ligands to target their cell surface receptors. However, the binding kinetics between the aptamers and their cell surface receptors, and more importantly, the potential regulation of the aptamers and associated stimuli-responsive surfaces on the downstream signaling cascades remains largely unknown and requires in-depth investigation and demonstration.

References

Akiyama, Y., A. Kikuchi, M. Yamato and T. Okano. 2004. Ultrathin poly(N-isopropylacrylamide) grafted layer on polystyrene surfaces for cell adhesion/detachment control. Langmuir 20(13): 5506–5511.

Auernheimer, J., C. Dahmen, U. Hersel, A. Bausch and H. Kessler. 2005. Photoswitched cell adhesion on surfaces with RGD peptides. J. Am. Chem. Soc. 127(46): 16107–16110.

Ballauff, M. and O. Borisov. 2006. Polyelectrolyte brushes. Current Opinion in Colloid & Interface Science 11(6): 316–323.

Bian, Q., W.S. Wang, S.T. Wang and G.J. Wang. 2016. Light-triggered specific cancer cell release from cyclodextrin/azobenzene and aptamer-modified substrate. Acs Applied Materials & Interfaces 8(40): 27360–27367.

Bock, L.C., L.C. Griffin, J.A. Latham, E.H. Vermaas and J.J. Toole. 1992. Selection of single-stranded-DNA molecules that bind and inhibit human thrombin. Nature 355(6360): 564–566.

Bordoni, A.V., M.V. Lombardo and A. Wolosiuk. 2016. Photochemical radical thiol-ene click-based methodologies for silica and transition metal oxides materials chemical modification: a mini-review. Rsc Advances 6(81): 77410–77426.

Bošković, F., J. Zhu, K. Chen and U.F. Keyser. 2019. Monitoring G-quadruplex formation with DNA carriers and solid-state nanopores. Nano Letters 19(11): 7996–8001.

Burkert, S., E. Bittrich, M. Kuntzsch, M. Muller, K.J. Eichhorn, C. Bellmann, P. Uhlmann and M. Stamm. 2010. Protein resistance of PNIPAAm brushes: application to switchable protein adsorption. Langmuir 26(3): 1786–1795.

Cheng, H.B., S.C. Zhang, J. Qi, X.J. Liang and J. Yoon. 2021. Advances in application of azobenzene as a trigger in biomedicine: molecular design and spontaneous assembly. Advanced Materials 33(26).

Chiang, E.N., R. Dong, C.K. Ober and B.A. Baird. 2011. Cellular responses to patterned poly(acrylic acid) brushes. Langmuir 27(11): 7016–7023.

Conde, J., E.R. Edelman and N. Artzi. 2015. Target-responsive DNA/RNA nanomaterials for microRNA sensing and inhibition: The jack-of-all-trades in cancer nanotheranostics? Advanced Drug Delivery Reviews 81: 169–183.

Ellington, A.D. and J.W. Szostak. 1990. *In vitro* selection of Rna molecules that bind specific ligands. Nature 346(6287): 818–822.

Esmaeilzadeh, P. and T. Groth. 2019. Switchable and obedient interfacial properties that grant new biomedical applications. Acs Applied Materials & Interfaces 11(29): 25637–25653.

Feng, L.Y., W. Li, J.S. Ren and X.G. Qu. 2015. Electrochemically and DNA-triggered cell release from ferrocene/beta-cyclodextrin and aptamer modified dual-functionalized graphene substrate. Nano Research 8(3): 887–899.

Friedl, J.D., V. Nele, G. De Rosa and A. Bernkop-Schnurch. 2021. Bioinert, stealth or interactive: how surface chemistry of nanocarriers determines their fate *in vivo*. Advanced Functional Materials 31(34).

Gao, X., Q. Li, F.C. Wang, X.H. Liu and D.B. Liu. 2018. Dual-responsive self-assembled monolayer for specific capture and on-demand release of live cells. Langmuir 34(28): 8145–8153.

Gomes, B.S., B. Simoes and P.M. Mendes. 2018. The increasing dynamic, functional complexity of bio-interface materials. Nature Reviews Chemistry 2(3).

Gosai, A., X. Ma, G. Balasubramanian and P. Shrotriya. 2016. Electrical stimulus controlled binding/unbinding of human thrombin-aptamer complex. Scientific Reports 6.

Gragoudas, E.S., A.P. Adamis, E.T. Cunningham, M. Feinsod, D.R. Guyer and V.I.S.O. Neova. 2004. Pegaptanib for neovascular age-related macular degeneration. New England Journal of Medicine 351(27): 2805–2816.

Hai, J., H. Wang, P. Sun, T. Li, S. Lu, Y. Zhao and B. Wang. 2019. Smart responsive luminescent aptamer-functionalized covalent organic framework hydrogel for high-resolution visualization and security protection of latent fingerprints. ACS Applied Materials & Interfaces 11(47): 44664–44672.

Hianik, T., V. Ostatna, Z. Zajacova, E. Stoikova and G. Evtugyn. 2005. Detection of aptamer-protein interactions using QCM and electrochemical indicator methods. Bioorganic & Medicinal Chemistry Letters 15(2): 291–295.

Hianik, T., V. Ostatna, M. Sonlajtnerova and I. Grman. 2007. Influence of ionic strength, pH and aptamer configuration for binding affinity to thrombin. Bioelectrochemistry 70(1): 127–133.

Homann, M. and H.U. Goringer. 1999. Combinatorial selection of high affinity RNA ligands to live African trypanosomes. Nucleic Acids Research 27(9): 2006–2014.

Huang, S.X. and Y. Chen. 2008. Ultrasensitive fluorescence detection of single protein molecules manipulated electrically on Au nanowire. Nano Letters 8(9): 2829–2833.

Kaiser, W. and U. Rant. 2010. Conformations of end-tethered DNA molecules on gold surfaces: influences of applied electric potential, electrolyte screening, and temperature. Journal of the American Chemical Society 132(23): 7935–7945.

Kroener, F., A. Heerwig, W. Kaiser, M. Mertig and U. Rant. 2017. Electrical actuation of a DNA origami nanolever on an electrode. Journal of the American Chemical Society 139(46): 16510–16513.

Krol, P. and P. Chmielarz. 2014. Recent advances in ATRP methods in relation to the synthesis of copolymer coating materials. Progress in Organic Coatings 77(5): 913–948.

Kuroki, H., I. Tokarev and S. Minko. 2012. Responsive surfaces for life science applications. Annual Review of Materials Research 42: 343–372.

Lahann, J., S. Mitragotri, T.N. Tran, H. Kaido, J. Sundaram, I.S. Choi, S. Hoffer, G.A. Somorjai and R. Langer. 2003. A reversibly switching surface. Science 299(5605): 371–374.

Langer, A., W. Kaiser, M. Svejda, P. Schwertler and U. Rant. 2014. Molecular dynamics of DNA–protein conjugates on electrified surfaces: solutions to the drift-diffusion equation. The Journal of Physical Chemistry B 118(2): 597–607.

Li, L., Y. Jiang, C. Cui, Y. Yang, P.H. Zhang, K. Stewart, X.S. Pan, X.W. Li, L. Yang, L.P. Qiu and W.H. Tan. 2018. Modulating aptamer specificity with pH-responsive DNA bonds. Journal of the American Chemical Society 140(41): 13335–13339.

Liu, D.B., Y.Y. Xie, H.W. Shao and X.Y. Jiang. 2009. Using azobenzene-embedded self-assembled monolayers to photochemically control cell adhesion reversibly. Angewandte Chemie-International Edition 48(24): 4406–4408.

Ma, X. and P. Shrotriya. 2012. Molecular dynamics simulation of electrical field induced conformational transition and associated frictional performance of monomolecular films. Journal of Physics D-Applied Physics 45(37).

Ma, X. and P. Shrotriya. 2015. Molecular dynamics simulation of conformational transition and frictional performance modulation of densely packed self-assembled mono layers based on electrostatic stimulation. Langmuir 31(24): 6729–6741.

Ma, X., A. Gosai, G. Balasubramanian and P. Shrotriya. 2017. Aptamer based electrostatic-stimuli responsive surfaces for on-demand binding/unbinding of a specific ligand. Journal of Materials Chemistry B 5(20): 3675–3685.

Ma, X., A. Gosai, G. Balasubramanian and P. Shrotriya. 2019. Force spectroscopy of the thrombin-aptamer interaction: Comparison between AFM experiments and molecular dynamics simulations. Applied Surface Science 475: 462–472.

Ma, X., A. Gosai and P. Shrotriya. 2020. Resolving electrical stimulus triggered molecular binding and force modulation upon thrombin-aptamer biointerface. Journal of Colloid and Interface Science 559: 1–12.

Martínez-Jarquín, S., A. Begley, Y.-H. Lai, G.L. Bartolomeo, A. Pruška, C. Rotach and R. Zenobi. 2022. Aptapaper—An aptamer-functionalized glass fiber paper platform for rapid upconcentration and detection of small molecules. Analytical Chemistry 94(14): 5651–5657.

McCorry, M.C., K.F. Reardon, M. Black, C. Williams, G. Babakhanova, J.M. Halpern, S. Sarkar, N.S. Swami, K.A. Mirica, S. Boermeester and A. Underhill. 2023. Sensor technologies for quality control in engineered tissue manufacturing. Biofabrication 15(1).

Mendes, P.M. 2008. Stimuli-responsive surfaces for bio-applications. Chemical Society Reviews 37(11): 2512–2529.

Nakanishi, J., Y. Kikuchi, S. Inoue, K. Yamaguchi, T. Takarada and M. Maeda. 2007. Spatiotemporal control of migration of single cells on a photoactivatable cell microarray. J. Am. Chem. Soc. 129(21): 6694–6695.

Nandivada, H., A.M. Ross and J. Lahann. 2010. Stimuli-responsive monolayers for biotechnology. Progress in Polymer Science 35(1-2): 141–154.

Pal, S., A. Naik, A. Rao, B. Chakraborty and M.M. Varma. 2022. Aptamer-DNA origami-functionalized solid-state nanopores for single-molecule sensing of G-quadruplex formation. ACS Applied Nano Materials 5(7): 8804–8810.

Peng, S.J. and B. Bhushan. 2012. Smart polymer brushes and their emerging applications. Rsc Advances 2(23): 8557–8578.

Qu, Y.C., K.Y. Lu, Y.J. Zheng, C.B. Huang, G.N. Wang, Y.X. Zhang and Q. Yu. 2022. Photothermal scaffolds/surfaces for regulation of cell behaviors. Bioactive Materials 8: 449–477.

Rant, U., K. Arinaga, S. Fujita, N. Yokoyama, G. Abstreiter and M. Tornow. 2004. Dynamic electrical switching of DNA layers on a metal surface. Nano Letters 4(12): 2441–2445.

Rant, U., K. Arinaga, S. Scherer, E. Pringsheim, S. Fujita, N. Yokoyama, M. Tornow and G. Abstreiter. 2007. Switchable DNA interfaces for the highly sensitive detection of label-free DNA targets. Proceedings of the National Academy of Sciences of the United States of America 104(44): 17364–17369.

Rant, U. 2012. Sensing with electro-switchable biosurfaces. Bioanalytical Reviews 4: 97–114.

Reinstein, O., M. Yoo, C. Han, T. Palmo, S.A. Beckham, M.C.J. Wilce and P.E. Johnson. 2013. Quinine binding by the cocaine-binding aptamer thermodynamic and hydrodynamic analysis of high-affinity binding of an off-target ligand. Biochemistry 52(48): 8652–8662.

Shim, J.W., Q. Tan and L.-Q. Gu. 2009. Single-molecule detection of folding and unfolding of the G-quadruplex aptamer in a nanopore nanocavity. Nucleic Acids Research 37(3): 972–982.

Si, W., H. Yang, J. Sha, Y. Zhang and Y. Chen. 2019. Discrimination of single-stranded DNA homopolymers by sieving out G-quadruplex using tiny solid-state nanopores. Electrophoresis 40(16-17): 2117–2124.

Tu, J., R.M. Torrente-Rodríguez, M. Wang and W. Gao. 2020. The era of digital health: a review of portable and wearable affinity biosensors. Advanced Functional Materials 30(29): 1906713.

Tuerk, C. and L. Gold. 1990. Systematic evolution of ligands by exponential enrichment - Rna ligands to bacteriophage-T4 DNA-polymerase. Science 249(4968): 505–510.

Willner, I. and E. Katz. 2005. Bioelectronics: From Theory to Applications. Wiley-VCH Verlag GmbH & Co.

Wong, I.Y., M.J. Footer and N.A. Melosh. 2008. Electronically activated actin protein polymerization and alignment. Journal of the American Chemical Society 130(25): 7908–7915.

Wong, I.Y. and N.A. Melosh. 2009. Directed hybridization and melting of DNA linkers using counterion-screened electric fields. Nano Letters 9(10): 3521–3526.

Wong, I.Y. and N.A. Melosh. 2010. An electrostatic model for DNA surface hybridization. Biophysical Journal 98(12): 2954–2963.

Wong, I.Y., B.D. Almquist and N.A. Melosh. 2010. Dynamic actuation using nano-bio interfaces. Materials Today 13(6): 14–22.

Wu, C., T. Chen, D. Han, M. You, L. Peng, S. Cansiz, G. Zhu, C. Li, X. Xiong, E. Jimenez, C.J. Yang and W. Tan. 2013. Engineering of switchable aptamer micelle flares for molecular imaging in living cells. ACS Nano 7(7): 5724–5731.

Yeung, C.L., P. Iqbal, M. Allan, M. Lashkor, J.A. Preece and P.M. Mendes. 2010. Tuning specific biomolecular interactions using electro-switchable oligopeptide surfaces. Advanced Functional Materials 20(16): 2657–2663.

Zhang, Y.F., Z.Y.H. Wang and Y.C. Chen. 2021. Biological tunable photonics: Emerging optoelectronic applications manipulated by living biomaterials. Progress in Quantum Electronics 80.

Zhu, J., T. Nguyen, R.J. Pei, M. Stojanovic and Q. Lin. 2012. Specific capture and temperature-mediated release of cells in an aptamer-based microfluidic device. Lab on a Chip 12(18): 3504–3513.

Chapter 6

Smart Polymers
Advances and Applications in Wound Healing

Premlata Ambre,* Abhishek Mishra, Aniket Kushare and
Gopal Agrawal

Introduction

The economic burden of acute and chronic wounds is escalating due to various factors, including the mounting healthcare costs, a rapidly aging population, and an upsurge in comorbidities such as diabetes. Consequently, the prevalence of these wounds is becoming a significant global concern (Wang et al. 2023). Additionally, the healing process of both acute and chronic wounds is prolonged due to repetitive infections, inflammation, poor nutrition, and inadequate blood supply. Therefore, there is a need for the development of advanced wound care technologies or precise wound dressings that can address the challenges by protecting the wound against infection, controlling exudate levels, enabling gas exchange, providing thermal insulation, and effectively delivering therapeutic compounds.

In the past, traditional wound treatments involved passive methods like applying medicated gauzes, plasters, bandages, natural ointments, and lint to relieve pain, prevent infection, which aid the wound closure (Saghazadeh et al. 2018). However, these methods have been found inadequate for effective wound healing. The traditional wound dressings struggle to maintain the optimal moisture level on the wound surface and can lead to secondary infections due to strong adhesion while changing the dressings.

Bombay College of Pharmacy, Kalina, Santacruz (E), Mumbai 400 098, India.
* Corresponding author: premlata.ambre@bcp.edu.in

An ideal wound dressing plays a crucial role in wound management by creating a protective barrier against bacteria and trauma (Derakhshandeh et al. 2018). It also maintains a moist environment that minimizes scarring and facilitates epithelization and cell migration into the wound. These dressings should possess high absorbency to manage significant amounts of wound exudate and low adherence to the wound, protecting newly formed tissue during dressing changes. Additionally, active intervention throughout the healing process, such as the release of medications or bioactive components incorporated into the dressings, is vital.

Significant efforts have been dedicated in the development of active treatment protocols to overcome these limitations, resulting in the introduction of smart dressings (Derakhshandeh et al. 2018, Van et al. 2021, Wang et al. 2023). These innovative smart dressings can consist of three key components: (i) Smart polymers which are designed to respond to specific wound conditions, such as temperature, pH, reactive oxygen species, or enzymes. They act as drug carriers and are commonly formulated as topical hydrogels, hydrofibers, foams, hydrocolloids, or films. (ii) Biosensors are utilized to detect various indicators including enzymes, temperature, pH, uric acid, or glucose, providing real-time monitoring of wound parameters. (iii) Additionally, wearable Bluetooth devices such as mobile phones or smartwatches can be used to connect to the smart dressings, enabling remote monitoring and data collection. Such "smart dressings" have the potential to revolutionize wound care management, offering advanced capabilities and tailored treatment approaches.

In this chapter, several aspects are explored, starting with the structure and functions of human skin, followed by an examination of wound healing processes and different types of wounds. The chapter also delves into the targets and biomarkers used for wound assessment. Additionally, there is a comprehensive discussion on the classification and significance of smart polymers in wound healing, particularly as active drug delivery systems. Furthermore, advanced topical formulations for topical dressings are extensively covered. Lastly, the chapter outlines the design and preparation of smart dressings for wound healing treatment, addressing the advantages and disadvantages and highlighting current issues in the field.

Human skin and its functions

Human skin constitutes roughly 8% of the average body mass and is considered the largest organ in the body. Numerous processes, including fluid hemostasis and sensory detection, heavily rely on the skin (Mayet et al. 2014). Moreover, skin permits the establishment of a self-repairing and self-renewing contact between the body and its environment.

The skin is composed of three main layers such as epidermis, dermis, and hypodermis. Each layer plays a crucial role in healing various types of wounds (Mayet et al. 2014). The outermost layer is epidermis, possessing a thin but densely cellular structure. It acts as a barrier with high impermeability, limiting water loss and protecting against potentially harmful external stimuli. The dermis is located beneath the epidermis and is separated by a membrane consisting of extracellular matrix (ECM). This ECM is characterized by a high concentration of collagen, elastin, fibroblasts, and glycosaminoglycans (Zhong et al. 2010). It provides support to the skin by housing large vascular and nerve bundles, and lymphatic system. Additionally, it contributes flexibility and physical strength to the skin. The dermis contains network of nerve fibers that serve sensory function in the skin and influence immune and inflammatory responses. Below the dermis is hypodermis, which consists of a significant number of adipose tissues. This layer is well vascularized and contributes to the skin's thermoregulatory and mechanical properties (Diegelmann and Evans 2004, Cohen 1983).

Damage to the normal anatomical structure and function of skin is called a "wound". This could be a minor rupture in the integrity of the skin's epithelium or a deeper wound affecting subcutaneous tissue, potentially damaging tendons, muscles, arteries, nerves, parenchymal organs, and even bone (Diegelmann and Evans 2004, Velnar et al. 2009). The healing process is complex, and it involves the coordinated interactions between several biological and immunological systems (Alonso et al. 1996, Labler et al. 2006, Rivera and Spencer 2007).

Wound healing processes

The wound healing process is a time-dependent and well-organized sequence that occurs in four distinct stages (Attinger et al. 2007, Vanwijck 2001).

a. Coagulation and hemostasis

A cascade of wound healing events begins with coagulation and hemostasis processes that ultimately lead to restoration of normal function and tissue repair (George et al. 2006, Jespersen 1998). At the initial stage, blood oozes from the damaged skin area that triggers the release of clotting components from platelets, leading to the formation of a blood clot. The clot is composed of fibrin, fibronectin, thrombospondin and vitronectin (refer Fig. 1a) (Pool 1997, Lawrence 1998, Skover 1991). Platelets also contain growth factors and cytokines, such as epidermal growth factor (EGF), transforming growth factor-alpha/beta (TGF-α/TGF-β), platelet derived growth factor (PDGF), Vascular Endothelial Growth Factor (VEGF), and insulin-like growth factors (Richardson 2004). These cells act as promoters in the cascade of wound healing and subsequently recruit macrophages, endothelial cells, and fibroblasts. Upon release from the blood clot, these substances increase vascular permeability, induce vasodilation, and cause fluid extravasation in the tissue, resulting in edema (Velnar et al. 2009).

b. Inflammatory Phase

The second stage of wound healing process is an inflammatory phase. It develops the immune defense against invasive microbes. It is subclassified into two phases such as early inflammatory phase and late inflammatory phase (Lawrence 1998, Hart 2002).

Early inflammatory phase- commences during the later stages of coagulation, triggering molecular processes that enable the infiltration of neutrophils into the wound site. Figure 1b illustrates the pivotal role of neutrophils in combating infections and initiating the phagocytosis process. Phagocytosis is aimed at

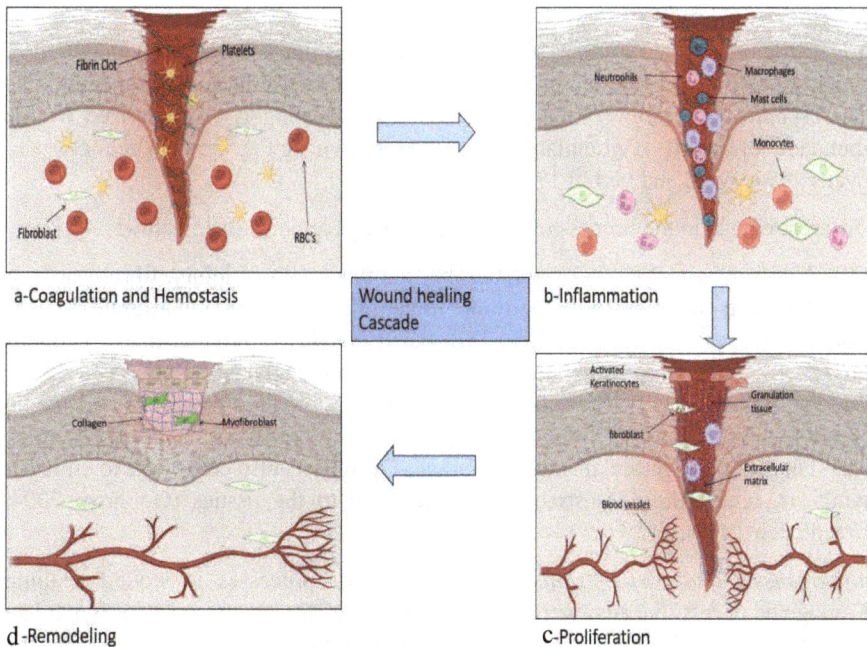

Fig. 1. Wound healing cascade.

eliminating bacteria, foreign objects, and damaged tissue (Robson 2001). Within 1–2 days of the injury, neutrophil migrate towards the chemotactic and chemokine substances like TGF-β, complement elements - C3a and C5a, formyl methionyl peptides (produced by bacteria), and platelet products to the wound site (Robson 2001, Velnar et al. 2009, Broughton et al. 2006a, b). A stronger adhesion mechanism is activated by endothelial cell-secreted chemokines through the action of integrins. When neutrophils enter the wound environment, they phagocytose bacteria and foreign objects, killing them by releasing protease enzymes and oxygen-derived free radical species (Hunt et al. 2000, Ramasastry 2005).

Late Inflammatory Phase- Macrophages emerge in the wound and continue the process of phagocytosis during the late inflammatory phase, which lasts for 2–3 days after injury (Hart 2002, Hunt et al. 2000). As depicted in Fig. 1b, the macrophages are drawn to the wound site by the variety of substances such as cytokines like tumor necrosis factor alpha (TNF-α), TGF-β, fibroblast growth factor (FGF), interferon gamma (IFNγ), interleukin-1 beta (IL-1 β), platelet-derived growth factor (PDGF), platelet factor IV, clotting factors, complement components, leukotriene B4, and lastly, breakdown products of elastin and collagen (Ramasastry 2005, Pierce et al. 1991a). These cells play a crucial regulatory role in the late stages of the inflammatory response, acting as a rich source of powerful tissue growth factors, particularly TGF-β, as well as other mediators that activate keratinocytes, fibroblasts, and endothelial cells (Ramasastry 2005, Witte and Barbul 1997).

c. Proliferative stage

It is distinguished by fibroblast migration and deposition of newly produced ECM, which takes place in the temporary network made of fibrin and fibronectin. The macroscopic observation of this stage of wound healing is characterized by the abundant formation of granulation tissue, as shown in Fig. 1c (Diegelmann and Evans 2004, Witte and barbul 1997).

Different events in the process of proliferative stage are discussed below:

Fibroblast migration process begins in the first three days of injury, fibroblasts and myofibroblasts in the surrounding tissue undergo proliferation and migrate into the wound. They release substances like TGF-β and PDGF, and produce matrix proteins such as procollagen, hyaluronan, fibronectin, and proteoglycans (Ramasastry 2005, Witte and Barbul 1997). These fibroblasts transform into myofibroblasts and extend pseudopodia that bind to fibronectin and collagen in the ECM, supported by thick actin bundles beneath the plasma membrane. Fibroblasts also contribute to collagen synthesis, which provides strength and integrity to the tissues (Goldman 2004, Greenhalgh 1998).

Angiogenesis and tissue granulation are essential processes in wound healing. Angiogenic agents like angiogenin, VEGF, FGF, PDGF, TGF-α, and TGF-β regulate the growth and differentiation of resident endothelial cells, thereby promoting the formation of new blood vessels (Pierce et al. 1991b, Takeshita et al. 1994). This angiogenesis process brings nutrients and oxygen to the growing tissue, supporting its development (Ribatti et al. 1991, Oike et al. 2004).

Protrusion is facilitated by actin polymerization at the leading edge of cell locomotion, where the concentration of chemoattractant is highest and adhesion is facilitated by integrins where cell migration occurs. Integrins act as receptors for ECM proteins, enabling cell motility and participating in signal transduction (Li et al. 2005, Giannone et al. 2004, Herman 1993).

Traction, or contractile force generation is achieved through integrin-cytoskeletal linkages. Myosin motor proteins connected to contractile actin bundles generate force, pushing the cell body forward. However, excessive traction forces at contact sites can deform the ECM (Young and Herman 1985, Bokel and Brown 2002, Baum and Arpey 2005).

d. Remodeling stage

During the final stage of wound healing, the production of new epithelium and the formation of complete scar tissue take place. As the granulation tissue grows, the synthesis of the ECM begins during the proliferative and remodeling stages. Collagen bundles increase in size, and hyaluronic acid along with fibronectin are degraded. The tensile strength of the wound gradually improves as collagen accumulates (English et al. 1999, Clark 1993). Initially, the formation of collagen bundles is disorganized, but over time, the newly formed collagen matrix becomes more organized and cross-linked. Myofibroblast interacts with the ECM, causing the underlying connective tissue to shrink and bringing the wound edges closer together (Ramsastry 2005, Baum and Arpey 2005).

Figure 1 illustrates the cascade of wound healing events, categorized into four parts: a. Coagulation and Hemostasis; b. Inflammation; c. Remodeling; and d. Proliferation.

Classification of wounds

Wounds are broadly classified into two classes such as acute and chronic wound.

a. Acute wounds

Acute wounds heal naturally and efficiently through a prompt and organized healing process, resulting in the restoration of both function and anatomy. The typical duration of recovery ranges from 6–8 weeks. In certain cases, like surgical procedure to remove a soft tissue and underlying tissue tumor from the skin may result in a large incision that cannot heal through primary treatment due to the tissue damage or traumatic wounds which may involve only soft tissue or be associated with bone fractures (Lararus et al. 1994, Szycher and Lee 1992).

b. Chronic wounds

Chronic wounds do not heal in a timely and organized manner and therefore take longer than usual time to heal. Several factors can contribute to the prolonged duration of wound healing, including hemostasis, inflammation, proliferation, and remodeling. These factors encompass tissue hypoxia, excessive exudate, presence of necrotic tissue, infection, and an overproduction of inflammatory cytokines. The persistent state of inflammation within the wound triggers a cascade of tissue reactions that collectively impede the healing process, resulting in compromised physiological outcomes. Various factors such as diabetes, blood pressure, arterial and venous insufficiency, burns, and vasculitis can contribute to the development of chronic wounds (Vanwijck 2001, Degreef 1998).

Table 1 presents a comprehensive overview of different types of acute and chronic wounds, highlighting wound characteristics, as well as the changes in temperature and pH within the wound environment (Degreef 1998). Different types of wounds express distinct wound environment. Therefore, healthcare professionals evaluate the wound and determine the optimal treatment approach by identifying key molecular targets or biomarkers involved in the process of wound healing.

Targets for wound healing treatment

The selection of target for wound healing depends on various factors such as the type and severity of the wound, the patient's overall health, and the specific requirements of the individual case.

a. Reactive oxygen species (ROS)

As a result of bacterial invasion, xenobiotics, cytokines, and mitochondrial oxidative metabolism, ROS are produced. These reactive molecules are oxygen superoxide

Table 1. Types of wounds, traits, and physiological changes in temperature and pH.

Sr No	Type of wound	Subclassification of wound	Traits of wounds	Physiological conditions	
				Temperature (°C)	pH
1	Acute Wound (Szycher and Lee 1992)	Frostbite	Frostbite is an injury when skin is exposed to extreme cold due to which the tissue freezes and vasoconstriction.	28–32	4.7–6.1
		Abrasion wound	Abrasions are injuries that occur when the skin scrapped or rubbed against the rough surface.	33–36	7.2–7.6
		Laceration wound	Laceration is cut in skin, which is caused due to sharp objects like blade, knife etc.	36–38	4.5–5.3
		Incision wound	Incision is specialized wound caused due to cutting of skin with scalpel for surgical purpose.	36–37.5	6.5–7.5
		Avulsions	Avulsions are an injury when part of body like muscle, tendon etc. is entirely torn out of body due to traumatic situation.	36–38	6.3–6.9
		Puncture wound	A puncture wound is an injury caused due to pointed objects.	36–38	6.2–7.2
		Pressure ulcer	A pressure ulcer is an injury which is caused due to prolonged pressure or friction.	34–36	6.8–8.0
2	Chronic wound (Evans 2019)	Arterial ulcer	Arterial ulcer is damage of tissue due to blockade caused in artery, causing shrinking of feet and open sore.	36.5–38	7.3–7.8
		Venous ulcer	A venous ulcer is a wound that occurs when there is reduced blood flow to veins in the leg, which causes open discomfort.	32–35	6.7–7.5

Stab wound	A stab wound is a wound caused due to forceful piercing of sharp objects like knife, arrow, etc. into skin leading to opening of skin.	38–40	7.2–8.6
Animal bite	Animal bite is a type of acute traumatic wound caused due to piercing of teeth of animal. Highly vulnerable to infectious diseases.	36–38	7–8.2
Vasculitis wounds	Vasculitis is a rare auto immune disorder wherein body's immune system attacks blood vessels leading to inflamed tissue and open skin surfaces.	36.5–37.5	7.5–8.5
Diabetic foot ulcer	Diabetic foot ulcer is injury which is caused due to nerve damage of leg diabetic patient leading to extreme wound condition like gangrene.	38–39	7–8.5
Diabetic wound	Diabetic wounds are chronic wounds that occur in diabetic patients due to nerve damage to organs. This wound shows slowest wound healing process.	36.5–38	7.5–8.87
Burn wound	A burn wound is an injury caused due to exposure to extreme heat or any hazardous chemical.	38–45	7.2–8.9

anions (O^{2-}), hydrogen peroxide (H_2O_2), and hydroxyl radicals ($OH^{\cdot}$). Mitochondria generate these molecules as byproducts of cellular metabolism (Dunnill et al. 2017).

ROS activates pro-inflammatory signaling pathways, recruit immune cells, and stimulate cytokine and growth factor production. However, the resulting inflammation may hinder healing and cause tissue damage. In angiogenesis and tissue repair, ROS regulates the expression of angiogenic factors like VEGF, promoting endothelial cell proliferation, migration, and capillary tube formation (Mittal et al. 2014). Nonetheless, excessive ROS disrupts the balance of angiogenic factors, inhibiting new blood vessel formation hence compromising blood supply to the wound site, hindering the delivery of oxygen and nutrients needed for healing. ROS participates in modeling of ECM through the activation of matrix metallo-proteinase (MMP) enzymes, which is vital for removing damaged tissue components (Dunnill et al. 2017). It also contributes to fibroblast activation and collagen synthesis, promoting tissue re-epithelialization and wound contraction. However, excessive ROS can degrade ECM components, particularly collagen, disrupting the synthesis of new ECM proteins. These compromises wound closure and tissue integrity. Moreover, ROS can hinder fibroblast function, impeding the formation of granulation tissue and overall healing (Kuroki et al. 1996). Therefore, maintaining a balanced level of ROS during wound healing is crucial to harness their beneficial effects while minimizing the harmful consequences of oxidative stress (Clempus et al. 2006).

b. Matrix Metallo-Proteinase (MMP) enzymes

These enzymes are involved in remodeling of the ECM, which is essential for tissue repair. In acute wound healing, MMPs are responsible for the initial breakdown and removal of damaged ECM components (Armstrong and Jude 2002). During the inflammatory phase of wound healing, immune cells release MMPs to degrade the existing matrix, allowing for the clearance of debris and the formation of a temporary matrix. This process promotes the migration of immune cells, fibroblasts, and endothelial cells into the wound site (Stadelmann et al. 1998). As the healing progresses, MMPs are also involved in the reorganization and remodeling of the ECM. Additionally, MMPs contribute to angiogenesis by degrading ECM components and releasing angiogenic factors, promoting the formation of new blood vessels necessary for adequate tissue perfusion and oxygenation (Stadelmann et al. 1998).

However, in chronic wounds, there is often dysregulation in the activity of MMPs, leading to excessive degradation of ECM components and impaired healing (Martis 2013). Chronic wounds typically have elevated levels of MMPs, which can lead to prolonged inflammation, delayed re-epithelialization, and impaired formation of granulation tissue (Menke et al. 2008). This imbalance in MMP activity can result in an abnormal wound healing environment that hinders proper tissue repair and leads to chronicity (Kuenzli and Saurat 2003). Therefore, the regulation of MMP activity is crucial for effective wound healing.

c. Peroxisome proliferator activated receptor (PPAR)

PPARs are a class of nuclear receptors that are involved in various cellular processes, including inflammation, metabolism, and cell differentiation. PPAR-α, PPAR-β/δ, and PPAR-γ are the three known subtypes of PPARs (Kuenzli and Saurat 2003).

PPAR-γ plays a crucial role in wound healing by regulating inflammation, angiogenesis, ECM remodeling, cell proliferation, and oxidative stress. PPAR-α and PPAR-γ improve angiogenesis process by regulating the expression of angiogenic factors and promoting the growth of new blood vessels in the wound area. They also stimulate the proliferation and differentiation of various cell types involved in wound healing. Additionally, PPARs possess antioxidant properties, helping to regulate the balance between reactive oxygen species (ROS) and antioxidant defense mechanisms, thereby protecting cells from further damage during the wound healing process (Sertznig et al. 2008).

Overall, PPARs play a multifaceted role in wound healing, and further research is needed to fully understand their mechanisms. PPAR-γ agonists and selective PPAR-γ modulators (SPPARMs) have shown potential in promoting a healing-associated macrophage phenotype, even in diabetic wound environments.

d. Advanced glycation end products (AGE)

AGEs impair the wound healing process by prolonging inflammation, disrupting collagen structure, hindering angiogenesis, and promoting oxidative stress. AGEs are formed through glycation, a natural process in which sugars react with proteins or lipids. However, conditions like hyperglycemia, oxidative stress, and inflammation accelerate AGE formation, leading to detrimental effects on wound healing (Qing 2017).

AGEs inhibit the recruitment and function of immune cells, prolonging the inflammatory phase and impeding subsequent repair stages. Additionally, they cross-link with collagen fibers, making them stiffer and less flexible. This disrupts collagen synthesis and remodeling, resulting in delayed wound closure and reduced tensile strength. AGEs also contribute to oxidative stress by promoting ROS production and inhibiting natural antioxidant defenses (Peppa et al. 2009). Cells involved in wound healing, such as fibroblasts, endothelial cells, and immune cells, are damaged by this oxidative stress, which impairs their capacity for migration, proliferation, and differentiation (Ahmed 2005).

e. Tumor Necrosis Factor - Alpha (TNF-α)

Excessive production of TNF-α involved in the immune response and inflammation by recruiting and activating immune cells, such as neutrophils and macrophages, lead to an extended inflammatory phase and disrupting subsequent healing stages (Bradley 2008). Moreover, TNF-α inhibits the proliferation and migration of various cells critical for wound healing. It suppresses the growth of fibroblasts responsible for producing collagen and ECM components necessary for tissue repair. Additionally, TNF-α impairs the migration of endothelial cells involved in forming new blood vessels during healing (Messadi 2008).

Furthermore, TNF-α negatively affects ECM remodeling, inhibiting collagen synthesis, and disrupting collagen fiber organization. Consequently, weak and less functional scar tissue may form (Banno et al. 2004). Additionally, TNF-α affects the function of keratinocytes, inhibiting their migration and proliferation, which is necessary for the process of formation of a new outer layer of skin to cover the wound.

f. Cellular senescence

Cellular senescence refers to a state in which cells experience permanent growth arrest. In the context of wound healing, cellular senescence can have both positive and negative impacts. Senescent cells play a beneficial role by initiating tissue repair processes and promoting the inflammatory response. They secrete various bioactive molecules, including growth factors, cytokines, and matrix metalloproteinases, which recruit immune cells and aid in clearing damaged tissue (Wilkinson and Hardman 2020).

Nevertheless, the persistent existence of senescent cells can lead to changes in gene expression patterns, resulting in the inhibition of cell proliferation. This, in turn, can disrupt the process of angiogenesis, tissue deposition and remodeling of the ECM (Wilkinson and Hardman 2020).

Assessment of wound status using biomarkers

Biomarkers play a significant role in the assessment of both acute and chronic wound healing processes. They help to unravel the complex interrelationships between several cellular and molecular components, such as the inflammatory response, angiogenesis, extracellular matrix remodeling, and cell proliferation (Van et al. 2021). Biomarkers, as depicted in Table 2, offer valuable insights into the underlying biological processes that initiate wound healing. They are used for diagnosis, monitoring, and forecasting therapy outcomes. With the help of biomarkers, researchers are able to design novel therapeutic approaches by better understanding the mechanics behind wound healing processes.

Table 2 given below discusses the significance of key biomarkers, and their role in assessing acute and chronic wound healing.

Smart Polymers

Researchers are investigating novel drug delivery carriers that can respond to altered physiological conditions in acute or chronic wounds. These systems aim to release drugs in response to specific stimuli, enhancing wound healing, improving patient compliance, and reducing therapy costs and duration. These drug delivery systems are described as "stimuli responsive" or "smart polymers" as they respond to external stimuli such as temperature, pH, light, enzymes, and ROS (Patel 2020). The responsive behavior has provided new opportunities for the development of smart drug delivery formulations, high-tech dressings, and AI-based biomedical devices targeted at managing wound care and accelerating the healing process.

Table 2. Key biomarkers and their role in wound assessment.

Sr. no.	Biomarkers	Healthy Skin	Acute/chronic wound infected skin	Role in assessment of wound	Reference
1	Temperature	Skin surface 32–35°C	32–39°C (3°C to 4°C higher than normal skin)	Provides information on infection, inflammation, oxygenation in the wound environment	(Van et al. 2021)
2	pH	Acidic (4–6)	Acute wound - (6.5–8.5) Chronic wound – (4.5–8.5)	pH that is too high or low could indicate a sign of infection	(Van et al. 2021)
3	Uric acid	220 to 750 × 10⁻⁶ M	Normal to higher than 200 × 10⁻⁶ M	Elevated levels of uric acid in wound indicates inflammation or tissue injury	(Gallani et al. 2020)
4	Glucose	$4.6–9.9 \times 10^{-3}$ M	Fluid range between 0 and 1.2×10^{-3} M reported for chronic wound	Elevated blood glucose levels impair the synthesis of hypoxia-inducible factor-1 which is a vital transcription factor in early wound healing. It controls cytokine activity, cellular oxygen levels, and the formation of blood vessel networks.	(Trenove et al. 1996)
5	ROS	100–250 μm	> 250 μm	The optimal level of oxygen indicates the rate of angiogenesis	(Safaee et al. 2021)
6	Moisture	Qualitative process; low moisture levels to be maintained and measured using bioimpedance method; optimal requirement is 15–50 kΩ	High moisture (1.4–15 kΩ) or dry wound (> 200 kΩ)	Optimal moisture in the wound bed improves healing by reducing inflammation, promoting dermal repair, and enhancing revascularization. Conversely, too much wound fluid raises the danger of bacterial infection.	(McColl et al. 2007)
7	Enzymes growth factors (EGF), and matrix metalloproteinases (MMPs)	EGF MW = 7.34 kDa EGF-MMP, 8.162 kDa	Elevated levels seen in acute wounds compared to chronic wounds.	EGF is used for distinguishing the acute and chronic wound	(Kim et al. 2017)

In response to stimuli, smart polymers have the ability to undergo a phase transition that alters their physical properties within a solvent. This transition typically involves a change from a soluble, liquid solution state to a gel-like phase (Roy and Gupta 2003). This quality is especially beneficial for targeted drug delivery. By harnessing the unique capabilities of smart polymers, innovative strategies can be developed to address the complexities of wound healing and tailor treatments to specific anatomical locations or conditions. These advancements have the potential to revolutionize wound care and improve therapeutic outcomes by enabling precise and controlled drug release in response to specific stimuli.

Classification of Smart Polymers

The classification of smart polymers is based on their responsiveness to specific stimuli, which enables researchers to tailor their properties and actively target the encapsulated drug at the site of action, tissue engineering, biosensing, and controlled release systems (Patel 2020). The classification of smart polymers and their responsive characteristics in wound healing is illustrated in Fig. 2 below.

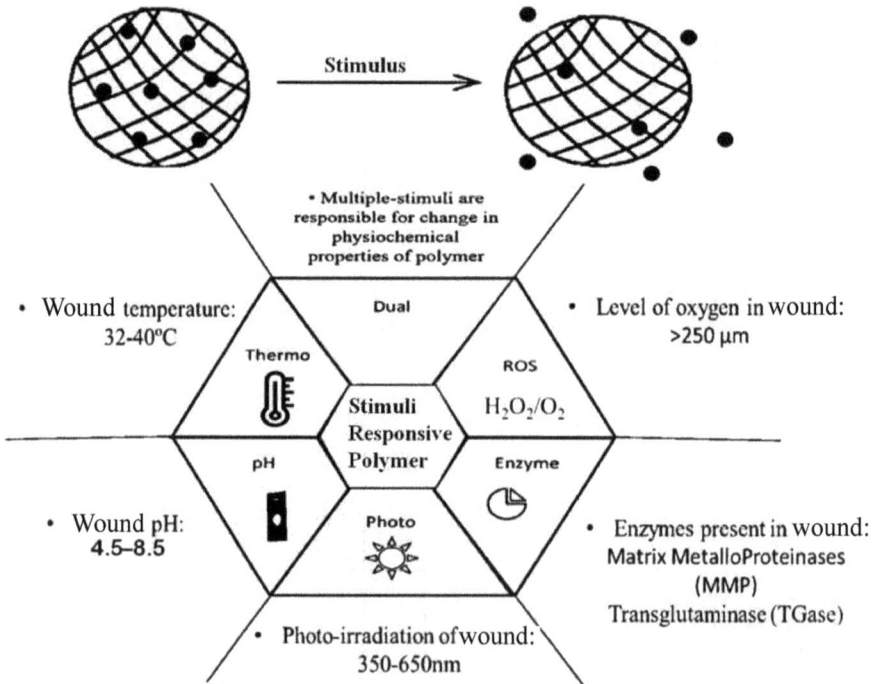

Fig. 2. Classification of smart polymers and their responsive characteristics in wound healing.

i. Thermoresponsive polymers

In response to temperature changes, these polymers experience reversible alterations in their physical characteristics. It exhibits two distinct parameters related to temperature-dependent changes in their solubility or phase behavior known as Lower Critical Solution Temperature (LCST) and Upper Critical Solution Temperature (UCST) (Gandhi et al. 2015).

LCST signifies the temperature at which a transition occurs in the polymeric structure, causing it to shift from being hydrophilic to hydrophobic (Schild 1992). Below the LCST, polymer dissolves completely and forms a solution. However, as the temperature rises and reaches the LCST, the polymer undergoes a phase transition, thereby resulting in the formation of aggregates, a gel-like structure, or precipitation. This transition leads to significant changes in the physical properties of polymer, such as its conformation, viscosity, or ability to interact with other molecules (Hoffman 1987, Gandhi et al. 2015).

Hoffmann and Schild (Schild 1992, Hoffman 1987) have reported the significance of phase transition behavior in thermoresponsive polymer such as PNIPAM. The polymer PNIPAM is soluble in water at room temperature below LCST and it aligns itself in a specific orientation to form hydrogen bonds between water molecule and amide group of PNIPAM. When the ambient temperature rises, and hydration structure breaks down and the water molecules try to reorient around hydrophobic isopropyl group of PNIPAM. This phenomenon is known as the hydrophobic effect and it is the entropy driven process.

On the other hand, UCST is the temperature below which a thermoresponsive polymer experiences a phase transition from a soluble state to an insoluble state in a solvent, but in the opposite manner compared to LCST. Below the UCST, the polymer is insoluble and forms aggregates or a gel-like structure. As the temperature increases above the UCST, the polymer becomes soluble and dissolves in the solvent (Seuring and Agrawal 2012, Gandhi et al. 2015).

Thermoresponsive polymers play a crucial role in targeted drug delivery systems for wound healing treatment. They can utilize the LCST behavior to facilitate drug release specifically at acute or chronic wound sites by modifying the solubility of the responsive polymer. Additionally, temperature-induced phase transitions in these polymers can be effectively employed to develop scaffolds, hydrogels, and responsive surfaces as dressing material for further enhancing their potential in wound healing applications (for details refer Table 3).

Role of thermoresponsive polymers in wound healing: The inflammatory stage of the wound healing process is mainly responsible for the rise in temperature. During inflammation, the body releases chemicals that promote dilation of blood vessels thereby increasing blood flow to the wound area, leading to a rise in local temperature. Additionally, the presence of bacterial toxins, metabolic byproducts, and foreign debris can trigger an immune response, which further impacts to elevate the wound site temperature. Typically, wound temperatures range between 36–39°C (Gandhi et al. 2015, Dethe et al. 2022), indicating an elevation compared to normal skin temperature (32–35°C) (Gandhi et al. 2015, Patel 2020). This temperature elevation can be leveraged for formulating drug delivery systems using smart thermoresponsive

Table 3. Advancements in thermoresponsive polymer formulations for wound healing applications.

Sr. no.	Formulation	Thermoresponsive Polymers	Drug	Responsive temperature in LCST (°C)	Reference
1	Hydrogel	P(N-isopropylacrylamide-co-acrylic acid)/PNIPAM	Diclofenac sodium	40	(Lin et al. 2020)
		P (Methacrylate arginine-co-NIPAM)	Chlorhexidine diacetate (CHX) and Poly hexamethylene guanidine phosphate (PHMG)	37–40	(Chi et al. 2020)
		Polycaprolactam and polyethylene glycol	-----	37	(Dethe et al. 2022)
		Chitosan-poly vinyl alcohol and chitosan-gelatin	Ketoprofen trometamol	38	(Castillo-Henriquez et al. 2021)
		PNIPAM, sodium alginate and methylcellulose	Octenidine dihydrochloride	37	(Niziol et al. 2021)
		PNIPAM, sodium alginate and calcium chloride	Mupirocin	34	(Chen et al. 2022)
2	Microneedle Array	Chitosan-co-PNIPAM	Vascular endothelial growth factor (VEGF)	36–40	(Chi et al. 2020)
3	Hydrofibers	PNIPAM, poly(l-lactic acid-co-ε-caprolactum)	Ciprofloxacin	above 32	(Li et al 2017)
4	Films	PNIPAM and pullulan	Silver Nanoparticles	35–36	(Paneysar et al. 2022)
		Chitosan with PNIPAM	Gentamycin	36–38	(Qureshi and Khatoon 2015)

polymers. To be more precise, smart polymers having LCST between 36–39°C can be used to release the encapsulated drug for treating acute or chronic wounds.

Table 3 provides a compilation of several thermoresponsive polymers utilized in diverse formulations for wound healing applications.

ii. pH-responsive polymers

Polymers that are pH-sensitive demonstrate a distinctive ability to respond to small changes in environmental pH by undergoing disproportionately large changes in size, shape, hydrophobicity, and degradation rate. pH-responsive polymers often incorporate ionizable functional groups, such as carboxylic acid (-COOH) or amino groups (-NH$_2$) (Gandhi et al. 2015, Patel 2020). These groups undergo protonation or deprotonation in response to pH changes. The alteration in pH affects the ionization state of these functional groups, leading to changes in the polymer's solubility or hydrophilicity. This solubility change drives the process of phase inversion, facilitating drug release (Van et al. 2021).

Polyacids or polyanions, which contain multiple ionizable acidic groups like carboxylic or sulfonic acid groups, are examples of pH-responsive polymers. Polyanionic polymers do not swell at low pH levels because the acidic groups are protonated and the repulsion due to the charges is abolished. However, the negatively charged polymer expands as the pH rises. Polymers that are polybasic or polycationic, however, behave differently. As the pH drops, their basic groups become more ionized (Patel 2020).

Role of pH responsive polymers in wound healing: The healthy and undamaged skin surface has a naturally acidic pH range from 4 to 6 because of the organic acid secreted by keratinocytes (Kuo et al. 2020). The acidic environment is considered to be good for stimulating fibroblast proliferation, encouraging epithelization and angiogenesis, regulating bacterial colonization, and making it easier for oxyhemoglobin to release oxygen. On the other hand, alkalinity may damage the wound tissue by depleting it with oxygen and promoting the right conditions for bacterial growth. It is evident from this observation that even slight fluctuations in pH can induce significant shifts in the wound condition. Nearly all of the surface wounds that were investigated had alkaline pH readings of around 6.8–8.5 (Kuo et al. 2020). The pH has an effect on all biochemical processes associated with the healing process of wounds, based on the phase of healing and the length of time.

pH responsive polymers play a significant role in the healing process of wounds by providing tailored and controlled drug delivery at the wound site. These polymers exhibit a response to changes in pH, which is a characteristic feature of wound environments. Chronic wounds frequently have a slightly alkaline pH which aids the pH responsive polymers for releasing therapeutic chemicals only in this acidic or alkaline range (Vanwijck 2001, Degreef 1998).

The use of pH-responsive polymers in wound healing has many benefits. By preventing sensitive medications from degrading in the hostile wound environment, they can guarantee their stability and effectiveness. These polymers provide sustained and localized medication release, which lessens the systemic negative effects.

Table 4. Examples of pH responsive polymer formulations for wound healing applications.

Sr. no.	Formulation	pH responsive Polymers	Drug	pH responsive characteristics	Reference
1	Hydrogel	Aminoethyl methacrylate hyaluronic acid and methacrylate methoxy polyethylene glycol	Chlorhexidine diacetate	5 to 7	(Zhu et al. 2018)
		Quaternized chitosan and benzaldehyde terminated Pluronic@F127	Curcumin	5 to 7	(Qu et al. 2018)
		NIPAAm-co-Acrylic acid	EGF and VEGF	6.7–7.9	(Banerjee et al. 2012)
		Chitosan and poly (N-vinyl-2-pyrolidone)	Silver sulfadiazine	8	(Rasool et al. 2019)
		Crosslinked - Methacrylic acid and acrylamide with N, N′-methylenebisacrylamide	Silver Nanoparticles	4	(Haidari et al. 2021)
		NIPAM cross-linked with Acrylic acid	Silver Nanoparticles	5.5–7.4	(Haidari et al. 2022)
2	Nano-vesicle	Oleyl amine based zwitterionic lipid with chitosan	Vancomycin	6	(Hassan et al. 2020)
3	Nanocomposite Film	Chitosan-PEG containing zeolite imidazolate framework	Cephalexin	5–7	(Mazloom-Jalali et al. 2020)
		Chitosan and polyphenolic tannic acid	Neomycin	5.5–7.5	(Chowdhury et al. 2022)

pH responsive polymers can enhance wound healing by regulating the local pH, which influences cell proliferation, migration, and tissue regeneration.

To achieve targeted and controlled drug delivery by facilitating the effective wound healing process, many researchers are investigating use of the pH responsive polymers in wound dressings, hydrogels, or other delivery methods (refer Table 4 for details). The potential for these polymers to provide individualized and successful wound care management is increased by their capacity to react to the particular pH values at the wound site.

Table 4 below offers several examples of different approaches undertaken by researchers for drug release using pH-responsive polymers in wound healing applications.

iii. Photoresponsive polymers

These polymers have emerged as crucial elements in the field of wound healing treatment, utilizing light as an external stimulus to achieve controlled and targeted drug release, support tissue regeneration, and enhance wound closure (Patel 2020).

The best example in this class is azobenzene containing polymers (Di Martino et al. 2023). It can undergo reversible photoisomerization in response to UV or visible light enabling the control on cell adhesion or drug release in the wound bed.

Role of photoresponsive polymers in wound healing process: When the wound area is exposed to light of the appropriate wavelength, it interacts with the light responsive polymers. This interaction can be achieved using light-emitting diodes (LEDs), lasers, or other light sources that emit specific wavelengths (Patel 2020, Di Martino et al. 2023).

Photoresponsive polymers have been used to monitor cell behavior and tissue regeneration during wound healing processes. By incorporation of specific peptides or proteins into the polymer matrix, photoresponsive polymers have the ability to activate or deactivate bioactive molecules in response to light-induced changes in polymer properties. This modulation of bioactive molecules influences crucial cellular responses including proliferation, migration, and differentiation, thereby facilitating tissue regeneration and expediting the wound healing process (Di Martino et al. 2023). Additionally, some photoresponsive polymers can generate heat when exposed to light. This localized heating effect can be utilized to enhance wound healing by promoting blood circulation, increasing collagen synthesis, and facilitating the eradication of bacteria (Van et al. 2021, Patel 2020, Di Martino et al. 2023).

Overall, photoresponsive polymers offer a versatile and targeted approach in wound healing treatment by utilizing light as an external stimulus to trigger specific responses, such as drug release or tissue regeneration, ultimately enhancing the healing process and improving patient outcomes.

Table 5 given below presents examples of research initiatives utilizing photoresponsive polymers in the field of wound healing applications.

Table 5. Advancements in photoresponsive polymer formulations for wound healing applications.

Sr. no.	Formulation	Photo responsive Polymers	Drug	Photo responsive characteristics (nm)	Reference
1	Hydrogel	Silver and graphene oxide fabrication via crosslinking with PNIPAM & N, N'-Methylene bisacrylamide	Polydopamine	-	(Huang et al. 2021)
		Metal-organic framework and Zeolite imidazolate framework mixed with nitrobenzaldehyde	Rifampicin	365	(Song et al. 2018)
2	Nano-fiber	AgNO₃, Poly-caprolactone	Silver metal	405	(Ballesteros et al. 2020)
3	Foam	Stearyl trimethyl ammonium chloride and polyvinyl alcohol	Methylene Blue	635–660	(He et al. 2020)

iv. Enzyme-responsive polymers

Enzyme-responsive polymers demonstrate the ability to undergo changes in their structure or behavior upon activation by specific enzymes. To achieve this responsiveness, these polymers are typically synthesized by incorporating enzyme-cleavable linkers, functional groups, or sequences into their structure (Patel 2020). These components are carefully designed to be recognized and cleaved by the target enzyme, allowing for selective release of the desired drug. Within the polymer matrix, the therapeutic drug is loaded or encapsulated. When the enzyme-responsive polymer encounters the target enzyme, it binds to the specific enzyme-cleavable sites or sequences embedded within its structure. As a result of this binding, the polymer chains are broken at predetermined places by an enzymatic cleavage reaction. Thus, the polymer undergoes degradation or fragmentation or disassembles by hydrolysis depending on its chemical composition and design. As a result, the loaded drug in the polymer matrix is released, and hence displays therapeutic activity (Stadelmann et al. 1998).

Role of enzyme responsive polymers in wound healing process: Enzyme-responsive polymers have been utilized for targeted drug delivery of growth factors, antibiotics, or anti-inflammatory medications, to the site of a wound. These polymers are made to react with certain enzymes found in the wound microenvironment, such as MMPs, and release the medications they are encapsulating. MMPs are enzymes that break down ECM as a part of the healing process for wounds, and their elevated levels at the wound site might cause the drug release from the polymer matrix (Patel 2020). This monitored drug delivery aids in preserving the optimal drug concentrations at the location of the injury and facilitating the healing process.

Enzyme-responsive polymers can also be designed to release growth factors, such as TGF-β or PDGF, in response to specific enzymes. These growth factors are vital for tissue regeneration because they influence cell migration, proliferation, and ECM deposition (Taniyama and Griendling 2003). Enzyme-responsive polymers can improve the capacity of wounds for regeneration and hasten the healing process by providing growth factors in a regulated and targeted manner.

In conclusion, enzyme responsive polymers can give specialized therapeutic interventions, strengthen tissue repair processes, and enhance overall wound healing process by making use of the enzymatic activity that is already present at the wound site. Enzyme-responsive polymers have been extensively investigated, with smart peptides playing a significant role in this field.

Table 6 below provides few examples of different approaches undertaken by researchers using enzyme responsive polymers in wound healing applications.

v. Reactive oxygen species (ROS)

ROS-responsive polymers offer a flexible design framework for creating stimuli-responsive materials that may be customized for preparation of smart active wound healing dressings. By harnessing the oxidative environment associated with various disease conditions or biological processes, these polymers offer the potential for targeted and controlled delivery of therapeutics, on-demand release of bioactive molecules, and responsive tissue engineering scaffolds. In the field of drug delivery, researchers have explored ROS-responsive materials incorporating following moieties—thioether, thioketal, diselenide, selenium, tellurium, polysaccharide, aminoacrylate, boronic ester, peroxalate ester, and polyproline (Liang et al. 2016).

Polymers containing thioketal linkages can undergo selective degradation in the presence of ROS, leading to the release of encapsulated therapeutics or modulation of material properties (Wilson et al. 2010), whereas polymers containing diselenide bonds exhibit a dual redox-responsive behavior. They can be oxidized in the presence of ROS, leading to the formation of selenic acid. Conversely, they can also be reduced to selenol by a reducing agent. On the other hand, polymers incorporating tellurium offer higher sensitivity due to their lower electronegativity and reduced toxicity compared to selenium. These properties make tellurium-containing polymers appealing as drug carriers (Ma et al. 2010, Fang et al. 2015).

Reactive oxygen species (ROS) are produced as a result of inflammation, and these ROS can interact with peroxalate ester bonds to cleave acetal connections. This cleavage facilitates the release of the encapsulated drug within the peroxalate ester-based system (Kwon et al. 2013). Similarly, in an acidic environment with escalating levels of hydrogen peroxide (H_2O_2), polymers incorporating ortho ester and boronic ester groups undergo degradation through a combined mechanism involving ortho ester hydrolysis and boronic ester oxidation (Song et al. 2013). Moreover, polyproline sequences can also be selectively cleaved by ROS, allowing for controlled release of therapeutics or triggering specific responses (Gupta et al. 2015).

Role of ROS in wound healing: As discussed in the previous sections, ROS plays a crucial role in inflammation, angiogenesis, cell proliferation, and ECM remodeling during the wound healing stages. Additionally, ROS has antimicrobial properties

Table 6. Advancements in enzyme responsive polymer formulations for wound healing applications.

Sr. no.	Formulation	Enzyme responsive polymers	Drug	Responsive enzymes or enzymes used for degradation	Reference
1	Nano capsule	Sulforhodamine 101 labeled hyaluronan nano capsule containing polyhexanide biguanide	Polyhexanide biguanide	Cell adhesion molecules enzyme immunoassays (EIAs) Enzyme linked immunosorbent assays (ELISA)	(Grützner et al. 2015)
2	Nano particles	poly-L-lactic acid with octenidine	Octenidine		
3	Film formulation	polyurethane and poly(glutamic acid) peptide	Cinnamic acid, p-coumaric acid, ferulic acid and chlorhexidine	Protease	(Ramezani and Monroe 2023)
4	Fiber patches	Genetically engineered epidermal growth factor containing a matrix metalloproteinase is covalently conjugated to a nonwoven poly(ε-caprolactone)	Growth factor	Metalloproteinase	(Kim et al. 2017)
5	Electrospun patch	Crosslinked hydrogel of allyl glycidyl ether, carboxymethyl chitosan and MMP-2 substrate peptide - CPLGLAGC loaded with polyplexes made of TGF-β1 siRNA and polycaprolactone	TGF-β1 siRNA polyplexes	MMP-2	(Cia et al. 2021)
6	Hydrogel	Cysteine containing peptides LIVAGKC and LK6C formed disulfide bonds	Peptides	---	(Seow et al. 2016)
7	Fibrillar hydrogel	Amphiphilic short peptide I_3QGK (80) sensitive to transglutaminase (TGase)	Short peptide I_3QGK	Transglutaminase (TGase)	(Chen et al. 2016)
		Ac-I_3SLKG-NH2 and Ac-I3SLGK-NH2	Peptide	Selectively peptide made of Ac-I_3SLKG-NH$_2$ sequence degrades in response to MMP-2 enzymes	(Chen et al. 2017)
8	Biomimetic hydrogel scaffolds	Ac-ILVAGK-NH2 peptide	Peptide	---	(Loo et al. 2014)

Table 7. Advancements in ROS responsive polymer formulations for wound healing applications.

Sr. no.	Formulation	ROS-responsive Polymers	Drug	Responsive characteristics	Reference
1	Hydrogel	Polyvinyl alcohol (PVA) crosslinked with ROS-responsive crosslinker N1-(4-boronobenzyl)-N3 -(4-boronophenyl)-N1,N1,N3,N3 -tetramethylpropane-1,3-diaminium (TSPBA)	Metformin and FGF21	Cleavage of TSPBA-PVA linkage	(Zhu et al. 2022)
2	Microsphere	Curcumin encapsulated in poly (propylene sulfide)	Curcumin	Cleavage of polymeric thioether linkage	(Poole et al. 2015)
3	Micelles	PEG-polyurethane (PU)-SeSe-PEG polymer using doxorubicin (DOX)	Doxorubicin	Cleavage of polymeric selenium linkage	(Ren et al. 2012)
4	Complex	Cationic poly(amino thioketal) and negatively charged DNA	DNA	Cleavage of polymeric thioketal linkage	(Shim and Xia 2013)
5	--	1,1,3,3-tetramethylguanidine promoted poly esterification for PEGylated ROS-responsive polymeric prodrug of cinnamaldehyde	Cinnamaldehyde	Cleavage of polymeric thioacetal linkage	(Lu et al. 2023)

and helps to combat invading microorganisms in the wound. They can directly kill bacteria and enhance the immune response against pathogens.

Table 7 given below discusses few examples belonging to ROS responsive polymers in drug delivery systems.

vi. Multi-responsive polymers

Certain smart polymers exhibit multi-responsive behavior, enabling them to respond to multiple stimuli within a single polymeric system. This characteristic is particularly advantageous when encapsulating two or more drugs within the same polymer. Such materials have shown promising potential in wound healing, as they can effectively target multiple pathways within the wound healing cascade to address different aspects and severities of the wound. These polymers act as facilitators in treating complex wound healing scenarios by simultaneously targeting multiple cascades.

The following Table 8 provides various examples of dual-responsive polymers used in drug delivery systems, along with their corresponding mechanisms.

Table 8. Advancements in multiple responsive polymer formulations for wound healing applications.

Sr. no.	Formulation	Thermo-responsive polymers	Drug	Responsive characteristics	Mechanism of targets	Reference
ROS and photo responsive polymers						
1	Micelles	Triblock copolymer (PEG-Polyurethane SeSe-PEG) with different PEG lengths with propylene sulfide, a porphyrin derivative 9,10-anthracenedipropionic acid for loading Doxorubicin	Doxorubicin	ROS and photo responsive	Active drug delivery mechanism	(Han et al. 2013)
2	Nanoparticles	Pheophorbide (photosensitizer) - decorated on the backbone of chondroitin sulfate (CS)	Doxorubicin	Laser light and hypoxia	Chondroitin sulfate is a ROS degradable carrier under hypoxic conditions, hence used in active targeted drug delivery	(Park et al. 2016)
ROS and Enzyme responsive polymers						
1	Micellar nano particles	The hydrophobic part of block polymer contains inactive MMP-2 component linked to the polymer backbone through a boronic ester linkage whereas the hydrophilic block of polymer comprises MMP-2 peptide substrate (GPLGLAGGERDG) that is loaded with a synthesized MMP-2 inhibitor (PY-2).	Synthesized MMP-2 inhibitor (PY-2)	ROS cleaves the phenolate linkage and releases the MMP inhibitor. On the other hand, MMP-2 enzyme cleaves the peptide substrate	Targeted for malignant cells where levels of ROS and MMP-2 are upregulated	(Daniel et al. 2016)
2	Gold Nanoparticles	PEG with a photosensitizer (PpIX) and linked to β-Cyclodextrin-SS and formulated as AuNPs using an MMP-2 responsive peptide linker (PLGVR). The AuNPs attached to drug through a ROS-responsive thioketal linker.	Doxorubicin	Photosensitizer and ROS	Active targeted drug delivery	(Han et al. 2015)

ROS and pH responsive polymers

1	Nanoparticles	Cy3-labelled pH-responsive N-palmitoyl chitosan linked with a polythioketal and loaded with therapeutic agent curcumin	Curcumin	Cy3-labelled pH responsive functionalized chitosan and polythioketal linker cleavage	It downregulates proinflammatory cascades.	(Pu et al. 2014)
2	Nano particles	Poly (vanillin oxalate) conjugated with vanillin	Vanillin	Acid cleavable acetal linkage makes the polymer pH sensitive and cleave the oxalate ester linkage due to presence of ROS	Antioxidant property is due to the presence of pH responsive polymer towards acidic pH and ROS.	(Kwon et al. 2013)

pH and thermo-responsive polymers

1	Hydrogel	Poly (ethylene glycol)-co-Poly (sulfamethazine) ester urethane)	Sulfamethazine	It releases drug by converting the gel to sol at pH-8.5, 23 °C. Assembly is stable gel at pH-7.4 and 37 °C temperature	Bioinspired adhesive hydrogel used as controlled delivery systems	(Fiaz et al. 2016)
2	Hydrogel micelles	Poly((propylenesulfide)-block-(N,N-dimethylacylamide)-block-PNIPAM) - triblock polymer loaded with hydrophobic drug	Nile red dye	NIPAAM is temperature responsive at its LCST and propylenesulfide is responsive towards ROS	-	(Gupta et al. 2014)

pH and Enzyme responsive copolymers

1	Hydrogel	Sodium alginate/ poly (N-vinyl caprolactam) incorporated tannic acid.	Tannic acid	Responsive at pH- 5.5-7.4	-	(Ninan et al. 2016)

Wound healing dressings

Smart polymers have a variety of uses in targeted wound healing since they may be included into traditional gels and creams as well as more contemporary hydrogels, nanoparticles, films, patches, and nanofibers. These contemporary/advanced dressings provide several benefits over conventional gauze dressings, which have been shown to be quite irritating. Smart dressings, on the other hand, made of smart polymers have superior qualities and act gently on wounds. However, development of smart polymers for wound healing applications poses a significant challenge in achieving an optimal balance between stimuli responsiveness, biocompatibility, and mechanical properties. The design and selection of smart polymers must consider factors such as cytotoxicity, degradation rates, and mechanical strength to ensure their suitability and sustainability.

a. Ideal properties of wound dressing

A wound dressing needs to have a few key components in order to be most successful in accelerating the healing process. First and foremost, it needs to be able to keep the area around the wound at the ideal moisture level to foster a healing environment (Vowden and Vowden 2017). Furthermore, the dressing should have great gas permeability, allowing for appropriate exchange of oxygen and moisture vapors. The capacity of the dressing to shield the wound from infections and microbial contamination, lowering the risk of consequences, is another critical component. Additionally, to avoid the growth of dead tissue, the dressing should reduce surface necrosis. To protect the wound from trauma or outside pressures, mechanical protection is necessary. It should also be simple to take off and replace the dressing. It is imperative to make sure that materials are non-toxic, biodegradable, biocompatible and elastic (Vowden and Vowden 2017). By incorporating these characteristics, wound dressings can dramatically improve the speed at which wounds heal.

b. Advanced wound healing dressings using smart polymers
   (Farahani and Shafiee 2021)

Compared to conventional dressings, advanced wound healing dressings using smart polymers offer several advantages. They provide active and targeted therapies, reducing the need for frequent dressing changes and minimizing wound disturbance. Additionally, smart dressings can enhance the healing process by promoting factors like angiogenesis, cell proliferation, and tissue regeneration.

The responsive polymers used in smart wound dressings are able to detect and react to particular triggers or alterations in the environment around the wound, such as changes in pH, temperature, moisture, or the presence of particular biomarkers. This enables the dressings to actively adapt, offer targeted medicines, or improve wound healing conditions. As an illustration, stimuli-responsive dressings can release bioactive molecules in response to the presence of bacteria or inflammatory signals, such as growth factors or anti-inflammatory or antibiotic drugs.

The inclusion of sophisticated sensing technology in smart dressings represents another important development. The pH, oxygenation, temperature, and wetness of the wound can all be monitored using sensing technologies. With the help of timely intervention and individualized treatment methods, it may enable real-time monitoring of these parameters and useful information on the development of the wound.

c. Formulations of advanced and smart dressings for wound healing treatment

i. Hydrocolloids

Hydrocolloid dressings are specialized wound dressing consisting of two layers: an outer impermeable layer and an inner colloidal layer. The inner layer contains carboxymethylcellulose or gelatin, or pectin; upon contact with wound exudate, it forms a gel (Thomas 2008). This gel formation creates a moist healing environment, promoting faster wound healing and reducing scar formation. The outer layer is a waterproof polyurethane film that provides protection against bacteria, foreign bodies, and debris. Most hydrocolloids are transparent or translucent substances that are applied to wounds with minimal to moderate exudate flow. These hydrocolloid dressings are biodegradable, non-toxic, breathable, and most patient-compliant dressings with good skin adhesion qualities. They are accessible in the form of semipermeable and impermeable sheets. These dressings can be used for up to a week, reducing the need for frequent changes. They are particularly suitable for pediatric patients as hydrocolloid dressings are non-adhesive and painless to remove (Thomas 2008). Dressings made of hydrocolloid are mostly used to treat non-infected wounds, including venous and diabetic ulcers. These dressings are not suitable for deeper wounds, especially those with infections that require oxygen to speed up the healing process of the wound because this restricted permeation impedes the free flow of oxygen (Gupta and Edwards 2009).

ii. Foam dressing

Foam dressings are specialized dressings composed of polyurethane sheets as outermost layer that adhere to the skin surface. Upon contact with wound exudate, the polyurethane films within the dressing create pores, facilitating drug release. The permeability of the film backing plays a crucial role in regulating gas exchange and water evaporation, thus preventing maceration, bacterial contamination, and maintaining a dry wound contact layer.

The foam structure in the dressing is prepared by using gelling and blowing methods. The blowing reaction produces carbon dioxide gas, which causes the foam to grow, while the gelling reaction creates urethane links, which helps to increase the strength of the foam (Trucillo and Di Maio 2021). The inner matrix of foam dressing is made up of polyols which makes the system enriched with water molecules. Also, hydrophilic natural polymers, like polysaccharides, cellulose, polyacrylates are often used in foam formulations to enhance water absorption and moisture protection (Trucillo and Maio 2021). Due to their hydrophilic composition and open-cell structure, these dressings have great absorbent qualities and can retain sizable volumes of exudate. Absorbent behavior and moisture vapor transport (MVT) are the two physical mechanisms that contribute to the fluid-handling properties of foam dressings (Zehrer et al. 2014). The porosity of the foam, or the volume fraction of the pores, has a significant role in determining absorbent behavior, whereas the permeability of the backing foam determines the MVT rate.

Foam dressings are often applied to treat chronic wounds such as diabetic foot ulcer and surgical wounds where the exudate is moderate to high exudate (Fogh and Nielsen 2015).

iii. Hydrofibers

Hydrofiber dressings are prepared by using a fibrous scaffold of synthetic polymers such as polycaprolactone or poly (L-lactic acid) or poly (L-lactic acid-co-glycolic acid) or sodium carboxymethylcellulose, or by using natural polymers such as gelatin, chitosan, and collagen (Barnea et al. 2010). Hydrofibers undergo transformation into a gel-like substance upon coming into contact with wound fluid. This gel-forming action promotes an effective absorption of exudates. The gel formation and fluid absorption protects the wound from dryness, supports clean healing, and immobilizes microorganisms and proteolytic enzymes. Hydrofiber dressings are suitable for treating moderately to highly exuding chronic and acute wounds, including burns, ulcers, and surgical wounds (Richetta et al. 2011). They prevent maceration, reduce the risk of infection, and have superior fluid retention compared to gauze or alginate dressings. The payload is released from the hydrofibers using swelling and diffusion mechanisms as they transform into a gel-like form.

iv. Film dressing

Films are transparent and flexible wound coverings generally prepared using polyurethane, a synthetic polymer (Savencu et al. 2021). However, there is a growing interest in the development of films using natural materials, including polysaccharides such as pectin, chitosan, and pullulan. These natural polysaccharides have the ability to enhance the activity of growth factors, facilitating cell proliferation and expediting the wound healing process. Film dressings create a moist environment for granulating wounds and facilitate a natural process of removing dead or necrotic tissue from wound surface. They absorb limited exudate while preventing excess fluid loss and maintaining the optimal moisture level. Although they do not have high water absorption capabilities, they can release small amounts of moisture through moisture vapor transpiration (MVT) process. Therefore, film dressings are not suitable for wounds with significant exudate but used for treating mild and superficial wounds with minimal discharge (Meuleneire 2014). Often film formulations are used in combination with absorbent dressings such as absorbent pad which helps to drain the excess of wound exudate.

v. Smart hydrogels

Hydrogels are porous, three-dimensional polymeric matrices. It is made up of a crosslinked network of hydrophilic polymers that can absorb and hold enormous volumes of water or biological fluids (Lima and Passos 2021). It is soft, flexible and retains high water contents between 90% to 99%. Hydrogels are particularly helpful as wound care dressings since they create and maintain a moist environment regardless of external conditions. By promoting cell proliferation, vascularization, and host integration, they offer an environment that is advantageous for wound healing. In addition, hydrogels offer beneficial properties such as effective oxygen transport, biocompatibility, biodegradability, high swelling capacity for absorbing wound exudate, antibacterial activities, and non-toxicity (Rasool et al. 2019). The mechanisms by which the drug-loaded hydrogel releases the drug are diffusion controlled, swelling controlled, chemical controlled and hydrolytic or enzyme controlled.

Traditional hydrogel dressings have the ability to encapsulate and distribute payload (drug) consistently, but they lack the capability to regulate drug release based on the specific disease environment. In recent years, the amount of research being done to create "smart" hydrogels has significantly increased in recent years. Smart hydrogels can respond to various stimuli, resulting in diverse drug delivery applications. These dressings are prepared by incorporating smart polymers into the hydrogel polymer backbone, enabling them to actively interact with the dynamic changes at the site of wound. These stimuli can include temperature, pH, light, oxygen level, glucose, ionic charge, or antigens (Seuring et al. 2023, Wang et al. 2023, Derakhshandeh et al. 2018).

One kind of smart hydrogels are "self-healing hydrogels", and as the name implies, these hydrogels have the capacity to do so by creating new bonds when the existing bonds on hydrogel become compromised or damaged (Zhang et al. 2021). The self-healing hydrogels can undergo multiple cycles of bond breaking and reforming of the covalent bonds. Such weak and labile covalent bonds are known as dynamic covalent bonds. This dynamic behavior enables the hydrogel to experience changes in its structure and properties, facilitating self-repair and allowing for adaptive responses. This feature may in turn improve the lifetime of hydrogel and enhance its robustness. Self-healing hydrogels demonstrate superior properties such as non-adhesiveness, retain high moisture, permeablity to gas and are biocompatible. They are used as a wound dressing to fill wound surfaces with uneven shapes, providing a bed to help in the usual healing process.

Table 9 gives examples of the advanced formulations of wound healing dressings, anticipated chances of wound maceration or infection, and their benefits and drawbacks.

Smart bandages or smart dressings

Most existing wound healing dressings are passive and do not actively respond to changes in the wound environment. While some passive dressings can release drugs with anti-inflammatory, antibiotic, or antibacterial properties, more advanced dressings release growth factors and drugs to aid tissue healing in a passive manner. However, a significant limitation of the current treatment process is the lack of information on the wound bed status and healing progress. Passive wound care dressings also struggle to distinguish between different stages of wound healing (Pang et al. 2023). As a result, patients require frequent screenings to assess the healing process and identify potential infections. This increased frequency of medical visits for monitoring the healing process contributes to the overall cost of treatment.

Table 9. Types of advanced dressings used for wound healing treatment.

Sr. No.	Dressing type	Fluid management @	Prevents maceration #	Advantages	Drawbacks	Marketed preparation (company name)	Reference
1	Hydrocolloids	Moderate absorption	No	Easy to use, waterproof, creates barrier against bacterial infection	May adhere to the wound bed as it prevents gaseous exchange, therefore difficult to remove.	**Tegaderm** (3M Company US) **Granuflex** (ConvaTec) **Comfeel plus** (Coloplast)	(Boateng et al. 2008, Finnie 2002)
2	Foams	Moderate to high absorption	Yes	Gives moist healing environment, convenient to apply over bony prominence or within exudative cavities.	Foams adhere to the wound surface, hence complicate the healing process. Contraindicated for dry, non-exudating wounds, and heavy bleeding wounds.	**Mepilex** (Molnlycke) **Allevyn** (Smith & Nephew healthcare) **Betafoam** (Betafoam corporation)	(Boateng et al. 2008)
3	Hydrofibers	Moderate to high absorption	Yes	Transform to gel on absorbing wound exudate. Durable and long-lasting dressing, easily removed	Non-adherent and need a secondary dressing to secure it. Contraindicated for dry, non-exudating wounds, and heavy bleeding wounds.	**Aquacel**(Convatec) **Versiva®** XC (Convatec)	(Boateng et al. 2008, Barnea et al. 2010)
4.	Film dressing	Very poor absorption	No	Transparent films are easy to monitor the healing wound. Can be used as a primary and secondary dressing cover.	Film has a poor capacity to absorb the wound exudate	**Cutifilm plus** (Smith & Nephew) **Hyalosafe** (KIBOU Pharma) **OpSite** (Smith & Nephew)	(Ahovan et al. 2022, Boateng et al. 2008)

| 5 | Hydrogels | Donate moisture | No | Suitable for dry wounds. Patient compliant and easily removable. | It needs to be clubbed with secondary dressing because of high water content. Not suitable for high exudating wound. | **AquaDerm** (DermaRite Industries) **SuprasorbG** (Lohman & Rauscher Global) **Neoheal Hydrogel** (Kikgel) | (Boateng et al. 2008) |

\# : Maceration is a process that involves softening and breaking of skin when it remains in contact with excessive moisture or wound exudate for extended period of time.

@ Fluid management:

-Moderate to high absorption means approx 15% to 50% of moisture absorption of their original weight

-Poor absorption means these dressings are non-adherent or non-absorbent and they do not actively absorb wound exudate.

-Donate moisture means they help to rehydrate the wound

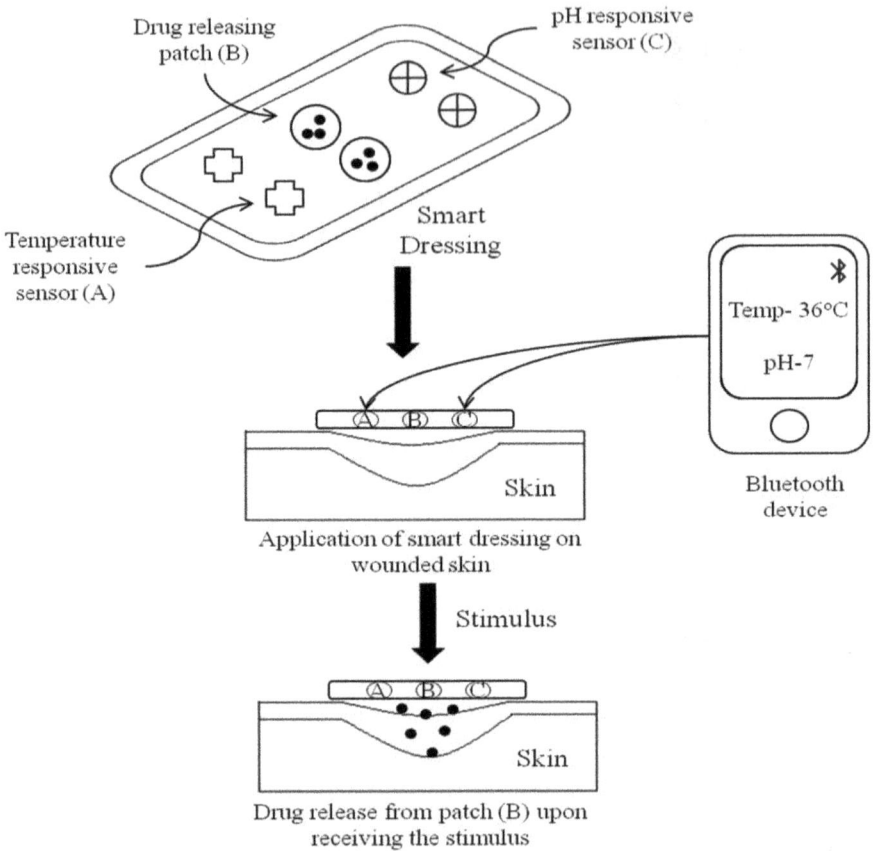

Fig. 3. AI-driven smart dressing prototype.

Researchers are currently investigating the development of active wound healing treatment through the integration of suitable drug delivery systems and wearable sensors in the form of smart dressings (Pang et al. 2023). These dressings are designed to monitor multiple wound biomarkers and detect physiological changes throughout the wound healing process. Additionally, they have the capability to deliver tailored therapy based on the specific needs of the wound. This cutting-edge technology, often referred to as "smart sense and treat dressings" or "all-in-one smart wound dressings", integrates the capability to monitor numerous biomarkers, enabling the detection of physiological changes throughout the wound healing process. Figure 3 showcases a prototype of innovative smart dressing. These advanced smart dressings have the capacity to administer precise therapy tailored to the unique requirements of the wound (Pang et al. 2023, Derakhsandeh et al. 2018). The emergence of smart dressings has revolutionized traditional clinical approaches to diagnosis and treatment, expediting decision-making processes and improving overall wound management outcomes. These advanced platforms have the capacity to integrate various types of (bio)sensors, enabling real-time monitoring and active wound healing treatment.

In 2020, Pang et al. (2021) demonstrated a proof of concept for an all-in-one advanced smart wound dressing system. They successfully integrated an ultraviolet (UV)-responsive antibacterial hydrogel and UV-LEDs into the dressing, allowing for controlled release of antibiotics directly at the wound site. The system also featured an integrated sensor that continuously monitored the temperature of the wound. This information was transmitted wirelessly using Bluetooth to portable devices such as smartphones. When the monitored wound temperature surpassed a predetermined threshold, the integrated UV-LEDs were triggered and prompted the controlled release of antibiotics directly at the wound site. A double-layer dressing was developed, comprising an upper layer consisting of a flexible electronic apparatus made of polyimide. The system's top layer included a Bluetooth device, four UV light-emitting diodes, a temperature sensor, and power management components. A UV-responsive antibacterial hydrogel made up the lower layer in contrast. Through the use of an o-nitrobenzyl (ONB) linkage that is UV cleavable, the antibiotic Gentamicin was covalently joined to the hydrogel's polyethylene glycol-based network. The integrated system made it possible to wirelessly transmit data from the wound site and monitor the temperature in real-time. By causing the localized release of antibiotics, this allowed for the early detection and on-demand treatment of bacterial infections.

Another study was conducted by researchers Xu et al. (2021) on a battery-free, wireless smart dressing patch for wound healing treatment. The patch consisted of two layers, each serving a specific purpose. The lower layer contained a stretchable electrode array made of polyimide (PI), which included a pH sensor, a uric acid sensor, and an electrode for controlled delivery of cefazolin, an antibiotic. The upper layer featured a flexible circuit board equipped with near-field communication (NFC) capabilities, as well as a temperature sensor and a drug delivery controller. By connecting to an NFC-enabled smartphone, the patch could receive power wirelessly, transmit monitored data, and regulate the release of drugs. This allowed for simultaneous monitoring of three essential biomarkers associated with bacterial infection and provided an accurate assessment of wound conditions during the healing process. The prototype device underwent *in vitro* and *in vivo* testing, confirming the efficacy of this AI-based wound dressing system.

Through the utilization of AI tools, Khatib et al. (2022) undertook a study on an electronic skin (e-skin) that possessed impressive self-healing capabilities and was specifically developed for monitoring temperature, pH, and pressure. The e-skin consisted of multiple layers, including a top layer with sensors and intermediate layer for detecting structural damage, and a bottom layer containing heater arrays for repairing extensive damage. Results revealed the rapid repair of superficial cuts within 30 seconds due to integrated array of silver nanowires (AgNW) based heater. The performance of sensors was fully recovered which indicated suitability of the device for long-term application at wound sites. Future technology will be able to incorporate advanced smart dressings for detecting damaged skin and healing processes by including wearable gadgets, electronic skins for robotics and prosthetics (Pang et al. 2023).

Challenges and future directions

Conventional wound healing treatment typically relies on a passive mechanism, wherein the body's natural healing processes are allowed to repair and regenerate the damaged tissue over time. This approach is suitable for acute or less severe wounds, such as minor cuts, scratches, or abrasions, where active drugs or external factors are not necessary for healing. Nevertheless, conventional treatment approaches for chronic wounds are frequently ineffective due to persistent inflammation, recurrent bacterial infections, the formation of drug-resistant microbial biofilms, and the lack of response from dermal and/or epidermal cells to healing signals. Therefore, comprehensive wound monitoring and the use of advanced dressings are necessary for the successful treatment of chronic wounds (Frykberg and Banks 2015, Lima and Passos 2021, Derakhsandeh et al. 2018).

In order to address these challenges, it is necessary to employ innovative strategies, such as integrating stimuli-responsive or smart polymers into topical dressing formulations. These smart polymers can respond to external stimuli, thereby allowing controlled release of loaded drugs and promoting effective wound healing process (Van et al. 2021). Another approach is by using natural or bioactive polymers for wound healing processes. Due to their inherent antibacterial properties, bioactive polymers effectively promote a balanced immune response and reduce excessive inflammation. Moreover, these polymers play a crucial role in supporting tissue formation by guiding cell growth and facilitating regeneration. However, they can be customized with the help of smart polymers for active healing processes. Such customized bioactive polymers show suitable mechanical properties, degradation rates, and drug release kinetics. Smart bioactive polymers can also serve as carriers for therapeutic agents, such as antibacterial or anti-inflammatory agents or growth factors (Ribeiro and Flores-Sahagun 2020).

These customized smart polymers have been formulated as advanced wound healing dressings, such as hydrogels, patches, hydrofibers, hydrocolloids, or films. Amongst these, smart hydrogels have been found to be more patient compliant as compared to dermal patches, hydrofibers and hydrocolloids in clinical trials (Ahovan et al. 2022) whereas topical films suffer from poor MVT; hence, they are combined with other advanced dressings (Meuleneire 2014). The limitation of conventional/advanced dressings is their inability to provide data about the healing progress of wound environment with respect to the altered pH, temperature, available enzyme concentration, ROS, etc.

The use of integrated dressing with smart polymers and AI technology using certain sensors have been the upcoming areas under the active wound healing process. The combination of sensors such as pH sensor, temperature sensor and glucose sensor (Derakhsandeh et al. 2018), etc. with smart polymers gives an important information to decision making progress in wound care and on demand release of bioactive agents which helps to heal the wound faster and also minimizes the healthcare cost and time of hospitalization. Along with that, the advanced dressings facilitate good oxygenation and gas exchange at the site of wound is essential for cellular respiration and promoting angiogenesis (Derakhsandeh et al. 2018). The advanced dressings also protect the wound bed from external contaminants such as bacteria, dirt, and debris, and stabilize the wound bed by preventing interference between the developed tissue and minimizing the risk of trauma or injury. In the case of chronic wounds, the advanced dressing also gives significant pain management by incorporating pain relieving agents such as analgesic or local anesthetics (Wang et al. 2023).

However, the challenge in AI-integrated smart wound dressings lies in finding the right balance between drug delivery functions and monitoring the sensor outputs. Integrating these functions may compromise the sensitivity and precision of the sensors. Additionally, there is a risk of molecule leakage from the wearable sensors into the wound, which can hinder the drug release and impact the healing process. Furthermore, incorporating multiresponsive polymers or multiple functions into the dressings can increase manufacturing complexity and cost. Therefore, it is recommended to consider independent monitoring and treatment modules using safer materials, as well as personalized dressings with specific functions, to optimize performance and reduce costs.

In summary, conventional wound healing approaches are suitable for treating acute wounds, while chronic wounds require the use of smart and bioactive polymers. Integrating these polymers with biomarker sensors in dressings can address the challenges of chronic wound healing. However, achieving the right balance is essential. Additional research is needed to optimize AI-based smart dressings and develop personalized dressings with suitable responsive polymers to ensure patient compliance.

References

Agarwal, A., J.F. McAnulty, M.J. Schurr, C.J. Murphy and N.L. Abbott. 2011. Polymeric materials for chronic wound and burn dressings. pp. 186–208. *In*: David Farrar (ed.). Advanced Wound Repair Therapies. Woodhead Press. Cambridge, Massachusetts.

Ahmed, N. 2005. Advanced glycation endproducts-role in pathology of diabetic complications. Diabetes. Res. Clin. Pract. 67: 3–21.

Ahovan, Z., Z. Esmaeili, B. Eftekhari, S. Khosravimelal, M. Alehosseini, G. Orive, M. Gholipourmalekabadi et al. 2022. Antibacterial smart hydrogels: New hope for infectious wound management. Mater. Today. Bio. 100499.

Alonso, J., J. Lee, A.R. Burgess and B.D. Browner. 1996. The management of complex orthopedic injuries. Surg. Oncol. Clin. N. 76: 879–903.

Alwan, H. and H. Kassab. 2021. Componental description of medicated foams: a review. Int. J. Pharm. Res. 13(1): 2183–2189.

Armstrong, D.G and E.B. Jude. 2002. The role of matrix metalloproteinases in wound healing. J. Am. Podiatr. Med. Assoc. 92: 12–18.

Attinger, C., J. Janis, J. Steinberg, J. Schwartz A. Al-Attar and K. Couch. 2006. Clinical approach to wounds: debridement and wound bed preparation including the use of dressings and wound-healing adjuvants. Plast. Reconst. Sur. 117: 72–109.

Ballesteros, C., D. Correa and V. Zucolotto. 2020. Polycaprolactone nanofiber mats decorated with photoresponsive nanogels and silver nanoparticles: Slow release for antibacterial control. Mater. Sci. Eng. 107: 110334.

Banerjee, I., D. Mishra, T. Das and T.K. Maiti. 2012. Wound pH-responsive sustained release of therapeutics from a poly (NIPAAm-co-AAc) hydrogel. J. Bio. Mater. Sci. Polym. 23: 111–132.

Banno, T., A. Gazel and M. Blumenberg. 2004. Effects of tumor necrosis factor-α (TNFα) in epidermal keratinocytes revealed using global transcriptional profiling. J. Bio. Chem. 279: 32633–32642.

Barnea, Y., J. Weiss and E. Gur. 2010. A review of the applications of the hydrofiber dressing with silver (Aquacel Ag) in wound care. Ther. 6: 21–27.

Baum, C. and C. Arpey. 2005. Normal cutaneous wound healing: clinical correlation with cellular and molecular events. Dermatol. Surg. 31: 674–686.

Boateng, J., K. Matthews, H.N. Stevens and G.M. Eccleston. 2008. Wound healing dressings and drug delivery systems: a review. J. Pharm. Sci. 97: 2892–2923.

Bökel, C and N. Brown. 2002. Integrins in development: moving on, responding to, and sticking to the extracellular matrix. Dev. Cell. 3: 311–321.

Bradley, J. 2008. TNF-mediated inflammatory disease. J. Pathol. 214: 149–160.

Broughton, G., J. Janis and C. Attinger. 2006a. Wound healing: an overview. Plast. Reconstr. Surg. 117: 1e-S–32e-S

Broughton, G., E. Jeffrey and E. Christopher. 2006b. The basic science of wound healing. Plast. Reconstr. Surg. 117: 12S–34S.

Cabral-Pacheco, G.A., I. Garza-Veloz, C. Castruita-De la Rosa, J.M. Ramirez-Acuna, B.A. Perez-Romero, J.F. Guerrero-Rodriguez et al. 2020. The roles of matrix metalloproteinases and their inhibitors in human diseases. Int. J. Mol. Sci. 21: 9739.

Cai, C., W. Wang, J. Liang, Y. Li, M. Lu, W. Cui et al. 2021. MMP-2 responsive unidirectional hydrogel-electrospun patch loading TGF-β1 siRNA polyplexes for peritendinous anti-adhesion. Adv. Funct, Mater. 31: 2008364.

Castillo-Henríquez, L., P. Sanabria-Espinoza, B. Murillo-Castillo, G. Montes de Oca-Vásquez, D. Batista-Menezes, B. Calvo-Guzmán et al. 2021. Topical chitosan-based thermo-responsive scaffold provides dexketoprofen trometamol controlled release for 24 h use. Int. J. Pharm. 13: 2100.

Chai, Q., Y. Jiao and X. Yu. 2017. Hydrogels for biomedical applications: their characteristics and the mechanisms behind them. Polym. Gels. 3(1): 6.

Chamkouri, H. and M. Chamkouri. 2021. A review of hydrogels, their properties and applications in medicine. Am. J. Biomed. Sci. Res. 11: 485–493.

Chen, C., Y. Zhang, R. Fei, C. Cao, M. Wang, J. Wang et al. 2016. Hydrogelation of the short self-assembling peptide I3QGK regulated by transglutaminase and use for rapid hemostasis. ACS. Appl. Mater. Interfaces 8: 17833–17841.

Chen, C., Y. Zhang, Z. Hou, X. Cui, Y. Zhao and H. Xu. 2017. Rational design of short peptide-based hydrogels with MMP-2 responsiveness for controlled anticancer peptide delivery. Biomacromolecules 18: 3563–3571.

Chen, G., Y. Zhou, J. Dai, S. Yan, W. Miao and L. Ren. 2022. Calcium alginate/PNIPAAm hydrogel with body temperature response and great biocompatibility: Application as burn wound dressing. Int. J. Bio. Macromol. 216: 686–697.

Chi, J., X. Zhang, Chen, C. Shao, Y. Zhao and Y. Wang. 2020. Antibacterial and angiogenic chitosan microneedle array patch for promoting wound healing. Bioact. Mater. 5: 253–259.

Chong, E.J., T.T. Phan, I.J. Lim, Y.J. Zhang, B.H. Bay, S. Ramakrishna et al. 2007. Evaluation of electrospun PCL/gelatin nanofibrous scaffold for wound healing and layered dermal reconstitution. Acta. Biomater. 3: 321–330.

Chowdhury, F., S. Ahmed, M. Rahman, M.A. Ahmed, M.D. Hossain, H.M. Reza et al. 2022. Chronic wound-dressing chitosan-polyphenolic patch for pH responsive local antibacterial activity. Mater. Today. Commun. 31: 103310.

Clark, R.A. 1993. Regulation of fibroplasia in cutaneous wound repair. Am. J. Med. 306: 42–48.

Clempus, R.E. and K.K. Griendling. 2006. Reactive oxygen species signaling in vascular smooth muscle cells. Cardiovasc Res. 71(2): 216–25.

Cohen, S. 1983. The epidermal growth factor (EGF). Cancer. Res. 51: 1787–17.

Daniel, K.B., C.E. Callmann, N.C. Gianneschi and S.M. Cohen. 2016. Dual-responsive nanoparticles release cargo upon exposure to matrix metalloproteinase and reactive oxygen species. Chem. Commun. 52: 2126–2128.

Degreef, H.J. 1998. How to heal a wound fast. Dermatologic Clinics 16: 365–375.

Derakhshandeh, H., S.S. Kashaf, F. Aghabaglou, I.O. Ghanavati and A. Tamayol. 2018. Smart bandages: the future of wound care. Trends in Biotechnol, 36: 1259–1274.

Dethe, M.R., A. Prabakaran, H. Ahmed, M. Agrawal, U. Roy and A. Alexander. 2022. PCL-PEG copolymer based injectable thermosensitive hydrogels. J. Control. Release. 343: 217–236.

Di Martino, M., L. Sessa, R. Diana, S. Piotto and S. Concilio. 2023. Recent progress in photoresponsive biomaterials. Molecules 28: 3712.

Diegelmann, R.F. and M.C. Evans. 2004. Wound healing: an overview of acute, fibrotic and delayed healing. Front Biosci. 9: 283–289.

Dunnill, C., T. Patton, J. Brennan, J. Barrett, M. Dryden, J. Cooke et al. 2017. Reactive oxygen species (ROS) and wound healing: the functional role of ROS and emerging ROS-modulating technologies for augmentation of the healing process. Int. Wound. J. 14: 89–96.

English, D., A.T. Kovala, Z. Welch, K.A. Harvey, R.A. Siddiqui, B.N. Brindley et al. 1999. Induction of endothelial cell chemotaxis by sphingosine 1-phosphate and stabilization of endothelial monolayer barrier function by lysophosphatidic acid, potential mediators of hematopoietic angiogenesis. J. Hematother. Stem. Cell. Res. 8: 627–634.

Fang, R., H. Xu, W. Cao, L. Yang and X. Zhang. 2015. Reactive oxygen species (ROS)-responsive tellurium-containing hyperbranched polymer. Polym. 6: 2817–2821.

Farahani, M and A. Shafiee. 2021. Wound healing: From passive to smart dressings. Adv. Healthc. Mater. 10: 2100477.

Finnie, A. 2002. Hydrocolloids in wound management: pros and cons. Br. J. Community. Nurs. 7: 338–345.

Fogh, K. and J. Nielsen. 2015. Clinical utility of foam dressings in wound management: A review. Chronic Wound Care Manag. Res. 2: 31–38.

Frykberg, R.G and J. Banks. 2015. Challenges in the treatment of chronic wounds. Adv. Wound. Caref. 4: 560–582.

G. Richetta, A., C. Cantisani, V.W. Li, C. Mattozzi, L. Melis, F. De Gado et al. 2011. Hydrofiber dressing and wound repair: review of the literature and new patents. Recent. Pat. Inflamm. Allergy. Drug. Discov. 5: 150–154.

Galliani, M., C. Diacci, M. Berto, M. Sensi, V. Beni, M. Berggren et al. 2020. Flexible printed organic electrochemical transistors for the detection of uric acid in artificial wound exudate. Adv. Mater. Interfaces 7: 2001218.

Gandhi, A., A. Paul, S.O. Sen and K.K. Sen. 2015. Studies on thermoresponsive polymers: Phase behaviour, drug delivery and biomedical applications. Asian. J. Pharma. 10: 99–107.

Ghomi, E., S. Khalili, S. Nouri Khorasani, R. Esmaeely Neisiany and S. Ramakrishna. 2019. Wound dressings: Current advances and future directions. J. Appl. Polym. Sci. 136(27): 47738.

Giannone, G., B. Dubin-Thaler, H.G. Döbereiner, N. Kieffer, A.R. Bresnick and M.P. Sheetz. 2004. Periodic lamellipodial contractions correlate with rearward actin waves. Cell. 116: 431–443.

Goldman, R. 2004. Growth factors and chronic wound healing: past, present, and future. Adv. Skin. Wound. Care. 17: 24–35.

Greenhalgh, D. 1998. The role of apoptosis in wound healing. Int. J. Biochem. 30: 1019–1030.

Grützner, V., R.E. Unger, G. Baier, L. Choritz, C. Freese, T. Böse et al. 2015. Enzyme-responsive nanocomposites for wound infection prophylaxis in burn management: *in vitro* evaluation of their compatibility with healing processes. Int. J. Nanomed. 10: 4111.

Gupta, B.S. and J.V. Edwards. 2009. Textile materials and structures for wound care products. Adv. Text for Wound Care. 48–96.

Gupta, M.K., J.R. Martin, T.A. Werfel, T. Shen, J.M. Page and C.L. Duvall. 2014. Cell protective, ABC triblock polymer-based thermoresponsive hydrogels with ROS-triggered degradation and drug release. J. Am. Chem. Soc. 136: 14896–14902.

Gupta, M.K., S.H. Lee, S.W. Crowder, X. Wang, L.H. Hofmeister, C.E. Nelson et al. 2015. Oligoproline-derived nanocarrier for dual stimuli-responsive gene delivery. J. Mater. Chem. B. 3: 7271–7280.

Haidari, H., Z. Kopecki, A.T. Sutton, S. Garg, A.J. Cowin and K. Vasilev. 2021. pH-responsive "smart" hydrogel for controlled delivery of silver nanoparticles to infected wounds. J. Antibiot. 10(1): 49.

Haidari, H., K. Vasilev, A.J. Cowin and Z. Kopecki. 2022. Bacteria-activated dual pH- and temperature-responsive hydrogel for targeted elimination of infection and improved wound healing. ACS. Appl. Mater. Interfaces 14: 51744–51762.

Han, K., J.Y. Zhu, S.B. Wang, X.H. Li, S.X. Cheng and X.E. Zhang. 2015. Tumor targeted gold nanoparticles for FRET-based tumor imaging and light responsive on-demand drug release. J. Mater. Chem. B. 3: 8065–8069.

Han, P., S. Li, W. Cao, Y. Li, Z. Sun, Z. Wang et al. 2013. Red light responsive diselenide-containing block copolymer micelles. J. Mater. Chem. B. 1: 740–743.

Hart, J. 2002. Inflammation 1: its role in the healing of acute wounds. J. Wound. Care. 11: 205–209.

Hasatsri, S., A. Pitiratanaworanat, S. Swangwit, C. Boochakul and C. Tragoonsupachai. 2018. Comparison of the morphological and physical properties of different absorbent wound dressings. Dermatol. Res. Prac. 1–6.

Hassan, D., C.A. Omolo, V.O. Fasiku, C. Mocktar and T. Govender. 2020. Novel chitosan-based pH-responsive lipid-polymer hybrid nanovesicles (OLA-LPHVs) for delivery of vancomycin against methicillin-resistant *Staphylococcus aureus* infections. Int. J. Bimol. Macromol. 147: 385–398.

He, M., F. Ou, Y. Wu, X. Sun, X. Chen, H. Li et al. 2020. Smart multi-layer PVA foam/CMC mesh dressing with integrated multi-functions for wound management and infection monitoring. Mater. Des. 194: 108913.

Henríquez, C., J. Castro-Alpízar, M. Lopretti-Correa and J. Vega-Baudrit. 2021. Exploration of bioengineered scaffolds composed of thermo-responsive polymers for drug delivery in wound healing. International Journal of Molecular Sciences 22: 1408.

Herman, I.M. 1993. Molecular mechanisms regulating the vascular endothelial cell motile response to injury. J. Cardiovasc. Pharmacol. 22: S25–S36.

Hoffman, A.S. 1987. Applications of thermally reversible polymers and hydrogels in therapeutics and diagnostics. Journal of Controlled Release 6(1): 297–305.

Hu, H. and F.J. Xu. 2020. Rational design and latest advances of polysaccharide-based hydrogels for wound healing. Biomater. Sci. 8: 2084–2101.

Huang, H., D. He, X. Liao, H. Zeng and Z. Fan 2021. An excellent antibacterial and high self-adhesive hydrogel can promote wound fully healing driven by its shrinkage under NIR. Mater. Sci. Eng. 129: 112395.

Hunt, T.K., H. Hopf and Z. Hussain. 2000. Physiology of wound healing. Adv. Skin Wound. Care. 13: 6.

Jespersen, J. 1988. Pathophysiology and clinical aspects of fibrinolysis and inhibition of coagulation. Experimental and clinical studies with special reference to women on oral contraceptives and selected groups of thrombosis prone patients. Dan. Med. Bull. 35: 1–33.

Khatib, M., O. Zohar, W. Saliba and H. Haick. 2020. A multifunctional electronic skin empowered with damage mapping and autonomic acceleration of self-healing in designated locations. Adv. Mater. 32: 2000246.

Kim, S.E., P.W. Lee and J.K. Pokorski. 2017. Biologically triggered delivery of EGF from polymer fiber patches. ACS. Macro. Lett. 6: 593–597.

Kuenzli, S. and J.H. Saurat. 2003. Peroxisome proliferator-activated receptors in cutaneous biology. B. R. Dermatol. 149: 229–236.

Kuo, S.H., C.J. Shen, C.F. Shen and C.M. Cheng. 2020. Role of pH value in clinically relevant diagnosis. Diagnostics 10: 107.

Kuroki, M., E.E. Voest, S. Amano, L.V. Beerepoot, S. Takashima, M. Tolentino et al. 1996. Reactive oxygen intermediates increase vascular endothelial growth factor expression *in vitro* and *in vivo*. J. Clin. Invest. 98: 1667–1675.

Kwon, J., J. Kim, S. Park, G. Khang, P.M. Kang and D. Lee. 2013. Inflammation-responsive antioxidant nanoparticles based on a polymeric prodrug of vanillin. Biomacromolecules 14: 1618–1626.

Labler, L., L. Mica, L. Härter, O. Trentz and M. Keel. 2006. Influence of VAC-therapy on cytokines and growth factors in traumatic wounds. Zentralbl. Chir. 131: S62–S67.

Lawrence, W.T. 1998. Physiology of the acute wound. Clin. Plast. Surg. 25: 321–340.

Lazarus, G.S., D.M. Cooper, D.R. Knighton, D.J. Margolis, R.E. Percoraro, G. Rodeheaver et al. 1994. Definitions and guidelines for assessment of wounds and evaluation of healing. Wound. Repair. Regan. 2: 165–170.

Le, T.M.D., H.T.T. Duong, T. Thambi, V.H.G. Phan, J.H. Jeong and D.S. Lee. 2018. Bioinspired pH- and temperature-responsive injectable adhesive hydrogels with polyplexes promotes skin wound healing. Biomolecules 19(8): 3536–3549.

Li, H., G.R. Williams, J. Wu, H. Wang, X. Sun and L.M. Zhu. 2017. Poly (N-isopropylacrylamide)/poly (l-lactic acid-co-ε-caprolactone) fibers loaded with ciprofloxacin as wound dressing materials. Mater. Sci. Eng. 79: 245–254.

Li, S., N.F. Huang and S. Hsu. 2005. Mechanotransduction in endothelial cell migration. J. Cell. Bio. 96: 1110–1126.

Liang, X., R. Liu, C. Chen, F. Ji and T. Li. 2016. Opioid system modulates the immune function: a review. Translational Perioperative and Pain Medicine 1(1): 5.

Lima, T.D.P.D.L. and M.F. Passos. 2021. Skin wounds, the healing process, and hydrogel-based wound dressings: a short review. J. Biomater. Sci. Polym. Ed. 32(14): 1910–1925.

Lin, X., X. Guan, Y. Wu, S. Zhuang, Y. Wu, L. Du et al. 2020. An alginate/poly (N-isopropylacrylamide)-based composite hydrogel dressing with stepwise delivery of drug and growth factor for wound repair. Mater. Sci. Eng. 115: 111123.

Loo, Y., Y.C. Wong, E.Z. Cai, C.H. Ang, A. Raju, A. Lakshmanan et al. 2014. Ultrashort peptide nanofibrous hydrogels for the acceleration of healing of burn wounds. Biomaterials 35(17): 4805–4814.

Lu, Y., P. Shan, W. Lu, X. Yin, H. Liu, X. Lian et al. 2023. ROS-responsive and self-amplifying polymeric prodrug for accelerating infected wound healing. Chem. Eng. J. 463: 142311.

Luo, Y., H. Diao, S. Xia, L. Dong, J. Chen and J. Zhang. 2010. A physiologically active polysaccharide hydrogel promotes wound healing. J. Biomed. Mater Res. A 94(1): 193–204.

Ma, N., Y. Li, H. Xu, Z. Wang and X. Zhang. 2010. Dual redox responsive assemblies formed from diselenide block copolymers. J. Am. Chem. Soc. 132(2): 442–443.

Martins, V.L., M. Caley and E.A. O'Toole. 2013. Matrix metalloproteinases and epidermal wound repair. Cell and Tissue Research 351: 255–268.

Mayet, N., Y.E. Choonara, P. Kumar, L.K. Tomar, C. Tyagi, L.C. Du Toit et al. 2014. A comprehensive review of advanced biopolymeric wound healing systems. J. Pharm. Sci. 103(8): 2211–2230.

Mazloom-Jalali, A., Z. Shariatinia, I.A. Tamai, S.R. Pakzad and J. Malakootikhah. 2020. Fabrication of chitosan–polyethylene glycol nanocomposite films containing ZIF-8 nanoparticles for application as wound dressing materials. Int. J. Biol. Macromol. 153: 421–432.

McColl, D., B. Cartlidge and P. Connolly. 2007. Real-time monitoring of moisture levels in wound dressings. Int. J. Surg. 5(5): 316–322.

Menke, M.N., N.B. Menke, C.H. Boardman and R.F. Diegelmann. 2008. Biologic therapeutics and molecular profiling to optimize wound healing. Gynecol. Oncol. 111(2): S87–S91.

Messadi, D.V., J.S. Pober, W. Fiers, M.A. Gimbrone Jr and G.F. Murphy. 1987. Induction of an activation antigen on postcapillary venular endothelium in human skin organ culture. J. Immun. 139(5): 1557–1562.

Meuleneire, F. 2014. A vapour-permeable film dressing used on superficial wounds. Br. J. Nurs. 23(15): S36–S43.

Mittal, M., R. Siddiqui, S. Reddy and A. Malik. 2014. Reactive oxygen species in inflammation and tissue injury. Antioxidants & Redox Signaling 20(7): 1126–1167.

Ninan, N., A. Forget, V.P. Shastri, N.H. Voelcker and A. Blencowe. 2016. Antibacterial and anti-inflammatory pH-responsive tannic acid-carboxylated agarose composite hydrogels for wound healing. ACS Appl. Mater. Interfaces 8(42): 28511–28521.

Nizioł, M., J. Paleczny, A. Junka, A. Shavandi, A. Dawiec-Liśniewska and D. Podstawczyk. 2021. 3D printing of thermoresponsive hydrogel laden with an antimicrobial agent towards wound healing applications. Bioengineering 8(6): 79.

Oike, Y., Y. Ito, H. Maekawa, Y. Morisada, Y. Kubota, M. Akao et al. 2004. Angiopoietin-related growth factor (AGF) promotes angiogenesis. Blood 103(10): 3760–3765.

Paneysar, J., S. Barton, P. Ambre and E. Coutinho. 2022. Novel temperature responsive films impregnated with silver nano particles (Ag-NPs) as potential dressings for wounds. J. Pharm. Sci. 111(3): 810–817.

Pang, Q., F. Yang, Z. Jiang, K. Wu, R. Hou and Y. Zhu. 2023. Smart wound dressing for advanced wound management: Real-time monitoring and on-demand treatment. Mater. Des. 229: 111917.

Park, W., B. Bae and K. Na. 2016. A highly tumor-specific light-triggerable drug carrier responds to hypoxic tumor conditions for effective tumor treatment. Biomaterials 77: 227–234.

Patel, D. 2020. Bioresponsive Polymers: Design and Application in Drug Delivery. Apple Academic Press, New Jersey and Canada.

Peppa, M., P. Stavroulakis and S. Raptis. 2009. Advanced glycoxidation products and impaired diabetic wound healing. Wound Repair and Regeneration 17(4): 461–472.

Pierce, G., J. Berg, R. Rudolph, J. Tarpley and T. Mustoe. 1991a. Platelet-derived growth factor-BB and transforming growth factor beta 1 selectively modulate glycosaminoglycans, collagen, and myofibroblasts in excisional wounds. Am. J. Pathol. 138(3): 629.

Pierce, G., T. Mustoe, B. Altrock, T. Deuel and A. Thomason. 1991b. Role of platelet-derived growth factor in wound healing. J. Cell. Biochem. 45(4): 319–326.

Pool, J. 1977. Normal hemostatic mechanisms: a review. Am. J. Med. Technol. 43(8): 776–780.

Poole, K., C. Nelson, R. Joshi, J. Martin, M. Gupta, S. Haws et al. 2015. ROS-responsive microspheres for on demand antioxidant therapy in a model of diabetic peripheral arterial disease. Biomaterials 41: 166–175.

Prabaharan, M. and J. Mano. 2006. Stimuli-responsive hydrogels based on polysaccharides incorporated with thermo-responsive polymers as novel biomaterials. Macromol. Biosci. 6(12): 991–1008.

Pu, H., W. Chiang, B. Maiti, Z. Liao, Y. Ho, M. Shim et al. 2014. Nanoparticles with dual responses to oxidative stress and reduced pH for drug release and anti-inflammatory applications. ACS Nano 8(2): 1213–1221.

Qing, C. 2017. The molecular biology in wound healing & non-healing wound. Chin. J. Traumatol. 20(04): 189–193.

Qu, J., X. Zhao, Y. Liang, T. Zhang, P.X. Ma and B. Guo. 2018. Antibacterial adhesive injectable hydrogels with rapid self-healing, extensibility and compressibility as wound dressing for joints skin wound healing. Biomaterials 183: 185–199.

Qureshi, M. and F. Khatoon. 2015. *In vitro* study of temperature and pH-responsive gentamycin sulphate-loaded chitosan-based hydrogel films for wound dressing applications. Polym. Plast. Technol. Eng. 54(6): 573–580.

Ramasastry, S. 2005. Acute wounds. Clin. Plast. Surg. 32(2): 195–208.

Ramezani, M. and M. Monroe. 2023. Bacterial protease-responsive shape memory polymers for infection surveillance and biofilm inhibition in chronic wounds. J. Biomed. Mater. Res. Part A 111(7): 921–937.

Rasool, A., S. Ata and A. Islam. 2019. Stimuli responsive biopolymer (chitosan) based blend hydrogels for wound healing application. Carbohydr. Polym. 203: 423–429.

Ren, H., Y. Wu, N. Ma, H. Xu and X. Zhang. 2012. Side-chain selenium-containing amphiphilic block copolymers: redox-controlled self-assembly and disassembly. Soft Matter. 8(5): 1460–1466.

Ribatti, D., A. Vacca, L. Roncali and F. Dammacco. 1991. Angiogenesis under normal and pathological conditions. Haematologica. 76(4): 311–320.

Ribeiro, A. and T. Flores-Sahagun. 2020. Application of stimulus-sensitive polymers in wound healing formulation. Int. J. Polym. Mater. 69(15): 979–989.

Richardson, M. 2004. Acute wounds: an overview of the physiological healing process. Nurs. Times, 100(4): 50–53.

Rivera, A.E. and J.M. Spencer. 2007. Clinical aspects of full-thickness wound healing. Clin. Dermatol. 25(1): 39–48.

Robson, M.C. 1997. Wound infection: a failure of wound healing caused by an imbalance of bacteria. Surg. Clin. North Am. 77(3): 637–650.

Robson, M.C. 2001. Wound healing: biologic features and approaches to maximize healing trajectories. Curr. Probl. Surg. 38: 61–140.

Roy, I. and M.N. Gupta. 2003. Smart polymeric materials. Chemistry and Biology 10(12): 1161–1171.

Rundhaug, J.E. 2005. Matrix metalloproteinases and angiogenesis. J. Cell. AMol. Med. 9(2): 267–285.

Safaee, M., M. Gravely and D. Roxbury. 2021. A wearable optical microfibrous biomaterial with encapsulated nanosensors enables wireless monitoring of oxidative stress. Adv. Funct. Mater. 31(13): 2006254.

Saghazadeh, S., C. Rinoldi, M. Schot, S.S. Kashaf, F. Sharifi, E. Jalilian et al. 2018. Drug delivery systems and materials for wound healing applications. Adv. Drug. Deliv. Rev. 127: 138–166.

Savencu, I., S. Iurian, A. Porfire, C. Bogdan and I. Tomuţă. 2021. Review of advances in polymeric wound dressing films. React. Funct. Polym. 168: 105059.

Schild, H.G. 1992. Poly(N-isopropylacrylamide): Experiment, theory and application. Progress in Polymer Science 17(2): 163–249.

Seow, W., G. Salgado, E. Lane and C. Hauser. 2016. Transparent crosslinked ultrashort peptide hydrogel dressing with high shape-fidelity accelerates healing of full-thickness excision wounds. Sci. Rep. 6(1): 1–12.

Serpico, L., S. Dello Iacono, A. Cammarano and L. De Stefano. 2023. Recent advances in stimuli-responsive hydrogel-based wound dressing. Gels 9(6): 451.

Seuring, J. and S. Agarwal. 2012. Polymers with upper critical solution temperature in aqueous solution. Macromol. Rapid Commun. 33(22): 1898–1920.

Sertznig, P., M. Seifert, W. Tilgen and J. Reichrath. 2008. Peroxisome proliferator-activated receptors (PPARs) and the human skin: importance of PPARs in skin physiology and dermatologic diseases. Am. J. Clin. Dermatol. 9: 15–31.

Shim, M. and Y. Xia. 2013. A reactive oxygen species (ROS)-responsive polymer for safe, efficient, and targeted gene delivery in cancer cells. Angew. Chemie. 125(27): 7064–7067.

Skover, G. 1991. Cellular and biochemical dynamics of wound repair, wound environment in collagen regeneration. Clin. Podiatr. Med. Surg. 8: 723–756.

Song, C., R. Ji, F. Du, D. Liang and Z. Li. 2013. Oxidation-accelerated hydrolysis of the ortho ester-containing acid-labile polymers. ACS Macro Lett. 2(3): 273–277.

Song, Z., Y. Wu, Q. Cao, H. Wang, X. Wang and H. Han. 2018. pH-responsive, light-triggered on-demand antibiotic release from functional metal–organic framework for bacterial infection combination therapy. Adv. Funct. Mater. 28(23): 1800011.

Stadelmann, W., A. Digenis and G. Tobin. 1998. Physiology and healing dynamics of chronic cutaneous wounds. Am. J. Surg. 176(2): 26S–38S.

Stan, D., C. Tanase, M. Avram, R. Apetrei, N. Mincu, A. Mateescu et al. 2021. Wound healing applications of creams and "smart" hydrogels. Exp. Dermatol. 30(9): 1218–1232.

Szycher, M. and S. Lee. 1992. Modern wound dressings: a systematic approach to wound healing. J. Biomater. Appl. 7(2): 142–213.

Takeshita, S., L. Zheng, E. Brogi, M. Kearney, L. Pu, S. Bunting et al. 1994. Therapeutic angiogenesis. A single intraarterial bolus of vascular endothelial growth factor augments revascularization in a rabbit ischemic hind limb model. J. Clin. Investig. 93(2): 662–670.

Taniyama, Y. and K. Griendling. 2003. Reactive oxygen species in the vasculature: molecular and cellular mechanisms. Hypertension 42(6): 1075–1081.

Thomas, S. 2008. Hydrocolloid dressings in the management of acute wounds: A review of the literature. Int. Wound J. 5(5): 602–613.

Trengove, N., S. Langton and M. Stacey. 1996. Biochemical analysis of wound fluid from nonhealing and healing chronic leg ulcers. Wound Repair Regen. 4(2): 234–239.

Trucillo, P. and E. Di Maio 2021. Classification and production of polymeric foams among the systems for wound treatment. Polymers 13(10): 1608.

Van Gheluwe, L., I. Chourpa, C. Gaigne and E. Munnier. 2021. Polymer-based smart drug delivery systems for skin application and demonstration of stimuli-responsiveness. Polymers 13(8): 1285.

Velnar, T., T. Bailey and V. Smrkolj. 2009. The wound healing process: an overview of the cellular and molecular mechanisms. J. Int. Med. Res. 37(5): 1528–1542.

Vowden, K. and P. Vowden. 2017. Wound dressings: principles and practice. Surgery (Oxford) 35(9): 489–494.

Wang, S., W. Wu, J. Yeo, X. Soo, W. Thitsartarn, S. Liu et al. 2023. Responsive hydrogel dressings for intelligent wound management. Biomed. Mater. Eng. 1(2): e2021.

Wilkinson, H. and M. Hardman. 2020. Wound healing: Cellular mechanisms and pathological outcomes. Open Biol. 10(9): 200223.

Williams, C. 1996. Tegasorb hydrocolloid dressing: advanced formulation. Br. J. Nurs. 5(20): 1271–1272.

Wilson, D., G. Dalmasso, L. Wang, S. Sitaraman, D. Merlin and N. Murthy. 2010. Orally delivered thioketal nanoparticles loaded with TNF-α–siRNA target inflammation and inhibit gene expression in the intestines. Nat. Mater. 9(11): 923–928.

Witte, M. and A. Barbul. 1997. General principles of wound healing. Surg. Clin. North Am. 77(3): 509–528.

Wu, D., J. Zhu, H. Han, J. Zhang, F. Wu, X. Qin and J. Yu. 2018. Synthesis and characterization of arginine-NIPAAm hybrid hydrogel as wound dressing: *In vitro* and *in vivo* study. Acta Biomater. 65: 305–316.

Xu, G., Y. Lu, C. Cheng, X. Li, J. Xu, Z. Liu et al. 2021. Battery-free and wireless smart wound dressing for wound infection monitoring and electrically controlled on-demand drug delivery. Adv. Funct. Mater. 31(26): 2100852.

Young, W. and I. Herman. 1985. Extracellular matrix modulation of endothelial cell shape and motility following injury *in vitro*. J. Cell Sci. 73(1): 19–32.

Zehrer, C., D. Holm, S. Solfest and S. Walters. 2014. A comparison of the *in vitro* moisture vapour transmission rate and *in vivo* fluid-handling capacity of six adhesive foam dressings to a newly reformulated adhesive foam dressing. International Wound Journal 11(6): 681–690.

Zhang, X., M. Qin, M. Xu, F. Miao, C. Merzougui, X. Zhang et al. 2021. The fabrication of antibacterial hydrogels for wound healing. Eur. Polym. J. 146: 110268.

Zhong, S., Y. Zhang and C. Lim. 2010. Tissue scaffolds for skin wound healing and dermal reconstruction. Wiley Interdiscip. Rev. Nanomed. Nanobiotechnol. 2(5): 510–525.

Zhu, H., J. Xu, M. Zhao, H. Luo, M. Lin, Y. Luo et al. 2022. Adhesive, injectable, and ROS-responsive hybrid polyvinyl alcohol (PVA) hydrogel co-delivers metformin and fibroblast growth factor 21 (FGF21) for enhanced diabetic wound repair. Front. Bioeng. Biotechnol. 10: 968078.

Zhu, J., Li. Faxue, Xu. Wang, Yu. Jianyong and Wu. Dequn. 2018. Hyaluronic acid and polyethylene glycol hybrid hydrogel encapsulating nanogel with hemostasis and sustainable antibacterial property for wound healing. ACS Appl. Mater. Interfaces 10: 13304–13316.

Chapter 7

pH-Responsive Polymer Prodrug Micelles for Anti-cancer Therapy

Shyam Vasvani,[1] *Rizia Bardhan,*[2,3] *Saji Uthaman*[2,3,*]
and *In-Kyu Park*[1,*]

Introduction

According to predictions from the Global Cancer Observatory (GLOBOCAN), the worldwide number of cancer cases is increasing daily. The most significant challenge in cancer treatment is the delayed diagnosis due to the absence of symptoms. The three conventional methods of anticancer therapy include radiation (Michon et al. 2022), chemotherapy (Krukiewicz and Zak 2016), and surgery (Nahm et al. 2021). Radiation therapy employs high-energy beams to destroy cancer cells and shrink tumors, but it is not always effective in selectively killing tumor cells. Chemotherapy, which entails treatment with one or more anticancer drugs, is utilized for the treatment of many primary and invasive tumors. However, traditional chemotherapy has several disadvantages, including poor absorption, chemoresistance, and severe side effects. Following surgical excision, cancer cells can migrate throughout the body, and the remaining tumor cells often become active, resulting in tumor recurrence and metastasis. However, a combination of diagnostic and therapeutic components with intelligent delivery systems that can track disease progression and deliver therapeutic agents to the tumor site selectively can overcome the limitations of conventional anticancer therapies.

[1] Department of Biomedical Sciences and BioMedical Sciences Graduate Program (BMSGP), Chonnam National University Medical School, Gwangju 61469, Republic of Korea.
[2] Department of Chemical and Biological Engineering, Iowa State University, Ames, IA 50012, USA.
[3] Nanovaccine Institute, Iowa State University, Ames, IA 50012, USA.
* Corresponding author: sajiuthaman@gmail.com, pik96@jnu.ac.kr

The tumor microenvironment (TME) is a multifaceted and continually evolving entity, encompassing blood vessels that supply nutrients and various TME cells, such as tumor cells, immune cells, and fibroblasts, embedded in the extracellular matrix (ECM) (Eble and Niland 2019). The TME exhibits unique characteristics, such as the permeability and discontinuity of tumor endothelial cells within the vasculature, hypoxia, low pH, and high interstitial pressure (Schaaf et al. 2018). Due to the rapid growth of the tumor, the blood vessels and junctions are inadequately developed, resulting in leaky vasculature. Through this permeable tumor vasculature, long-circulating nanosized drugs can selectively accumulate within the tumor and be retained in the tumor bed by decreased lymphatic outflow. This phenomenon is referred to as the "enhanced permeability and retention" (EPR) effect (Wu 2021). The EPR effect varies depending on the tumor type and location, ranging from microscopic animal tumors to human malignancies. Therefore, nanomedicine (NM) designs must be improved to utilize the EPR effect more efficiently for the treatment of human cancers. Understanding tumor vascular transport control is critical, and the EPR effect is widely recognized as one of the universal pathophysiological characteristics of solid tumors. It also serves as a crucial premise for the development and design of tumor-targeted delivery of anticancer medications.

Nanomedicine (NMs) offer a promising alternative for anticancer therapy due to their unique properties. Their small size allows them to target specific sites, enhancing drug bioavailability and reducing hazardous side effects (Rasool et al. 2022). Moreover, NMs can exploit distinct cancer physiologies to increase drug uptake and specificity. NMs can accumulate within tumors through passive and active targeting. In passive targeting, NMs exploit changes in the cancer vasculature and preferentially accumulate within tumors through the enhanced permeability and retention (EPR) effect. This property provides advantages over small molecules, as NMs can passively accumulate in tumors. In active targeting, NMs are attached to a targeting moiety that recognizes specific changes in cancer biology, such as upregulated markers. NMs rely more on the EPR effect and perfusion than small-molecule pharmaceuticals, leading to increased drug availability and improved physical combination therapies. Additionally, NMs can be engineered to respond to specific endogenous and external stimuli, making them a promising approach in the field of cancer therapy (Rasool et al. 2022, Golombek et al. 2018).

Role of nanotechnology or nanomedicine in anti-cancer therapy

In current cancer treatment, a combination of surgery, radiation therapy, chemotherapy, and immunotherapy are used depending on factors such as tumor size, grade, and phase (Kumari et al. 2021). However, the heterogeneity and complexity of the tumor microenvironment (TME) continue to be significant obstacles in delivering effective drugs to most cancers (Khot et al. 2021). To address this issue, it is crucial to understand the TME and develop more potent targeted treatments. One promising avenue is the use of nanomedicines (NMs) that offer state-of-the-art therapeutic advances for cancer treatment.

Nanoformulations of modified polymers and drug combinations have been developed as one of the most productive drug carriers for cancer therapies, with improved systemic stability, lower early leakage, reduced systemic toxicity, and more regulated drug-release kinetics. Many studies have investigated the use of polymer nanoparticles (NPs) for the administration of various medications, including gemcitabine and betulinic acid loaded with poly(lactic-co-glycolic acid) with poly(ethylene glycol) (PLGA-PEG) polymer NPs for improved anticancer efficacies in solid tumor models (Saneja et al. 2019), docetaxel (DTX) and resveratrol (RES) via copolymer methoxy poly(ethylene glycol)-poly(dl-lactide) (mPEG-PDLA) for improving the treatment of drug-resistant tumors (Guo et al. 2019), transferrin-conjugated polymeric NPs composed of poloxamer 407 (F127) and poloxamer 123 (P123) for receptor-mediated delivery of doxorubicin (DOX) in DOX-resistant breast cancer cells (Soe et al. 2019), paclitaxel (PTX) delivery with polyglutamic acid and paclitaxel peptide conjugate PGA-PTX-E-[c(RGDfK)2] that targets tumors, and ormeloxifene delivered by a pluronic polymer to induce impactful effects in pancreatic cancer cells (Eldar-Boock et al. 2011, Chauhan et al. 2020). Additionally, oxaliplatin co-loaded with hollow polydopamine nanoshells has been reported to alter drug resistance to enhance chemotherapeutic effects (Zhang et al. 2021a), while sulforaphane, a nature-derived component, has been shown to mediate glutathione depletion via cisplatin-incorporating polymeric micelles (NC-6004), which is currently in phase III clinical trials in Asian countries (Xu et al. 2019).

Despite the potential of polymeric NPs, conventional monotherapy still needs to overcome a few obstacles, such as low biocompatibility and less-effective tumor targeting by the medication, to participate in clinical trials for refractory illnesses. Modern cancer monotherapies and combinations of different monotherapies can somewhat mitigate the side effects of conventional cancer treatments, including pain, nausea, systemic toxicity, and nonspecific drug distributions. Various modified polymeric NPs with some near-infrared absorbance also support conventional monotherapies, such as chemotherapy and radiotherapy (Gong et al. 2013).

Need for prodrugs anti-cancer therapy

A prodrug is a compound that requires biotransformation before exhibiting its pharmacological effects, as defined by the International Union of Pure and Applied Chemistry (IUPAC) (Wermuth et al. 1998). Prodrugs have a moiety that is covalently bonded to the active drug molecule and works to improve the physicochemical, biological, and pharmacokinetic characteristics of the drug. Depending on the cellular locations where they are converted into the final active drug forms, prodrugs are categorized as type I or type II. In type I prodrugs, the biotransformation occurs intercellularly, while in type II, they occur intracellularly. Statins, which decrease cholesterol, and antiviral nucleoside analogs are some examples of type I prodrugs. Investigational new drug (IND) applications for type II prodrugs, such as fexofenadine, often include two tracks of toxicity profile studies: one on the prodrug and the other on the active drug (Wu and Farrelly 2007). Prodrugs consist of an active drug molecule that is covalently bonded to a moiety, which enhances the physicochemical, biological, and pharmacokinetic characteristics of the drug. Two types of prodrugs are recognized based on the location of biotransformation: type I prodrugs undergo biotransformation intercellularly, while type II prodrugs undergo biotransformation intracellularly. Examples of type I prodrugs include statins and antiviral nucleoside analogs. Investigational new drug (IND) applications for type II prodrugs, such as fexofenadine, require toxicity profile studies for both the prodrug and the active drug. Molecular prodrugs are crucial in many chemotherapy regimens, and they increase drug solubility, prolong *in vivo* circulation, and reduce the adverse effects of various anticancer treatments (Huo et al. 2014).

By focusing on specific markers that are overexpressed on cancer cells, prodrugs have been developed to enhance the specificity and efficacy of cancer treatments. To achieve this, various approaches have been researched and developed. One promising strategy is the combination of targeted prodrugs with the fabrication of multipurpose nanoparticles (NPs) for targeted cancer treatments (Chen et al. 2021). The main objective of prodrug use is to exploit the biological differences between cancer cells and healthy cells. Polymer-based prodrug delivery is currently receiving increased attention from researchers due to advances in nanotechnology-based prodrug manufacturing practices and the use of polymers for diverse applications, such as polypeptide drugs and poly-prodrugs with various stimuli, as depicted in Fig. 1. Several prodrug delivery methods have been widely used with polymer-based drug-delivery systems (DDSs). In Fig. 2, the hypoxia-triggered prodrug tirapazamine

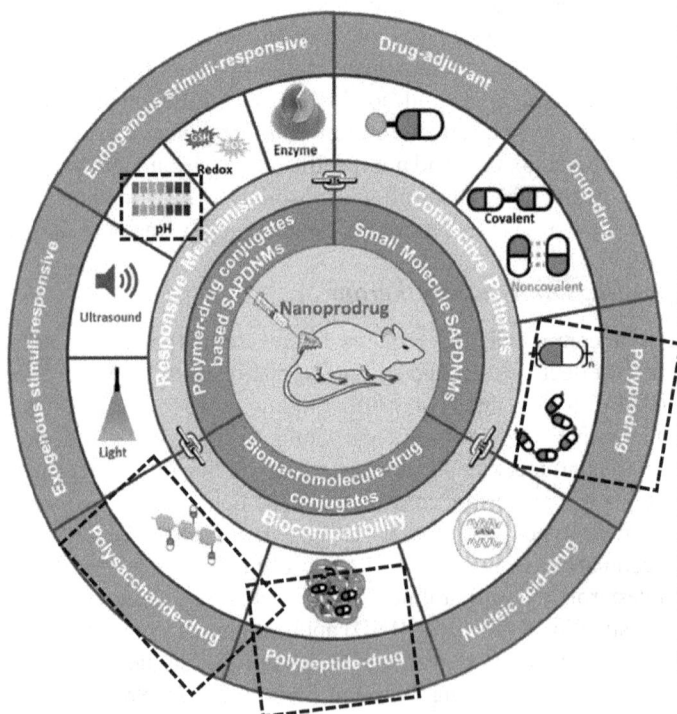

Fig. 1. Schematic representation for the fabrication of nano prodrug as newly developing cancer therapeutic formulations under various endogenous/exogenous stimuli and drug connectivity patterns. (Reprinted with permission from (Zhang et al. 2021) © 2021 The Authors. Advanced Science published by Wiley-VCH GmbH).

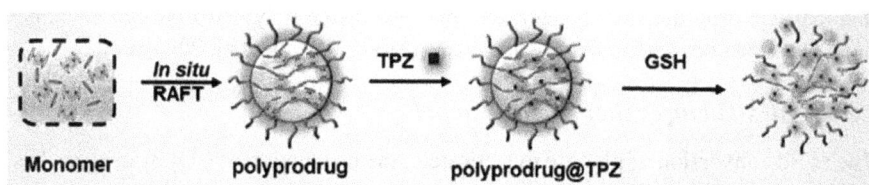

Fig. 2. The formation and drug release of the polyprodrug@TPZ nano gel (Reprinted with permission from (Guo et al. 2020)).

Table 1. FDA-approved anti-cancer prodrugs.

No.	Drug Name	Active Ingredient	Type of Cancer	Year of Approval	References
1	Isodax	Romidepsin	Lymphoma	2009	(Barbarotta and Hurley 2015)
2	Zytiga	Abiraterone acetate	Prostate cancer	2011	(Gardner et al. 2012)
3	Rituxan	Rituximab	Lymphoma	1997	(Grillo-Lopez et al. 2000)
4	Mylotag	Gemtuzumab ozogamicin	Acute myeloid leukemia	2018	(Norsworthy et al. 2018)
5	Erbitux	Cetuximab	Colorectal	2006	(Blasco et al. 2017)
6	NINLARO	ixazomib	Multiple myeloma	2015	(Shirley 2016)

(TPZ) was rationally designed for synergistic chemotherapy with platinum-based polyprodrug delivery, with reactive oxygen species (ROS) serving as the stimulus response.

This book chapter provides a comprehensive overview of the potential applications of prodrugs in targeted cancer therapy, with a particular focus on pH-based environments. The advantages of using prodrugs and the use of a polymeric nanomicelle-based approach are examined, along with current updates in anticancer research. Table 1 displays some FDA-approved anticancer prodrugs for various types of cancers and their active forms with the year of approval.

Chemistry behind pH-responsive drug delivery

Studies have shown that cancer cells tend to thrive in acidic environments and struggle to survive in typical, more alkaline environments. This link between pH levels and cancer has been demonstrated in several studies. Due to their abnormal metabolism compared to normal tissues, cancers are known to create extracellular microenvironments with pH levels of 6.5–7.0. Specific pH variations have been observed at the subcellular to tissue levels, highlighting the need for research on pH-responsive materials for various biological applications, such as sensing, diagnostics, tissue engineering, and targeted drug delivery for cancer and other inflammatory conditions (Liu and Thayumanavan 2017). Smart pH-responsive nanoparticle drug delivery systems and pH-responsive polymer-prodrug conjugates have also been developed for use in cancer therapy (Zhang et al. 2021b).

Chemical structures that respond to pH

The rapid conversion of glucose to lactic acid, the production of CO_2 by cancer cells, and the presence of abnormal vascular-type proton pumps in solid tumors are all known to contribute to the reduction of pH levels in tumor tissues compared to normal tissues (Zhang et al. 2021). Normal tissues typically have a pH level of approximately 7.4, while tumor tissues range from 5.5 to 7.0. The pH levels in subcellular organelles are even more acidic than those in tumor tissues. The cytoplasm has a pH of around 7.0, while the pH of endosomes and lysosomes ranges from 5–6 and 4–5, respectively (as shown in Fig. 3A) (Ding et al. 2022a). In such acidic environments, various chemical bonds, such as esters and amide linkers, can hydrolyze and activate the prodrug, eventually releasing the parent medication. On the other hand, pH-sensitive functional groups, such as pH sensitive biodegradable linkage, acid-labile chemical linkages (Fig. 3B), and acid-unstable inorganic materials are considered ideal catalysts for creating DDSs for cancer therapies. Due to the activation of the linkers by acidic environments and subsequent cleavage-cascaded chemical-structure modifications, orthoesters (Sartaj et al. 2021), hydrazone (Xu et al. 2021), acetal/ketal (Domiński et al. 2020), imine (Chen and Liu 2022), and cis-aconite (Peng et al. 2021) are some of the acid-labile chemical linkages used to generate pH-sensitive prodrugs (as summarized in Table 2). These acid-labile chemical linkages of prodrugs dissociate upon exposure to acidic environments, reverting them to their initial states. Below, we discuss some examples of

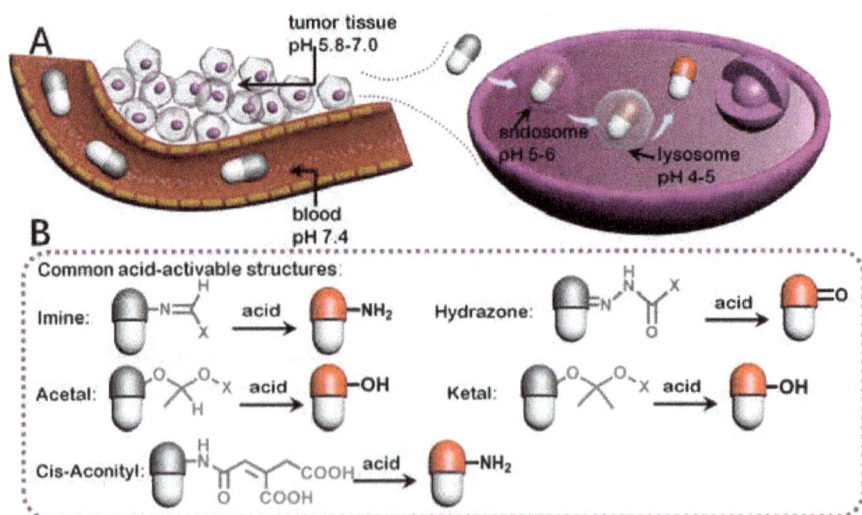

Fig. 3. Biodegradable bond-based polymers utilized in drug delivery systems are depicted in Figure. The degradability of the polymeric carrier is determined by biodegradable linkages. (Reprinted with permission from (Ding et al. 2022) American Chemical Society).

Table 2. Chemical bond activation with polymeric prodrug micelles.

Bond Activation	Parent Drug	Application	Reference
Imine hydrolysis	MTX	As an anti-cancer drug study in HeLa cells	(Wang et al. 2019b)
Hydrazone hydrolysis	MTX/DOX	Multiple drugs synergistic anti-cancer therapy	(Zhu et al. 2021)
Benzamide bonds	siVEGF & siPIGF	Lung metastasis (4T1) tumor	(Kumar and Rajnikanth 2020)
Ortho-ester hydrolysis	IND	Combination therapy	(Wang et al. 2019a)
Amide bonds	Oxaliplatin, NLG919	4T1 tumor	(Guo et al. 2022)
Metal–ligand bonds	OVA, CpG	B16 melanoma	(Bera et al. 2022)
Silyl ether hydrolysis	Gemcitabine	Acid-dependent prodrug activation	(Pandit and Royzen 2022)

acid-activated prodrugs to understand their pH-sensitive drug release with various chemical bonds.

Imine

To improve the specificity and effectiveness of cancer treatments, prodrug conjugates have been developed by linking a drug to a carrier molecule through a pH-sensitive linker. One such linker is the imine bond, which is formed between the amino group of the drug and the aldehyde group on the carrier molecule. The imine bond is pH-sensitive, meaning it can be cleaved in the acidic environment of tumor cells, releasing the active drug [as shown in Fig. 4 (Ahmad et al. 2019)]. Rolitetracycline is an example of an imine derivative used in intravenous administration. This book chapter also discusses the application of imine bond formation in the synthesis of a prodrug conjugate using DOX and a copolymer.

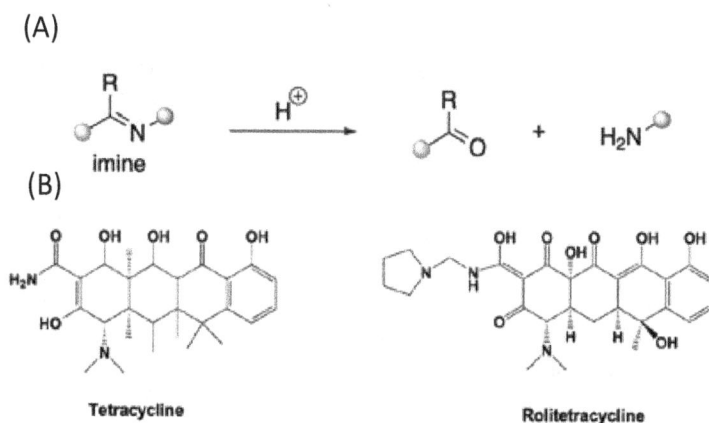

Fig. 4. (A) pH-responsive linkages and the corresponding hydrolyzed products. (B) Chemical structure of imine bond-based prodrug.

Acetal and ketal bond

Acetals are a type of geminal-diether aldehyde or ketone derivatives produced by the reaction of two equivalents (or more) of alcohol with the exclusion of water. Under basic conditions, acetals and ketals are known to be relatively stable, but in an acidic environment, they hydrolyze easily to the equivalent carbonyl compound (aldehyde and ketone) and alcohol, as shown in Fig. 5A.

Acetals and ketals have been identified as potential linkers for the development of acid-responsive prodrugs. For instance, the original drug Gemcitabine (GEM) was conjugated to the side chains of a poly(amino acid) using a cyclic acetal bond to produce PGEM (Takemoto et al. 2020). Acetals and ketals have gained significant interest as pH-sensitive linkers because they generate charge-neutral and perhaps harmless byproducts upon cleavage. Several series of acetal and ketal linkers have been designed and investigated for their relative pH sensitivities. As illustrated in Fig. 5B, both acetal- and ketal-based pH-sensitive polymeric prodrugs can deliver drug payloads at tumor-specific sites (Liu and Thayumanavan 2017).

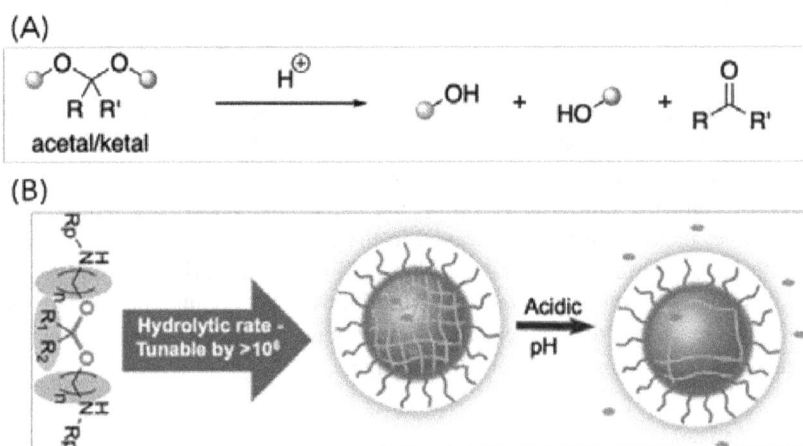

Fig. 5. (A) pH-responsive linkages and the corresponding hydrolyzed products and (B) Subsequent effects on the pH sensitivity of acetals and ketals and their correlation with encapsulation stability in polymeric nanogel (Reprinted with permission from (Liu and Thayumanavan 2017)).

Hydrazone bond

The hydrazone molecule can undergo decomposition to create a carbinolamine intermediate, which can further decompose into hydrazine and aldehyde/ketone moieties. Hydrazone linkages are sensitive to pH and are used in drug delivery systems because they hydrolyze more rapidly at acidic pH than at neutral physiological pH, as shown in Fig. 6A. Greenfield et al. (Sonawane et al. 2017) studied several hydrazone connections created by conjugating the ketone group of the anticancer medication DOX to hydrazides. The resulting double hydrazone bond of the unimolecular conjugate improves the pH sensitivity and drug loading of the conjugate-based micelles. As per Fig. 6B, researchers (Sonawane et al. 2017)

(A)

(B)

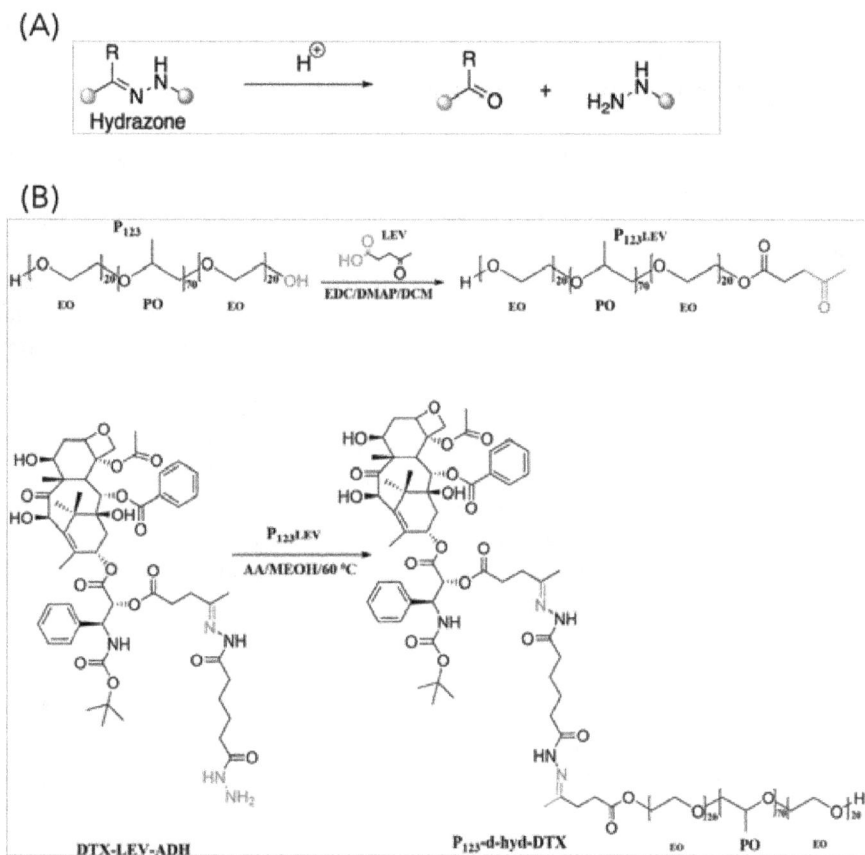

Fig. 6. (A) pH-responsive linkages and the corresponding hydrolyzed products and (B) The synthesis scheme of polymeric micelles conjugate containing double-hydrazone bond. (Reprinted with permission from (Sonawane et al. 2017) © 2016 Elsevier B.V.).

have developed a pH-responsive amphiphilic polymer-drug conjugate by connecting the hydrazide group of the mPEG portion of the copolymer to the ketone group of levulinic-acid-modified curcumin (mPEG-PLA-Hyd-Cur). Similar pH-sensitive linkers and bond cleavage applications are listed in Table 2.

pH-responsive polymeric prodrug micelles in drug delivery

The development of pH-sensitive polymeric prodrug micelles has gained significant attention due to the importance of pH stimuli at both physiological and subcellular levels. Research in this field has mainly focused on internal-stimulus-based drug delivery systems (DDSs) that utilize pH stimuli. These systems typically utilize bonds that can be easily broken in acidic environments, thereby facilitating rapid release of the therapeutic cargo (Vallet-Regí et al. 2022). Acid-labile linkages are particularly beneficial for effective drug and gene delivery since they allow the nanocarriers to escape from the endosomes (Zou et al. 2015). This escape is critical in preventing payload degradation in the lysosome or endosome, which have pH values of around 4.5 and 6.8, respectively. Several studies have demonstrated the design of various acid-labile nanoparticle delivery systems to achieve endosome escape via endosomal rupture, membrane instability, and membrane fusion.

The linker chemistry plays a crucial role in the design of polymeric prodrug combination therapy as it enables stimuli-based release of active components during the cleavage process. Insufficient linker stability can cause a conjugated drug to release too quickly and lose the benefits of its macromolecular carrier. On the other hand, failure to activate the linker as expected can result in insufficient therapeutic efficacy in the case of irregular drug release from the polymeric prodrug. Therefore, maintaining a pharmacological release profile is crucial for achieving extended therapeutic effectiveness. Despite the numerous and intricate challenges involved in the practical application of polymer therapeutics, the methods currently used to create polymer-drug conjugates are mostly empirical. To address these challenges and advance the therapeutic effect of prodrug carriers, a bottom-up design approach that starts with the active quantitative design requirements is necessary, particularly considering the poor predictability of pre-clinical models for clinical outcomes. The following sections of this chapter will discuss the mechanism of acid-triggered drug release in the tumor microenvironment (TME).

Acid-triggered drug release

In order to utilize cationic polymers and liposomes, Nie et al. (2011) developed a programmable lipo-polyplex system that tightly compacts DNA through the complexation of liposomes and polyethyleneimine (PEI), as shown in Fig. 7. To protect the liposomes, a combination of pH-cleavable PEG-hydrazone and cholesterol with the helper lipid dioleoyl phosphatidylethanolamine (DOPE) is employed. In the extracellular compartment under neutral physiological conditions, this lipo-polyplex is expected to remain stable. However, in environments with lower pH such as certain malignancies or endosomes, the pH-sensitive hydrazone linkage is broken, leading to the shedding of the PEG layer, and reinstating the endosome destabilizing function of the lipo-polyplex components, resulting in cytosolic release. A ligand can also be attached to the system for targeted receptor recognition and internalization. Several studies have demonstrated the potential of pH-sensitive lipo-polyplexes targeting the transferrin receptor for delivering therapeutic molecules such as the ribonucleotide reductase antisense oligonucleotide GTI-2040. In primary acute myeloid leukemia (AML) cells, these polyplexes enhanced the activity of cytarabine by 45% (Szulc et al. 2016), and similar improvements were observed with miR-29b administration. Other studies have investigated DOX-encapsulated Prussian blue nanoparticles (NPs) targeting dual cluster of differentiation 44 (CD44) and C-X-C chemokine receptor 4 (CXCR4), which showed increased cellular uptake in leukemic cells and reduced AML cell proliferation in a xenograft mice model. Given the potential of pH-responsive systems in cancer therapy, efforts are being made to develop novel pH-sensitive materials for drug delivery. However, many existing drug-delivery polymers are vulnerable to hydrolytic cleavage under acidic pH conditions, which can result in toxic byproducts. Therefore, research is ongoing to create biodegradable,

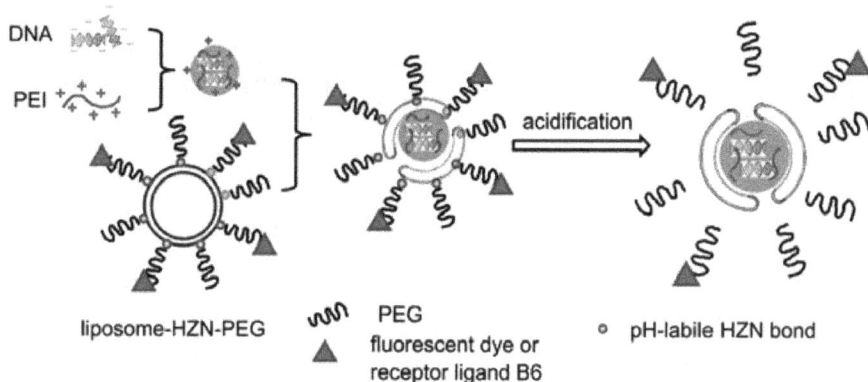

Fig. 7. The synthesis scheme of polymeric micelles conjugate containing double hydrazone bond (Reprinted with permission from (Nie et al. 2011)).

pH-responsive polymers that do not produce harmful substances. A PEG carrier is used to couple an inactive prodrug of unmodified human growth hormone (GH) using a hydrazone (HZN) linker to create long-circulating GH that is more manageable with a single weekly dose and provides an exposure equivalent to daily GH administration through physiological pH- and temperature-dependent hydrolysis. One potential strategy for achieving high local concentrations of active drugs and reducing their off-target toxicity is to trigger site-specific release using the chemical and biological specificities of the target indication. To achieve this, many linker chemistries have been developed, such as combining DNA and PEI to form complexes that are incubated with liposomes to generate cationic polyplexes (Jerzykiewicz and Czogalla 2022). These liposomes contain pyridyl-hydrazone-based cholesterol-polyethylene glycol (Chol-PEG), which allows dynamic deshielding in acidic environments, such as endosomes, and shielding in extracellular compartments.

Mechanism of pH-responsive polymeric prodrug micelles

Cancer cells typically endocytose NPs and ingest them into the acidic lysosomal compartments (early at ~ pH 6.5 and late at ~ pH 4.5) and fused at pH 4.5, containing hydrolytic lysosomal enzymes. One or more of the following pathways often accomplish endosomal escape, for example, the entry and escape of the drugs into the cytoplasm have been explored and shown in Fig. 8. The effective endosomal escape of pH-based NPs allows drug transfer into the cytosol. On the other hand, because they may fuse with cellular membranes directly, NPs enable the drug to be released directly into the cytosol without the need for endosomes, as shown in Fig. 8 (Shinn et al. 2022).

Micelles are stable colloids, generated when a hydrophilic material, with a hydrophobic chain, self-assemble in water. As the micelle core is naturally hydrophobic, hydrophobic drugs can be easily loaded into the core, shielding them from environmental media and supporting prolonged drug circulation in the body. Hence, polymeric micelles are potential nanocarriers owing to their high stability,

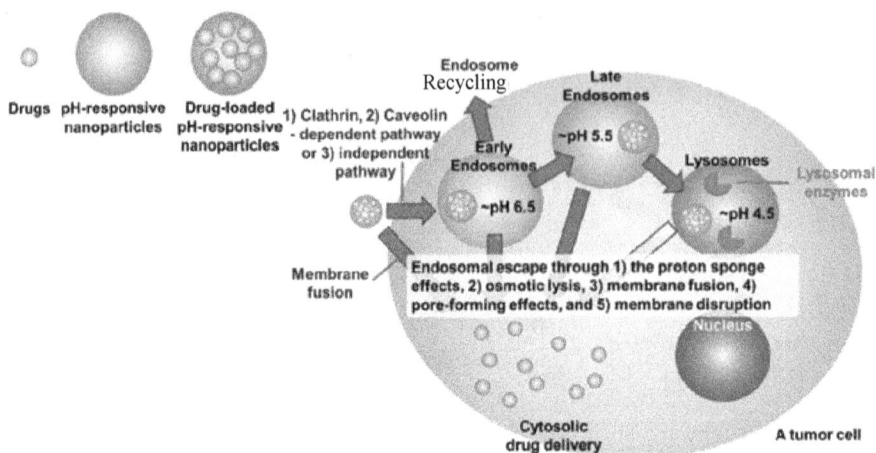

Fig. 8. Schematic mechanism for intracellular delivery of pH-responsive nanoparticles. (Reprinted with permission from (Mitchell et al. 2021, Shinn et al. 2022)).

higher drug-loading content, and prolonged circulation endurance. One of the means to address the issue of unwanted drug leakage during systemic circulation is through development of stimuli-responsive micelles. Among the stimuli-responsive micelles, pH is the stimulus that is used most frequently. Drug leakage can be reduced by securing drugs within micelles via different bonds, such as pH-based chemical bonding (Table 2). A pH-responsive drug or prodrug is a type of medication that exhibits changes in its properties, such as solubility, stability, or bioactivity, in response to changes in the pH (acidity or alkalinity) of its environment. This allows the drug to be selectively activated or released in specific areas of the body that exhibit characteristic pH levels, such as the gastrointestinal tract, tumor tissues, or intracellular compartments. In a tumor environment, micelles with pH-responsive drug or prodrug react differently to various pH levels. In physiological conditions, pH-sensitive micelles may release medications gradually or not at all, but they may release them quickly in an acidic environment. Currently, some of the most promising anticancer DDSs, with the pH-sensitive micelle, may adapt to the TME by reducing the systemic toxicity of the drug and increasing the drug bioavailability (Zhang et al. 2020b).

Polymeric micelles are formed by self-assembly of amphiphilic block copolymers in an aqueous environment due to hydrophobic and ionic interactions between the polymer blocks. They are being investigated as potential nanocarriers for drug delivery, offering several appealing features such as the ability to solubilize water-insoluble drugs in the hydrophobic core, high solubility, minimal toxicity, and passive tumor targeting via the EPR effect. Additionally, hydrophilic surfaces of the micelles can be modified with functional groups such as cell-penetrating peptides and specific ligands like monoclonal antibodies to enhance intracellular absorption. Polymeric micelles that are pH-sensitive can selectively release their payload in the acidic tumor microenvironment, which has a pH typically ranging from

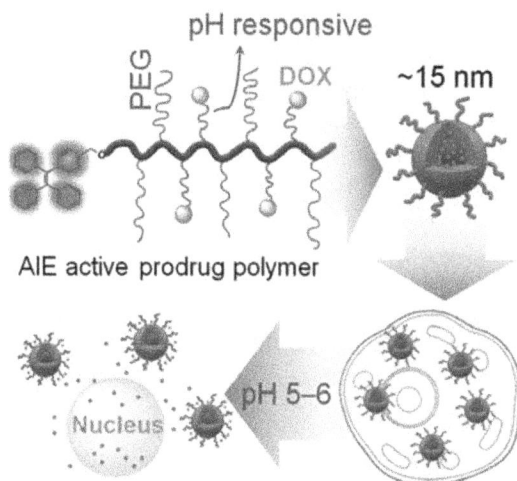

Fig. 9. pH responsive AIE based polymeric prodrug (Reprinted with permission from (Kalva et al. 2021)).

~ 5.5–7 (Cao et al. 2019). Cancer therapies increasingly rely on polymeric micelles that can provide simultaneous bioimaging, drug delivery tracking, and effective medicine administration to sites. In Fig. 9, ring-opening polymerization and click reactions were combined to synthesize a new biodegradable polycarbonate brush-like polymer that is pH-responsive and exhibits aggregation-induced emission (AIE). To covalently attach the pH-sensitive anticancer drug DOX to the polymer backbone, an acid-sensitive Schiff base linkage was used. Transmission electron microscopy and dynamic light scattering studies confirmed that the resulting prodrug polymer self-assembled into nanomicelles in an aqueous solution. The prodrug micelles were identified via fluorescence imaging due to their AIE characteristics. *In vitro* drug release experiments demonstrated that DOX was released more quickly in acidic conditions (pH 5.0) than in physiological conditions (pH 7.4). Finally, the integration of intracellular imaging and cancer therapy using these novel prodrug micelles shows considerable promise (Kalva et al. 2021).

The release of bioactive substances, such as drugs or prodrugs, from nanocomposites was evaluated at pH 4.8 (lysosomal) and pH 7.4 (blood). The initial burst release of the drug during the first few hours at pH 4.8 may be attributed to the partial breakdown of two potential hydrogen bonds between the drug and nanocarrier, which are associated with drug bound to the nanocomposite surface (Buskaran et al. 2021). The subsequent section of this article describes the mechanism underlying the release of the bioactive drug moiety and current applications of pH-responsive polymeric prodrugs in the tumor microenvironment.

Protonation and deprotonation

Several drug delivery systems (DDSs) contain functional groups capable of engaging in protonation activities to alter the hydrophilic-hydrophobic equilibrium at the tumor site. These functional groups often have low acidity or alkalinity. Figure 8 schematically illustrates some of the protonatable groups commonly found in literature, which will be briefly discussed below. Protonation is one of the most common mechanisms for pH-dependent drug release. The protonation and deprotonation of functional groups in polymers can change the physical properties or structure of the material, such as its stability, solubility, surface charge, and chain conformation (Ding et al. 2022b). Early pH-sensitive nanomaterials (NMs) were created by altering the ionization of carboxylic acid (-COOH) or amine (-NH_2) groups. In response to pH changes, these nanocarriers expand to release their medication. To create pH-sensitive DDSs, anionic polymers with -COOH groups are typically used. These -COOH groups would indeed result in the removal of charge from the group, rather than adding a charge to it. This can lead to a decrease in the overall polarity of the molecule, due to which it becomes less polar compared to the carboxylate anion (Ofridam et al. 2021), as explained below.

Amine (-NH_2) group: Amines are classified as primary, secondary, or tertiary based on the number of hydrogen atoms in the ammonia molecule that are replaced with alkyl groups. These functional groups often result in increased hydrophilicity of the molecule upon protonation. The rate of DOX release increases significantly when the pH drops from 7.4 to 5.0. In one study, researchers activated the amine group by ionizing tertiary amine groups in N,N-diethylaminomethyl methacrylate (DEAEMA) in the intermediate layer of the micelles. The *in vitro* cytotoxicity of DOX-encapsulated micelles to HepG2 cells (hepatocellular carcinoma) showed enhanced anticancer activity and bioavailability compared to free DOX, suggesting that a lower drug dose could be used for therapeutic purposes (Karimi et al. 2016).

Carboxyl (-COOH) group: Some DDSs contain carboxyl groups, which are another type of acid-responsive group that can promote drug release by altering the hydrophobic-hydrophilic balance of the system. At low pH conditions, the protonation of the carboxyl group leads to polarity changes in carboxyl ion, which mimics hydrophobic nature and causes the surface to become more hydrophobic. On the other hand, under neutral or high pH conditions, the carboxyl group undergoes deprotonation, resulting in hydrophilicity. This property can be useful when delivering nanomaterials by exploiting leaky vasculature to achieve selective tumor accumulation with nanoparticles. Under acidic conditions, the carboxyl group can be protonated, causing the polymer to form larger aggregates that enhance micelle retention in the tumor matrix and cancer cell uptake.

Polyacrylic acid (PAA) is a pH-responsive polymer that exhibits high hydrophobicity. In acidic conditions, it can self-interact and form precipitates, but in neutral or alkaline environments, it deprotonates and becomes hydrophilic. Charge exchange of a nanocarrier may lead to protonation or deprotonation of the nano DDS at various pH levels. The drug in the nanocarriers is released upon contact with tumor cells in the TME after the protective layer is removed, making it easier to target. Charged polymers are classified as cationic, anionic, or zwitterionic based on their electric charge. Recent research has focused on noncovalent and metal coordination bonds. While pH-triggered chemical bonding can sustain fundamental stability in neutral environments, hydrolytic fracture can hasten the release of medicines from nanoscale prodrugs in acidic environments. Most cationic polymers contain ionized

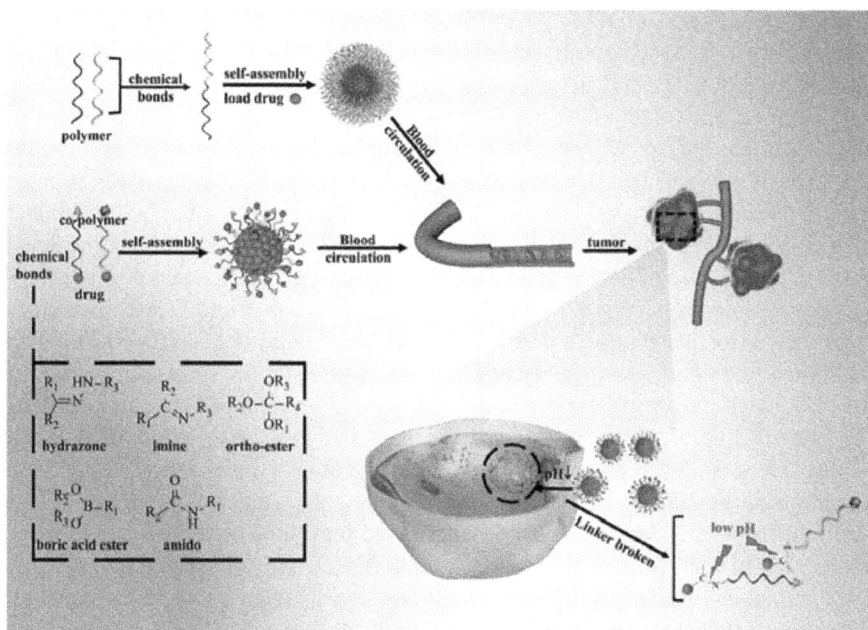

Fig. 10. Drug release mechanism after pH based chemical bond breakage. Reprinted with permission from (Ding et al. 2022).

amino groups, while ionized organic acids are present in anionic polymers, and both cationic and anionic groups are present in zwitterionic polymers (Ding et al. 2022). As illustrated in Fig. 10, negatively charged NPs are more compatible with cells than neutral or positively charged ones, but they may be eliminated from the bloodstream during systemic circulation upon interacting with blood serum components. In a pH 7.4 environment, charge-convertible polymers may still be negatively charged, which limits their ability to bind to certain serum components and prevents them from being scavenged by the reticuloendothelial system. The acidic environment of tumor tissue changes the negative charges or neutral nature of these polymers into positive charges, enhancing their absorption by tumor cells. Chitosan can protonate in low-pH solutions because its amine groups have a pKa value of 6.5.

Current application of pH responsive prodrug micelles

Acid-sensitive dimeric prodrug-based nano-systems (DPNS)

The acidic tumor microenvironment (TME) is a noteworthy target for tumor-specific treatment and significantly impacts tumor development and spread. Several recent examples of pH-based chemical couplings include the hydrazone bond (Yang et al. 2016), acetal bond (Roth and Lowe 2017), and silyl ether link (Kanamala et al. 2016). The hydrazone link is frequently used in the development of pH-sensitive drug delivery systems (DPNS). For instance, hydrazone bonds can be used to conjugate two molecules of the anticancer drug doxorubicin (DOX) to create an acid-labile dimer of DOX. When the DOX dimer is self-assembled into DPNS, it displays pH-responsive drug-release capabilities. Due to its exceptional cellular absorption efficiency and pH-triggered drug-release capabilities, the DOX dimer nanoassembly effectively kills cancer cells (Cao et al. 2019, Li and Liu 2019). In addition to the hydrazone bond, several dimeric prodrugs linked with ester bonds also showed pH-based drug-release properties. For example, irinotecan was coupled with the cytotoxic chemical enediyne to create an ester-bonded amphiphilic dimeric molecule. The amphiphilic dimer demonstrated a 100% bioavailability of drugs and pH-sensitive release of the drug characteristics, and it could self-assemble into DPNS without the aid of a carrier molecule. All-trans retinoic acid (ATRA) molecules were joined to phosphorylcholine to form an ester-bond-linked dimeric lipid in addition to the dimer structure. Acidic stimuli may trigger the release properties of the phosphorylcholine-drug dimer, resulting in the formation of liposomes. Most of these readily hydrolysable ester bonds were integrated into the structures of nonpolar conjugates, Despite the presence of pH-responsive drug release capabilities in these dimeric prodrugs linked by ester bonds, it is noteworthy that the majority of these easily hydrolyzed ester bonds were incorporated into the structure of amphipathic conjugates. In contrast, hydrolyzing ester bonds in hydrophobic conjugates proves to be quite challenging, leading to significantly slower drug release rates and suboptimal therapeutic efficacy. Since the ester bond is a pH-sensitive connection used to create hydrophobic dimeric conjugates, it is critical to ascertain if such medicines can be suitably activated from the terminal active sites of the dimers (Luo et al. 2016). The terminal active sites in pH-sensitive hydrophobic dimeric conjugates with ester bonds refer to the specific positions within the molecule where the ester bond is located. These sites play a critical role in ensuring the appropriate activation and release of the active drug from the conjugate when exposed to the appropriate pH conditions.

In drug delivery applications using hydrophobic dimeric conjugates, it is crucial to assess whether the pharmaceuticals can be effectively activated at their intended action sites when ester bonds are used as pH-sensitive links. This requires examining the stability of the ester bond across a range of pH values experienced by the conjugate *in vivo*, including physiological pH (around 7.4) and acidic pH (below 7.4). If the ester bond can be selectively broken at acidic pH while remaining stable at physiological pH, this property may be desirable for drug delivery to specific sites such as tumor cells, which exhibit an acidic microenvironment that triggers drug release.

To ensure successful drug activation and therapeutic efficacy at the intended sites, various factors such as ester hydrolysis rate, drug stability upon release, and pharmacokinetics of the conjugates should also be carefully evaluated. Therefore, in the manufacture of hydrophobic dimeric conjugates for drug delivery, the sensitivity of the ester bond to pH must be rigorously examined to ensure their efficacy and safety.

Multifunctional prodrug-based polymeric micelle

Mao and coworkers developed pH-responsive delivery of a block-copolymer-based prodrug poly (PEGMA-co-SEMA)-doxorubicin (BCP-DOX) conjugate by imine bond hydrolyzed at pH < 5.5 and release of DOX at the target site. This drug-delivery method can synergistically accomplish improved cellular uptake and regulate DOX release by combining the pH-induced charge conversion and pH-triggered breaking of chemical bonds (Mao et al. 2016).

In this study, the authors utilized pH-responsive nanocarriers that are sensitive to changes in charge, shifting from a negatively charged to a positively charged state in the slightly acidic extracellular environment of tumors (as shown in Fig. 11A). This charge conversion resulted in significantly improved cellular absorption efficiency and greater effectiveness in killing cancer cells. The authors demonstrated that positively charged nanocarriers exhibit a greater propensity to attach to negatively charged cell membranes, leading to increased internalization by the cells. As a result, this system can change the charge of the nanocarrier from negative to positive in response to the intracellular pH environment. Through a combination of pH-triggered cleavage and pH-induced charge conversion (as depicted in Fig. 11B), the authors achieved enhanced cellular uptake and controlled release of prodrugs using self-assembled polymeric micelles as the nanocarriers. These findings demonstrate the synergistic effects of pH-triggered mechanisms, highlighting the potential of pH-sensitive nanocarriers for drug delivery applications.

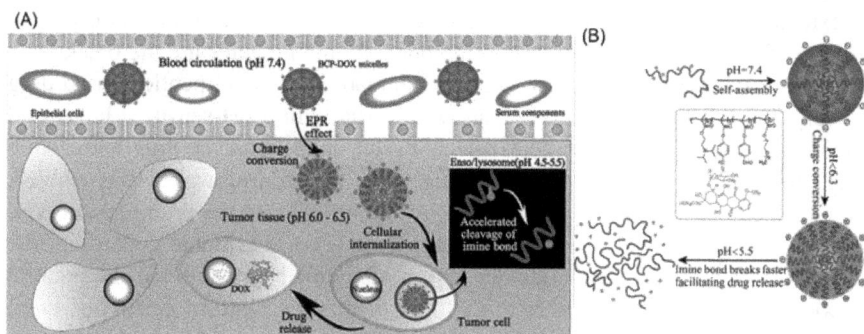

Fig. 11. (A) Scheme of dual-pH responsive feature to precisely deliver DOX to the cancer cells and (B) Structural mechanism of pH targeted charge conversion and drug release (Reprinted with permission from (Mao et al. 2016)).

Imaging application of pH-based polymer prodrug micelles

Lastly, after exploring multifunctional pH-based prodrug delivery, we consider the imaging application of pH-based polymeric micelles.

Fluorescence sensing technology is used for *in vivo* monitoring of tumor tissue growth due to its high sensitivity, resolution, simplicity, and noninvasiveness. Nanomaterials M can be used to deliver drugs to cancer-specific sites while also offering on-site fluorescent imaging and treatment. Carbon dots (CDs) have been successfully employed in drug delivery and fluorescence imaging applications due to

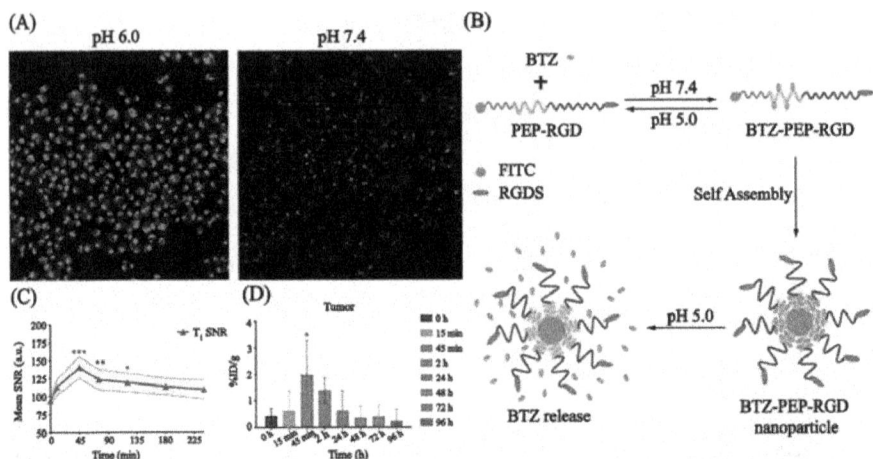

Fig. 12. (A) Biomedical imaging by DOX-CDs@LCP pH based Fluorescence confocal microscopy. (B) Structure of a peptide-based prodrug to target cancer. (C) Quantification data of NPs by SNR (Signal to noise ratio) and (D) ICP-MS. (Reprinted with permission from (Yao et al. 2021)).

their excellent photosensitive stability, biocompatibility, and low toxicity (Yao et al. 2018). The fluorescent dye 1,8-naphthylamine was delivered using CDs made from glutathione (FP) (Ansari et al. 2021), as illustrated in Fig. 12A.

Foreign polymers or inorganic nanoparticles administered to humans over an extended period may cause injuries or create serious adverse effects owing to buildup of the breakdown products in the body. As amino acids are the breakdown products of peptides, which have good biocompatibility, novel peptide-based drug delivery is used to overcome this issue. A prodrug with a pH-sensitive linker and peptide was joined to boronic acid–catechol ester bonds with the drug bortezomib (BTZ) to mark the bionic peptide. To create NPs, the prodrug was self-assembled. Fluorescence imaging was made possible by adding FITC to the prodrug, and BTZ was released in a pH-sensitive environment (Fig. 12B). Additionally, 45 min after injection, MnO_2 loaded BSA (Bovine serum albumin) NPs exhibited the greatest T1 weighted image (also referred to as T1WI or the "spin-lattice" relaxation time) magnetic resonance (MR) signal, which may be the ideal imaging time point to track the therapy response (Fig. 12C, D). The pH sensitivity of the prodrug made it easier for the peptide-based DDS to target cancer cells (Santhosh and Chandrasekar 2020). The pH sensitivity of the prodrug enabled targeting of cancer cells by exploiting the acidic tumor microenvironment. The pH-sensitive functional groups in the prodrug selectively activate and release the active drug in response to the acidic pH, which is characteristic of cancer cells. The peptide-based DDS facilitates the targeted delivery of the pH-sensitive prodrug to cancer cells, further enhancing the localization and effectiveness of the treatment.

Conclusion

In various pathological disorders, pH is an essential functional parameter that can be exploited for effective treatment. Nanomaterials can be easily designed to respond to intrinsic pH changes within tumors. Furthermore, active targeting moieties can be added to the surfaces of nanocarriers to reduce the adverse outcomes due to systemic toxicity. pH-responsive nanomicelles are an attractive option for delivering therapeutic and diagnostic compounds to diseased tissues. Several acid-sensitive chemical bonds, noncovalent bonds, and protonation and deprotonation groups have been utilized to generate pH-responsive nanocarriers. Hybrid nanoparticles with pH-responsive nanomicelles have been developed using organic, inorganic, or a combination of both components. Prodrugs based on drug-loaded pH-responsive nanomicelles have been used to diagnose and treat various disorders such as cancer, inflammation, antimicrobial infections, and neurological diseases. Table 3 summarizes the different types of prodrugs that are currently undergoing clinical

Table 3. Clinical update of prodrug with different cancer.

Study Title	NCT No.	Interventions	Cancer Type	Reference
Docetaxel-polymeric Micelles (PM) and Oxaliplatin for Esophageal Carcinoma	NCT03585673	Docetaxel-PM Oxaliplatin	Metastatic Cancer	(Oh 2021)
A Clinical Trial of Paclitaxel Loaded Polymeric Micelle in Patients With Taxane-Pretreated Recurrent Breast Cancer	NCT00912639	Polymeric micelle loaded with paclitaxel	Breast Cancer	(Group)
A Study of Docetaxel Polymeric Micelles for Injection in Patients with Advanced Solid Tumors	NCT05254665	Docetaxel Polymeric Micelles for Injection	Solid Tumors	(Atrafi et al. 2020)
Paclitaxel-Loaded Polymeric Micelle and Carboplatin as First-Line Therapy in Treating Patients with Advanced Ovarian Cancer	NCT00886717	carboplatin paclitaxel-loaded polymeric micelle	Ovarian Cancer	(Khalifa et al. 2019)
Study to Evaluate the Efficacy and Safety of Docetaxel Polymeric Micelle (PM) in Recurrent or Metastatic HNSCC	NCT02639858	Docetaxel-PM	Head and Neck Carcinoma	(Keam et al. 2019)
A Trial of Paclitaxel (Genexol®) and Cisplatin Versus Paclitaxel Loaded Polymeric Micelle (Genexol-PM®) and Cisplatin in Advanced Non-Small Cell Lung Cancer	NCT01426126	Paclitaxel (Genexol®)	Non-Small Cell Lung Cancer (NSCLC)	(Park et al. 2010)
Study of Genexol-PM in Patients with Advanced Urothelial Cancer Previously Treated with Gemcitabine and Platinum	NCT01426126	Genexol PM	Bladder Cancer, Ureter Cancer	(Ahn et al. 2014)

trials for cancer treatment, indicating the promising therapeutic potential of prodrug delivery based on polymeric nanomicelles.

pH-responsive nanomicelles have shown significant success in treating diseased tissues, but they face substantial challenges when dealing with complex diseases in the human body. To overcome these challenges, multifunctional pH-reactive nanomicelles have been developed by incorporating functional groups responsive to various stimuli, attaching multiple ligands, and encapsulating them in a different carrier. However, it should be noted that the effectiveness of drug transport or biocompatibility in these complex DDSs may sometimes be reduced, resulting in limited therapeutic benefits.

Therefore, drug distribution must be carefully considered to ensure that these multifunctional nanomedicines consistently provide greater therapeutic effects. One promising application of pH-responsive nano DDSs is the distribution of multiple drugs in a single DDS while coordinating their pharmacokinetics and mechanism of action to maximize therapeutic benefits. Another potential use is the combination of several medications into a single DDS with coordinated pharmacokinetic characteristics and mechanisms of action to increase therapeutic advantages.

References

Ahn, Hee Kyung, Minkyu Jung, Sun Jin Sym, Dong Bok Shin, Shin Myung Kang, Sun Young Kyung et al. 2014. A phase II trial of Cremorphor EL-free paclitaxel (Genexol-PM) and gemcitabine in patients with advanced non-small cell lung cancer. Cancer Chemotherapy and Pharmacology 74: 277–82.

Ansari, Anees A., Abdul K. Parchur, Nanasaheb D. Thorat and Guanying Chen. 2021. New advances in pre-clinical diagnostic imaging perspectives of functionalized upconversion nanoparticle-based nanomedicine. Coordination Chemistry Reviews 440: 213971.

Atrafi, Florence, Herlinde Dumez, Ron H.J. Mathijssen, Catharine W. Menke van der Houven, Cristianne J.F. Rijcken, Rob Hanssen et al. 2020. A phase I dose-escalation and pharmacokinetic study of a micellar nanoparticle with entrapped docetaxel (CPC634) in patients with advanced solid tumours. Journal of Controlled Release 325: 191–97.

Barbarotta, Lisa and Kristen Hurley. 2015. Romidepsin for the treatment of peripheral T-cell lymphoma. Journal of the Advanced Practitioner in Oncology 6: 22.

Bera, Sourabh, Hemanta Kumar Datta and Parthasarathi Dastidar. 2022. Nitrile-containing terpyridyl Zn (II)-coordination polymer-based metallogelators displaying helical structures: synthesis, structures, and "druglike" action against B16-F10 melanoma cells. ACS Applied Materials & Interfaces.

Blasco, Michael A., Peter F. Svider, S. Naweed Raza, John R. Jacobs, Adam J. Folbe, Pankhoori Saraf et al. 2017. Systemic therapy for head and neck squamous cell carcinoma: Historical perspectives and recent breakthroughs. The Laryngoscope 127: 2565–69.

Buskaran, Kalaivani, Saifullah Bullo, Mohd Zobir Hussein, Mas Jaffri Masarudin, Mohamad Aris Mohd Moklas and Sharida Fakurazi. 2021. Anticancer molecular mechanism of protocatechuic acid loaded on folate coated functionalized graphene oxide nanocomposite delivery system in human hepatocellular carcinoma. Materials 14: 817.

Cao, Zhiwen, Wen Li, Rui Liu, Xiang Li, Hui Li, Linlin Liu et al. 2019. pH-and enzyme-triggered drug release as an important process in the design of anti-tumor drug delivery systems. Biomedicine & Pharmacotherapy 118: 109340.

Chauhan, Neeraj, Amber Kruse, Hilary Newby, Meena Jaggi, Murali M. Yallapu and Subhash C. Chauhan. 2020. Pluronic polymer-based ormeloxifene nanoformulations induce superior anticancer effects in pancreatic cancer cells. ACS Omega 5: 1147–56.

Chen, Bin Bin, Meng Li Liu and Cheng Zhi Huang. 2021. Recent advances of carbon dots in imaging-guided theranostics. TrAC Trends in Analytical Chemistry 134: 116116.

Chen, W. and P. Liu. 2022. Dendritic polyurethane-based prodrug as unimolecular micelles for precise ultrasound-activated localized drug delivery. Materials Today Chemistry 24: 100819.

Ding, Chendi, Chunbo Chen, Xiaowei Zeng, Hongzhong Chen and Yanli Zhao. 2022a. Emerging strategies in stimuli-responsive prodrug nanosystems for cancer therapy. ACS Nano 16: 13513–53.

Ding, Haitao, Ping Tan, Shiqin Fu, Xiaohe Tian, Hu Zhang, Xuelei Ma et al. 2022b. Preparation and application of pH-responsive drug delivery systems. Journal of Controlled Release 348: 206–38.

Domiński, Adrian, Monika Krawczyk, Tomasz Konieczny, Maciej Kasprów, Aleksander Foryś, Gabriela Pastuch-Gawołek et al. 2020. Biodegradable pH-responsive micelles loaded with 8-hydroxyquinoline glycoconjugates for Warburg effect based tumor targeting. European Journal of Pharmaceutics and Biopharmaceutics 154: 317–29.

Eble, Johannes A. and Stephan Niland. 2019. The extracellular matrix in tumor progression and metastasis. Clinical & Experimental Metastasis 36: 171–98.

Eldar-Boock, Anat, Keren Miller, Joaquin Sanchis, Ruth Lupu, María J. Vicent and Ronit Satchi-Fainaro. 2011. Integrin-assisted drug delivery of nano-scaled polymer therapeutics bearing paclitaxel. Biomaterials 32: 3862–74.

Gardner, Thomas, Bennett Elzey and Noah M. Hahn. 2012. Sipuleucel-T (Provenge) autologous vaccine approved for treatment of men with asymptomatic or minimally symptomatic castrate-resistant metastatic prostate cancer. Human Vaccines & Immunotherapeutics 8: 534–39.

Golombek, Susanne K., Jan-Niklas May, Benjamin Theek, Lia Appold, Natascha Drude, Fabian Kiessling et al. 2018. Tumor targeting via EPR: Strategies to enhance patient responses. Advanced Drug Delivery Reviews 130: 17–38.

Gong, Hua, Liang Cheng, Jian Xiang, Huan Xu, Liangzhu Feng, Xiaoze Shi and Zhuang Liu. 2013. Near-infrared absorbing polymeric nanoparticles as a versatile drug carrier for cancer combination therapy. Advanced Functional Materials 23: 6059–67.

Grillo-Lopez, A.J., C.A. White, B.K. Dallaire, C.L. Varns, C.D. Shen, A. Wei et al. 2000. Rituximab the first monoclonal antibody approved for the treatment of lymphoma. Current Pharmaceutical Biotechnology 1: 1–9.

Group, Korean Breast Cancer Study. A clinical trial of paclitaxel loaded polymeric micelle (Genexol-PM®) in patients with taxane-pretreated recurrent breast cancer. NCT00912639: https://clinicaltrials. gov [accessed on April 2015].

Guo, Dongbo, Shuting Xu, Wumaier Yasen, Chuan Zhang, Jian Shen, Yu Huang et al. 2020. Tirapazamine-embedded polyplatinum (iv) complex: a prodrug combo for hypoxia-activated synergistic chemotherapy. Biomaterials Science 8: 694–701.

Guo, Hui, Fangzhe Liu, Enqi Liu, Shanshan Wei, Wenbo Sun, Baoqiang Liu et al. 2022. Dual-responsive nano-prodrug micelles for MRI-guided tumor PDT and immune synergistic therapy. Journal of Materials Chemistry B.

Guo, Xiong, Zhiyue Zhao, Dawei Chen, Mingxi Qiao, Feng Wan, Dongmei Cun et al. 2019. Co-delivery of resveratrol and docetaxel via polymeric micelles to improve the treatment of drug-resistant tumors. Asian Journal of Pharmaceutical Sciences 14: 78–85.

Huo, Meng, Jinying Yuan, Lei Tao and Yen Wei. 2014. Redox-responsive polymers for drug delivery: from molecular design to applications. Polymer Chemistry 5: 1519–28.

Jerzykiewicz, Julia and Aleksander Czogalla. 2022. Polyethyleneimine-based lipopolyplexes as carriers in anticancer gene therapies. Materials 15: 179.

Kalva, Nagendra, Saji Uthaman, Soo Jeong Lee, Yu Jeong Lim, Rimesh Augustine, Kang Moo Huh et al. 2021. Degradable pH-responsive polymer prodrug micelles with aggregation-induced emission for cellular imaging and cancer therapy. Reactive and Functional Polymers 166: 104966.

Kanamala, Manju, William R. Wilson, Mimi Yang, Brian D. Palmer and Zimei Wu. 2016. Mechanisms and biomaterials in pH-responsive tumour targeted drug delivery: a review. Biomaterials 85: 152–67.

Karimi, Mahdi, Masoud Eslami, Parham Sahandi-Zangabad, Fereshteh Mirab, Negar Farajisafiloo, Zahra Shafaei et al. 2016. pH-Sensitive stimulus-responsive nanocarriers for targeted delivery of therapeutic agents. Wiley Interdisciplinary Reviews: Nanomedicine and Nanobiotechnology 8: 696–716.

Keam, Bhumsuk, Keun-Wook Lee, Se-Hoon Lee, Jin-Soo Kim, Jin Ho Kim, Hong-Gyun Wu et al. 2019. A phase II Study of Genexol-PM and cisplatin as induction chemotherapy in locally advanced head and neck squamous cell carcinoma. The Oncologist 24: 751–e231.

Khalifa, Alaa M., Manal A. Elsheikh, Amr M. Khalifa and Yosra S.R. Elnaggar. 2019. Current strategies for different paclitaxel-loaded nano-delivery systems towards therapeutic applications for ovarian carcinoma: A review article. Journal of Controlled Release 311: 125–37.

Khot, Vishwajeet M., Ashwini B. Salunkhe, Sabrina Pricl, Joanna Bauer, Nanasaheb D. Thorat and Helen Townley. 2021. Nanomedicine-driven molecular targeting, drug delivery, and therapeutic approaches to cancer chemoresistance. Drug Discovery Today 26: 724–39.

Krukiewicz, Katarzyna and Jerzy K. Zak. 2016. Biomaterial-based regional chemotherapy: Local anticancer drug delivery to enhance chemotherapy and minimize its side-effects. Materials Science and Engineering: C 62: 927–42.

Kumar, Manish and P.S. Rajnikanth. 2020. A mini-review on HER2 positive breast cancer and its metastasis: resistance and treatment strategies. Current Nanomedicine (Formerly: Recent Patents on Nanomedicine) 10: 36–47.

Kumari, Smita, Dia Advani, Sudhanshu Sharma, Rashmi K. Ambasta and Pravir Kumar. 2021. Combinatorial therapy in tumor microenvironment: where do we stand? Biochimica et Biophysica Acta (BBA)-Reviews on Cancer 1876: 188585.

Li, Jiagen and Peng Liu. 2019. Self-assembly of drug–drug conjugates as drug self-delivery system for tumor-specific pH-triggered release. Particle & Particle Systems Characterization 36: 1900113.

Liu, Bin and Sankaran Thayumanavan. 2017. Substituent effects on the pH sensitivity of acetals and ketals and their correlation with encapsulation stability in polymeric nanogels. Journal of the American Chemical Society 139: 2306–17.

Luo, Cong, Jin Sun, Dan Liu, Bingjun Sun, Lei Miao, Sara Musetti et al. 2016. Self-assembled redox dual-responsive prodrug-nanosystem formed by single thioether-bridged paclitaxel-fatty acid conjugate for cancer chemotherapy. Nano Letters 16: 5401–08.

Mao, Jie, Yang Li, Tong Wu, Conghui Yuan, Birong Zeng, Yiting Xu et al. 2016. A simple dual-pH responsive prodrug-based polymeric micelles for drug delivery. ACS Applied Materials & Interfaces 8: 17109–17.

Michon, Simon, Francis Rodier and François T.H. Yu. 2022. Targeted anti-cancer provascular therapy using ultrasound, microbubbles, and nitrite to increase radiotherapy efficacy. Bioconjugate Chemistry.

Mitchell, Michael J., Margaret M. Billingsley, Rebecca M. Haley, Marissa E. Wechsler, Nicholas A. Peppas and Robert Langer. 2021. Engineering precision nanoparticles for drug delivery. Nature Reviews Drug Discovery 20: 101–24.

Nahm, S.H., Agata Rembielak, H. Peach and Paul C. Lorigan. 2021. Consensus guidelines for the management of melanoma during the COVID-19 pandemic: surgery, systemic anti-cancer therapy, radiotherapy and follow-up. Clinical Oncology 33: e54–e57.

Nie, Yu, Michael Günther, Zhongwei Gu and Ernst Wagner. 2011. Pyridylhydrazone-based PEGylation for pH-reversible lipopolyplex shielding. Biomaterials 32: 858–69.

Norsworthy, Kelly J., Chia-Wen Ko, Jee Eun Lee, Jiang Liu, Christy S. John, Donna Przepiorka et al. 2018. FDA approval summary: mylotarg for treatment of patients with relapsed or refractory CD33-positive acute myeloid leukemia. The Oncologist 23: 1103–08.

Ofridam, Fabrice, Mohamad Tarhini, Noureddine Lebaz, Emilie Gagniere, Denis Mangin and Abdelhamid Elaïssari. 2021. pH-sensitive polymers: Classification and some fine potential applications. Polymers for Advanced Technologies 32: 1455–84.

Oh, Sung Yong. 2021. Docetaxel-polymeric micelles (PM) and oxaliplatin for esophageal carcinoma (DOSE). Clinical Trial Citation.

Pandit, Bhoomika and Maksim Royzen. 2022. Recent development of prodrugs of Gemcitabine. Genes 13: 466.

Park, Choel-Kyu, Hyun-Wook Kang, Tae-Ok Kim, Ho-Seok Ki, Eun-Young Kim, Hee-Jung Ban et al. 2010. A case of ischemic colitis associated with paclitaxel loaded polymeric micelle (Genexol-PM®) chemotherapy. Tuberculosis and Respiratory Diseases 69: 115–18.

Peng, Shiyuan, Fusheng Zhang, Baihao Huang, Jufang Wang and Lijuan Zhang. 2021. Mesoporous silica nanoprodrug encapsulated with near-infrared absorption dye for photothermal therapy combined with chemotherapy. ACS Applied Bio Materials 4: 8225–35.

Rasool, Mahmood, Arif Malik, Sulayman Waquar, Mahwish Arooj, Sara Zahid, Muhammad Asif et al. 2022. New challenges in the use of nanomedicine in cancer therapy. Bioengineered 13: 759–73.

Roth, Peter J. and Andrew B. Lowe. 2017. Stimulus-responsive polymers. Polymer Chemistry 8: 10–11.

Saneja, Ankit, Robin Kumar, Mubashir J. Mintoo, Ravindra Dhar Dubey, Payare Lal Sangwan, Dilip M. Mondhe, Amulya K. Panda and Prem N. Gupta. 2019. Gemcitabine and betulinic acid co-encapsulated PLGA−PEG polymer nanoparticles for improved efficacy of cancer chemotherapy. Materials Science and Engineering: C 98: 764–71.

Santhosh, S.B. and M.J.N. Chandrasekar. 2020. Isoelectric point based dual sensitive peptide-drug conjugate prodrug to target solid tumors. International Journal of Peptide Research and Therapeutics 26: 2225–29.

Sartaj, Ali, Zufika Qamar, Farheen Fatima Qizilbash, Shadab Md, Nabil A. Alhakamy, Sanjula Baboota et al. 2021. Polymeric nanoparticles: exploring the current drug development and therapeutic insight of breast cancer treatment and recommendations. Polymers 13: 4400.

Schaaf, Marco B., Abhishek D. Garg and Patrizia Agostinis. 2018. Defining the role of the tumor vasculature in antitumor immunity and immunotherapy. Cell Death & Disease 9: 1–14.

Shinn, Jongyoon, Nuri Kwon, Seon Ah Lee and Yonghyun Lee. 2022. Smart pH-responsive nanomedicines for disease therapy. Journal of Pharmaceutical Investigation 52: 427–41.

Shirley, Matt. 2016. Ixazomib: first global approval. Drugs 76: 405–11.

Soe, Zar Chi, Jun Bum Kwon, Raj Kumar Thapa, Wenquan Ou, Hanh Thuy Nguyen, Milan Gautam et al. 2019. Transferrin-conjugated polymeric nanoparticle for receptor-mediated delivery of doxorubicin in doxorubicin-resistant breast cancer cells. Pharmaceutics 11: 63.

Sonawane, Sandeep J., Rahul S. Kalhapure and Thirumala Govender. 2017. Hydrazone linkages in pH responsive drug delivery systems. European Journal of Pharmaceutical Sciences 99: 45–65.

Szulc, Aleksandra, Lukasz Pulaski, Dietmar Appelhans, Brigitte Voit and Barbara Klajnert-Maculewicz. 2016. Sugar-modified poly (propylene imine) dendrimers as drug delivery agents for cytarabine to overcome drug resistance. International Journal of Pharmaceutics 513: 572–83.

Takemoto, Hiroyasu, Takanori Inaba, Takahiro Nomoto, Makoto Matsui, Xiaomeng Liu, Masahiro Toyoda et al. 2020. Polymeric modification of gemcitabine via cyclic acetal linkage for enhanced anticancer potency with negligible side effects. Biomaterials 235: 119804.

Vallet-Regí, María, Ferdi Schüth, Daniel Lozano, Montserrat Colilla and Miguel Manzano. 2022. Engineering mesoporous silica nanoparticles for drug delivery: where are we after two decades? Chemical Society Reviews.

Wang, Xin, Xu Cheng, Le He, Xiaoli Zeng, Yan Zheng and Rupei Tang. 2019a. Self-assembled indomethacin dimer nanoparticles loaded with doxorubicin for combination therapy in resistant breast cancer. ACS Applied Materials & Interfaces 11: 28597–609.

Wang, Yulong, Avik Khan, Yanxin Liu, Jing Feng, Lei Dai, Guanhua Wang et al. 2019b. Chitosan oligosaccharide-based dual pH responsive nano-micelles for targeted delivery of hydrophobic drugs. Carbohydrate Polymers 223: 115061.

Wermuth, C.G., C.R. Ganellin, Per Lindberg and L.A. Mitscher. 1998. Glossary of terms used in medicinal chemistry (IUPAC Recommendations 1998). Pure and Applied Chemistry 70: 1129–43.

Wu, Jun. 2021. The enhanced permeability and retention (EPR) effect: the significance of the concept and methods to enhance its application. Journal of Personalized Medicine 11: 771.

Wu, Kuei-Meng and James G. Farrelly. 2007. Regulatory perspectives of Type II prodrug development and time-dependent toxicity management: Nonclinical Pharm/Tox analysis and the role of comparative toxicology. Toxicology 236: 1–6.

Xu, Caidie, Long Xu, Renlu Han, Yabin Zhu and Jianfeng Zhang. 2021. Blood circulation stable doxorubicin prodrug nanoparticles containing hydrazone and thioketal moieties for antitumor chemotherapy. Colloids and Surfaces B: Biointerfaces 201: 111632.

Xu, Ying, Xuexiang Han, Yiye Li, Huan Min, Xiao Zhao, Yinlong Zhang et al. 2019. Sulforaphane mediates glutathione depletion via polymeric nanoparticles to restore cisplatin chemosensitivity. ACS Nano 13: 13445–55.

Yang, Kai, Liangzhu Feng and Zhuang Liu. 2016. Stimuli responsive drug delivery systems based on nano-graphene for cancer therapy. Advanced Drug Delivery Reviews 105: 228–41.

Yao, Jun, Pingfan Li, Lin Li and Mei Yang. 2018. Biochemistry and biomedicine of quantum dots: from biodetection to bioimaging, drug discovery, diagnostics, and therapy. Acta Biomaterialia 74: 36–55.

Yao, Y., X. Guo, W. Jiang, M. Jiang, J. Yang, Y. Li et al. 2021. Precise cancer anti-acid therapy monitoring using pH-Sensitive $MnO_2@$ BSA nanoparticles by magnetic resonance imaging. ACS Applied Materials & Interfaces 13: 18604–18.

Zhang, Min, Xinli Guo, Mingfu Wang and Kehai Liu. 2020. Tumor microenvironment-induced structure changing drug/gene delivery system for overcoming delivery-associated challenges. Journal of Controlled Release 323: 203–24.

Zhang, Yin-Ci, Cheng-Guang Wu, A-Min Li, Yong Liang, Dong Ma and Xiao-Long Tang. 2021a. Oxaliplatin and gedatolisib (PKI-587) co-loaded hollow polydopamine nano-shells with simultaneous upstream and downstream action to re-sensitize drugs-resistant hepatocellular carcinoma to chemotherapy. Journal of Biomedical Nanotechnology 17: 18–36.

Zhang, Yuezhou, Huaguang Cui, Ruiqi Zhang, Hongbo Zhang and Wei Huang. 2021b. Nanoparticulation of prodrug into medicines for cancer therapy. Advanced Science 8: 2101454.

Zhu, Dan, Bin Wang, Xiao-Hua Zhu, Hai-Liang Zhu and Shen-Zhen Ren. 2021. A MnO_2-coated multivariate porphyrinic metal–organic framework for oxygen self-sufficient chemo-photodynamic synergistic therapy. Nanomedicine: Nanotechnology, Biology and Medicine 37: 102440.

Chapter 8

Smart Hydrogels for Tissue Engineering of Hard and Soft Tissues in the Oral Cavity

Jan C. Kwan, Sangeeth Pillai, Uyen M. N. Cao and *Simon D. Tran**

Introduction

The overarching theme of regenerative medicine has always been to develop and apply novel therapies that can restore, repair, or replace damaged tissue or organ function and promote regeneration using the body's own healing mechanism (Dzobo et al. 2018). This has been accomplished through a variety of approaches in the field of regenerative medicine such as cellular or gene therapy, tissue engineering, and biomaterials. Tissue engineering falls under the theme of regenerative medicine but uniquely focuses on the generation of functional tissues that closely resemble native tissue in terms of structure and function (Atala 2004, Fisher and Mauck 2013, Sharma et al. 2019). The ultimate goal of tissue engineering is to develop tissue substitutes that can be implanted into patients to restore, repair, or replace damaged or lost tissues (Olson et al. 2011, Khademhosseini and Langer 2016). There are three general key components of tissue engineering: (1) cells that can be patient or donor-derived or from stem cells that can be grown in a laboratory and seeded onto scaffolds encouraging tissue formation; (2) scaffolds that can help provide structural support for cells to attach, proliferate, differentiate, and mimic the extracellular matrix (ECM) of native tissues, essentially guiding the organization of cells; (3) biochemical factors such as growth factors and bioactive

McGill Craniofacial Tissue Engineering and Stem Cell Laboratory, Faculty of Dental Medicine and Oral Health Sciences, McGill University, Montreal, QC H3A 0C7, Canada.
Emails: jan.kwan@mail.mcgill.ca, sangeeth.pillai@mail.mcgill.ca, uyen.cao@mail.mcgill.ca
* Corresponding author: simon.tran@mcgill.ca

molecules incorporated either at the cellular or scaffold level which help to regulate cell behavior, promote tissue growth, or direct cell differentiation (Furth et al. 2007, Olson et al. 2011, Griffith and Naughton 2002).

There have been significant advances in the field of tissue engineering with few existing FDA-approved tissue-engineered products for skin, cartilage, and bone (Ashammakhi et al. 2021, Hoffman et al. 2019). However, as the field continues to develop our understanding of tissue engineering, various unmet needs still require extensive research of these technologies to be effective and routine within a clinical setting. Examples of these challenges include tissue complexity, immune response and rejection, clinical outcomes, cost, and scalable manufacturing processes (Williams 2019a, Crupi et al. 2015, Place et al. 2009). To overcome many of these challenges in the field of tissue engineering, researchers have employed the use of polymers that have influenced scaffold design, surface modifications, drug delivery methods, biocompatibility, and hybrid materials (Dobrzyński and Pamuła 2021, Place et al. 2009, Hudiță and Gălățeanu 2023, Yousefzade et al. 2020). There are many different polymers that are available to choose from; however, whether they can be used for tissue engineering depends heavily on the specific tissue that is being targeted, desired scaffold properties, the intended clinical application, and their *in vivo* tolerance and interactions with cells and biological systems.

Hydrogels are an example of a three-dimensional polymer that can be used in tissue engineering, consisting of a cross-linked network of hydrophilic polymer chains that can both absorb and retain large amounts of water or biological fluids (Lee and Mooney 2001). Hydrogels are unique as their structure consists of a gel-like consistency maintaining both solid and flexible characteristics (Xu et al. 2022). Hydrogels also have tunable properties dependent on the desired clinical application, crosslinking can be achieved through hydrogen or covalent bonding to influence the mechanical strength or degradation of the hydrogel (Lee and Mooney 2001, Xu et al. 2022). Additionally, with their hydrophilic polymer chains, hydrogels naturally have a strong affinity for water and are highly tolerated biomaterials *in vivo* as they mirror the natural ECM found in tissues (Skardal et al. 2012). With their high-water content capabilities, hydrogels can also allow for the hydration of biological tissues aiding in the support of cell survival, proliferation, and migration (Caliari and Burdick 2016, Ahmed 2015). Hydrogels can also be composed of a variety of different polymers which opens the door for a wide range of hydrogel formulations. Additionally, researchers have also made advancements in hydrogel technology through the development of a subclass known as 'Smart Hydrogels'. Briefly, what differentiates a smart hydrogel is that it possesses an additional level of responsiveness to external stimuli which then can exhibit a physical or chemical change, typically in response to a specific cue in the surrounding environment for example a change in pH, temperature, or presence of certain biomolecules (Zhang and Huang 2021, Bustamante-Torres et al. 2021, Hu et al. 2022).

Smart hydrogels offer a dynamic approach that can be both controlled and predictable to enable hydrogels to adapt and respond to their surrounding environment (Mantha et al. 2019). This feature has led to the exploration of more tailored and targeted tissue engineering applications, controlled drug delivery systems, biosensors, multifunctional platforms, and improved scaffold functionality and bioactivity (Zhang and Huang 2021). Although smart hydrogels have shown promising potential in tissue engineering applications, there are still challenges to their clinical translation and regular use in a clinical setting (Xu et al. 2022). This chapter will explore a general view of how smart hydrogels are classified, fabricated, and characterized along with their unique properties, all in the context of tissue engineering and regenerative medicine. With the existing and more recent applications of smart hydrogels, this chapter will focus on their applications in the oral cavity for the regeneration and revitalization of hard and soft dental tissues. As smart hydrogels become more sophisticated, so too will the limitations and challenges of taking this technology into a clinical setting. This will provide insight into the current and potential limitations as well as the outlook of smart hydrogels for tissue engineering in general and within the field of dentistry.

Classification of hydrogels

Polymeric scaffolds alone have limitations in their ability to promote tissue regeneration and cell-mimicking properties. As a result, hydrogel-based scaffolds have allowed researchers to overcome many of the challenges encountered when using polymeric scaffolds. Hydrogels can act as a scaffold and mimic the properties of various tissues while providing structural and biomechanical support to nearby cells without dissolving in high concentrations of water. Hydrogels can be classified in several different ways that include physical or chemical interactions, preparation methods, or structural properties. These are not to be confused with smart hydrogels or stimuli-responsive hydrogels where they are typically categorized on the basis of physical or chemical responsiveness. Hydrogels can also be generally classified based on the makeup of their polymeric network which can be categorized between natural or synthetic polymers.

Synthetic and natural polymers

Synthetic polymer-based hydrogels are hydrogels that are composed of polymers chemically synthesized in the laboratory. The advantage of using a synthetic composition for hydrogels is that they can be easily synthesized and scaled up while being tailored to elicit specific physiochemical properties such as controlled degradation, elasticity, or strength (Ahmad et al. 2022). This also allows researchers complete control over the hydrogel's structure and functionality in terms of molecular weight and crosslinking ratios; however, some of the main limitations are their tolerance *in vivo*, toxicity, and their interactions with neighboring cells (Madduma-Bandarage and Madihally 2021). Natural polymer-based hydrogels are derived from naturally occurring polymers found in nature or living organisms. Researchers have highlighted their similarity to the body's native environment, to be well tolerated by living tissues, and less likely to provoke an immune response. Since many natural polymers possess sites that cells can recognize or cell adhesion domains, integrating these traits within a hydrogel can promote cell attachment, proliferation, and tissue formation (Silva et al. 2021, Mano et al. 2007). However, some general

Table 1. Different types of synthetic and natural polymers used in the formulation of hydrogels and their associated strengths and limitations.

Polymer Type	Composition	Strengths	Limitations	Ref.
Synthetic	Poly(ethylene glycol) (PEG)	-Anti-foul -Easily tunable -High reproducibility -Less expensive/scalable -Easily functionalized -Water soluble -FDA approved	-Weak mechanical properties -Lack bioactivity -Degradation particles (possible acute immune response) -Require functionalization	(Lin and Anseth 2009), (Tan et al. 2010), (Zhu 2010), (Lei et al. 2022), (Gopinathan and Noh 2018)
	Poly(vinyl alcohol) (PVA)	-High water content -High reproducibility -Less expensive/scalable -Easily functionalized -Water soluble -FDA approved	-Weak mechanical properties -Lack bioactivity -Degradation particles (acute immune response) -Require functionalization	(Bodugoz-Senturk et al 2009, Wang et al. 2021a, Chen et al. 2022), (Zhang et al. 2022, Li et al. 2015)
	Polyurethane (PU)	-Good mechanical properties (tensile and compression) -Water-resistant -High reproducibility -Less expensive/scalable -Easily functionalized	-Hydrophobic -Easily degraded *in vivo* -Weak thermal properties -Lack bioactivity	(Shen et al. 2022), (Petrini et al. 2003), (Liu et al. 2022b)

		Advantages	Disadvantages	References
Natural	Polydopamine	-Strong adhesive properties -Tolerated *in vivo* (low immune response) -Biodegradable -Bioactive properties	-High variability -Complex structure -High cost (extraction/production)	(O'Connor et al. 2020), (Li et al. 2021, Trinh et al. 2020)
	Collagen	-Good cell adhesion -Easily functionalized -Tolerated *in vivo* (low immune response) -Biodegradable	-Mechanical instability -High degradation rate -High variability -High cost and difficult to sterilize	(Zhang et al. 2023), (Holmes et al. 1998, Bienkowski et al. 1986).
	Elastin	-High elasticity and flexibility -Tolerated *in vivo* (low immune response) -Biodegradable	-Low solubility -Vulnerability to ultraviolet radiation -Limited capacity for repair and regeneration -High variability -Poor mechanical strength	(Baumann et al. 2021), (Heinz 2020, Stojic et al. 2021, Vazquez-Portalatin et al. 2020), (Heinz 2020), (Ravanetti et al. 2023, Shokri et al. 2022)
	Chitosan	-Antibacterial and antifungal properties Soluble in water -Abundant in nature -Tolerated *in vivo* (low immune response) -Biodegradable	-Weak mechanical properties -Slow gelation (hydrogel formulation) -High variability	(Kean and Thanou 2010, Guillen-Carvajal et al. 2023, Younes and Rinaudo 2015), (Trinh et al. 2020)
	Hyaluronic acid	-High water retention properties -Well tolerated *in vivo* (low immune response) -Easily functionalized -Fast gelation -Biodegradable	-High degradation (Short half-life) -Weak mechanical properties -High variability -High surface tension and viscosity	(Xu et al. 2012, Khaliq et al. 2023), (Shokri et al. 2022, Gao et al. 2023)

limitations when using these types of hydrogel are their batch-to-batch variability, lower mechanical strength compared to synthetic polymers, and challenges in cross-linking as some functional groups present in natural polymers may interfere with cross-linking reactions (Moon et al. 2023). Table 1 shows examples of the most common synthetic and natural polymers used to synthesize smart hydrogels.

With the many different synthetic and natural polymer-based hydrogels that are available, their clinical adoption is still limited especially in tissue engineering due to either their overall lack of cell-specific bioactivity, poor mechanical properties, *in vivo* degradation, or a combination of other factors that reduce the efficacy of these hydrogels. A common theme that can be observed among many hydrogel studies is the investigation of the tunability or functionalization physically or chemically prior to or post-gelation of the hydrogel to enhance either the mechanical properties, bioactivity, or responsiveness to external stimuli. Although many studies continue to report on innovative approaches and add to our current understanding of the ideal hydrogel formulation, there remains a lot more challenges before clinical testing of the technology (Hu et al. 2022). However, many studies still stress the importance of how polymer choice for the hydrogel can largely be attributed to its success *in vivo* over the short and long term.

Fabrication techniques

Hydrogels have great potential in the field of tissue engineering, many techniques have been scrutinized for the fabrication of novel gels. Scientists have synthesized hydrogels through traditional methods such as emulsification, lyophilization, gas foaming, and now more recently, electrospinning. Emulsification is a process of mixing an immiscible substance consisting of bioactive molecules in a second solution to form a uniform mixture. Emulsifying agents help to reduce the surface tension between two different solutions, usually between hydrophilic and hydrophobic ones, allowing them to interact and prevent separation (Miguel Ângelo Cerqueira 2017). Emulsion hydrogels or emulgels have been used for the delivery of insoluble or poor soluble substances to improve their stability and bioactivity *in vivo* (Liu et al. 2007). Another technique of scaffold fabrication is lyophilization or freeze-drying method. In regenerative medicine, this is a common process of removing water out of a soluble polymer scaffold through sublimation (Brougham et al. 2017). The gas foaming technique has been used to create a porous structure within materials by incorporating gas bubbles within a scaffold. The gas bubbles can be incorporated *in situ* by various approaches including a chemical reaction or physical pressurization (Dehghani and Annabi 2011). A more recent approach to fabricate scaffolds for tissue engineering is the technique of electrospinning. Microfibers are obtained from an extrusion of a liquid biomaterial out from the tip of a needle and collected on a collector plate by applying high voltage forces in the system. When the surface tension in the extrusion is overcome by the force of the electric field, the extrusion is distorted. The distortion leads to a charged jet ejection that moves towards the collector plate and generates ultrathin fibers (Kamali and Shamloo 2020, Song et al. 2023).

As tissue engineering draws more attention worldwide in recent decades, more approaches to fabricate regenerative scaffolds have been developed and explored, including photolithography, 3D printing, and advanced bioprinting techniques. Photolithography is a crucial method used in microfabrication and semiconductor devices. This technique utilizes light, mainly ultraviolet (UV) light, to transfer a 3D blueprint into a thin photoresist film connected to a silicon substrate (Sun et al. 2023). By controlling the extent of light exposure over the substrate, the geometric patterns, porosity of scaffold, and the gradient of biomolecules can be adjusted (Martin et al. 2011). While this technique can efficiently produce microstructures with high precision and repeatability, the cost of the instrument and required space are expensive. This technology is commonly used in the fields of semiconductors, microfluidics, and tissue engineering. Future research may consist of achieving higher resolutions, more cost-effective manufacturing, and a wider range of applications.

More recently, 3D printing techniques have become prominent in the fabrication of scaffolds in the field of tissue engineering and regenerative medicine. It is an additive method, in which the framework is constructed by depositing the printing material layer by layer following a pre-designed scaffold using computer-aided design (CAD) software (Martinelli et al. 2023). The printer can move in three dimensions (X, Y, Z) while releasing ink according to the design requirements. After printing, depending on the printing technique used, some post-fabrication steps including curing, sintering, and light or chemical treatments are employed to strengthen the mechanical properties. The final scaffolds can be seeded with cells for culturing or implanted *in vivo* and this technique is ideal for fabricating scaffolds due to its high precision, simplicity, and cost-effectiveness. Other methods have been widely employed for scaffold fabrication including stereolithography (SLA), selective laser sintering (SLS), fused deposition (FDM), extrusion printing, and inkjet printing (Huanbutta et al. 2023). With these techniques, multiple applications for tissue engineering have been successfully reported. A study by Yao et al. (2022) produced a 3D silk fibroin and hydroxyapatite scaffold that mimics the silkworm spinning structure for the regeneration of cortical and cancellous bone (Yao et al. 2022). In the same year, Liu and colleagues reported the application of cryogenic and coaxial 3D printing in the fabrication of PCL and alginate shell/core scaffolds for drug delivery (Liu et al. 2022a).

Among many enhancements of 3D printing technology, bioprinting is the most desirable for tissue engineering. The key difference between these methods is that the bioprinting method utilizes bioinks, in which the ink material is mixed with living cells before extrusion to construct a functional and viable bioproduct similar to tissues or organs. The procedure usually consists of three steps: pre-bioprinting, bioprinting, and post-bioprinting (Shafiee and Atala 2016). Parameters including ink viscosity, cell density, temperature, speed, nozzle diameter, and extrusion pressure, need to be controlled to obtain optimal viability of cells before and after printing (Shafiee and Atala 2016). Recently, Miao et al. (2023) reported a bioactive composite hydrogel scaffold using a 3D bioprinting technique to deliver mouse bone marrow stem cells for the regeneration of periodontal tissue. The authors transplanted the cell-embedded scaffolds into a dog with a periodontal defect and observed significant tissue reconstruction after eight weeks of transplantation (Miao et al. 2023). Another recent work has attempted to reconstruct a dermal model via bioprinting. Keratinocytes were mixed with sacrificial gelatin (4 wt%) bioink to create a uniform and stratified epidermal layer, which led to epidermal stratification. The results demonstrated that the bioprinting approach is a reliable technique for the mass production of full-thickness dermal models (Ahn et al. 2023). Both the use of 3D printing and bioprinting have tremendous potential in the field of tissue engineering and regenerative medicine, yet despite the work that has been done there are still major challenges to effectively and reliably scale and re-produce hydrogels with high bioactivity for their clinical application. This is further complicated by the cost factor due to hydrogel structure complexity and the efficacy of varying polymeric mixtures to formulate respective hydrogels. As a result, more research is required before its reliable use.

Properties of hydrogels

The method by which smart hydrogels are designed can be controlled at each step of the process during synthesis and fabrication. This can depend on the type and number of polymers, cross-linking method, fabrication strategy, and many other factors. This is further complicated by the desired therapeutic effect, type of microenvironment, and biocompatibility requirements. An important consideration throughout the entire process is the type of properties the smart hydrogel should exhibit based on its intended clinical use. Due to the gel-like structure of hydrogels, certain resulting properties may influence the clinical outcome of a hydrogel.

Mechanical properties

For hydrogels to resist shear and compressive forces, many studies have elucidated the use of different cross-linkers. An example would be the addition of a high enough concentration of glutaraldehyde in a cross-linked gelatin hydrogel, demonstrating its ability to withstand much larger strains of compression and shear stress (Yousefi-Mashouf et al. 2023). Other methods to increase the overall mechanical properties such as toughness and impact resistance can be to use fiber-reinforced polymer composites combined with the hydrogel that can bear external loads such as polycaprolactone (PCL), polylactic acid (PLA), gelatin, and silk (Teixeira et al. 2021, Lin et al. 2022). Double-network hydrogels are similar in the goal of improving the mechanical properties but differ as they are made up of two different types of polymer networks that are interwoven and layered with the more charged or dense polymers (polyelectrolyte) as the first layer to disperse external stress and the more neutral and soft polymers as the second layer to sustain deformation (Gu et al. 2018, Chen et al. 2015). Hydrogels inherently have high-water content, and as a result must maintain a level of elasticity; studies have shown that controlling the crosslinking density of a polymeric network can be a method of tuning elasticity, ranging from kilopascals (kPA) to megapascals (MPA), which can also be referred to Young's modulus (Min et al. 2022, Lv et al. 2019).

Degradation

The degradation rate is a critical factor that affects the lifespan and overall stability of a smart hydrogel. In the context of tissue engineering, it is important for the rate of degradation to be controlled to allow new tissue formation during the early stages of tissue development (Zhu and Marchant 2011). A rapid degradation rate may lead to the premature collapse of the scaffold, while too slow may hinder tissue growth. It is also important to keep in mind the degradation of by-products produced by smart hydrogels over time and limit their potential to induce systemic toxicity (Meyvis et al. 2000). This is true for injectable or implantable smart hydrogels that are bioresorbable and degrade naturally within patients mainly to reduce the need for an additional surgical procedure for removal after its initial function is fulfilled. Many studies have now focused on implementing naturally biodegradable ECM components that can be broken down via enzymatic digestion or hydrolysis (Brovold et al. 2018).

Rheology

The rheology of the smart hydrogel refers to viscosity or flowability and plays an important role in the smart hydrogel's ability to be injected and its ease of administration. This is relevant in the latest trend of using extrusion-based 3D-printing technology to fabricate smart hydrogel scaffolds as shape fidelity and porosity are important factors in the overall structure (Herrada-Manchón et al. 2023). Injectable smart hydrogels have great importance in the field of tissue engineering as drugs and cells can be easily incorporated into the gelling matrix. Some of the most common clinical applications are cartilage repair and soft tissue regeneration (Zhu et al. 2022, Bertsch et al. 2023). Using an injection method allows for homogenous cell distribution that can take the size or shape of a cavity or defect before gelation. As more novel smart hydrogels are created with different compositions, so does the ability to quantitatively characterize the rheological properties of smart hydrogels before its fabrication. These include measuring viscosity and shear stress at different shear rates and temperatures, which can be tested using flow curves, oscillatory, and recovery tests to help provide viscoelastic properties or flow behavior (Tan and Marra 2010, Stojkov et al. 2021).

Surface property

The surface of a hydrogel plays a critical role in cell behavior, particularly in the initial stages of adhesion and proliferation (Cui et al. 2021, Meier et al. 2019). The surface of smart hydrogels can be smooth or rough, but researchers have gone to the lengths of functionalizing the surface of smart hydrogels with growth factors, collagen, or fibronectin as a method to improve biocompatibility (Palomino-Durand et al. 2021, Cui et al. 2021, Nilasaroya et al. 2021). The complexity that arises when developing a smart hydrogel, especially for tissue engineering, is how each of the properties will be affected when an adjustment or modification is made for a desired outcome. To have a smart hydrogel with high strength and toughness is to potentially sacrifice the water content of the hydrogel; as a result, there needs to be a balance. Although significant progress has been made in the field of smart hydrogels, the incorporation of additive manufacturing (AM) such as 3D printing adds another component that needs to be taken into consideration (Li et al. 2020). Ensuring consistent characterization of smart hydrogels quantitatively concerning critical mechanical properties will help benefit the field in adding to the existing foundation of potential modification strategies.

Characterization of hydrogels

The connection between the synthesis of smart hydrogels and their practical clinical application is the characterization through quantitative means. This provides investigators with the necessary information to predict its behaviors in different environments for a desired functionality. Understanding how the properties of smart hydrogels behave before injection or implantation *in vivo* is a crucial step in its potential use clinically.

Crosslinking

A common theme amongst many of the aforementioned properties of smart hydrogels can be closely related to the type or degree of cross-linking within the hydrogel network. The most common type of method is through a chemical cross-link involving the use of cross-linking agents such as glutaraldehyde, carbodiimides, and epoxides that can be added through physical mixing, sequential addition, encapsulation in micro/nanoparticles, or a condensation polymerization reaction (Xue et al. 2022, Hu et al. 2019). Higher the degree of cross-linking in a hydrogel, the denser it becomes and in turn has a higher mechanical strength. A lower degree of cross-linking leads to a more open network but with greater swelling capacity. The degree of cross-linking can easily be characterized through scanning electron microscopy (SEM) or Fourier transform infrared spectroscopy (FTIR) (Martinez-Garcia et al. 2022). Other means can be through analytical methods such as the swelling ratio, referring to the change in hydrogel volume upon swelling, and the output of this ratio can then be used in the Flory-Rehner equation, which relates the equilibrium swelling ratio to the cross-link density (Borges et al. 2020). Studies have also used mathematical modeling such as finite-element modeling (FEM) which can be used to estimate and provide a prediction of the behavior and other mechanical properties of the hydrogel based on cross-linking density (Colter et al. 2018, Bisotti et al. 2022).

1. Swelling

Due to the high water content of hydrogels, swelling is an important factor that needs to be characterized. By the very definition of "smart hydrogel," they are designed to undergo significant changes in volume in response to changes in specific stimuli such as temperature, pH, or ions. Swelling is one of the key indicators of how smart hydrogels might respond to external cues. Swelling simply refers to the hydrogel's ability to absorb and retain water or other solvents within its structure which naturally leads to an increase in volume (Romischke et al. 2022). Swelling characterization can be performed by submerging a dry hydrogel in a specific solvent (water or buffer) over a given period and then measuring the change in mass or volume. This can be further characterized either by performing real-time monitoring of changes in volume through dynamic-swelling kinetic analysis or plotting the change in volume or mass of the hydrogel over time during swelling providing a visual representation of its swelling kinetics (Zhan et al. 2021).

2. Temperature

The average temperature of our internal body falls into a narrow range of around 37°C and as such smart hydrogels for biomedical applications need to account for this parameter within the overall design and structure (Geneva et al. 2019). The thermal behavior of a smart hydrogel can be commonly characterized using differential scanning calorimetry (DSC) that can detect phase transitions such as the glass transition temperature or melting point of a hydrogel (Hilmi et al. 2016). More importantly, DSC can also detect the lower critical solution temperature (LCST) of temperature-responsive hydrogels which refers to when the hydrogel undergoes a phase transition from a swollen and hydrated state to a collapsed or de-swollen state (Shi et al. 2020). Another method is thermogravimetric analysis (TGA) but unlike the DSC, it measures the change in weight of the hydrogel as it is being heated and can help characterize the thermal stability of the hydrogel including the temperature at which it degrades and the weight loss due to decomposition (Jiang et al. 2018).

3. pH

Similar to our internal body temperature, the pH of our internal tissues also falls into a narrow pH range of around 7.35–7.45 under normal physiological conditions (Cencer et al. 2014). However, the surrounding pH of certain internal tissues can fluctuate out of this normal range in cases where individuals suffer from inflammation or infection (Derwin et al. 2023). Smart hydrogels need to be able to adapt to these changes in pH and researchers need to ensure that the hydrogel can maintain a suitable pH range for optimal cell or tissue viability and growth over the short and long term. Cell and swelling behavior studies are simple methods to predict and characterize how a smart hydrogel may respond to changes in pH. Cell behavior can be evaluated by culturing tissue cells within the smart hydrogel submerged in different varying pH buffers and observing over time their attachment, proliferation, and differentiation under different pH conditions, leading to an understanding of the hydrogel's impact on cellular responses (Hilmi et al. 2016). Swelling behavior is a very similar concept but instead, the hydrogel's volume or weight in different pH levels would be monitored over time.

4. Porosity

Within the matrix of the hydrogel's structure, typically there will be interconnections of different-sized pores that help promote cell-matrix interactions, nutrient and oxygen diffusion, waste removal, and tissue remodeling. The size and degree of porosity can be controlled through the choice of polymer with respect to the degree of cross-linking density; as mentioned earlier, higher cross-linking density leads to a denser network with smaller pore sizes (Chavda and Patel 2011). Gas foaming is a technique that incorporates gas bubbles, into the hydrogel's precursor solution, prior to gelation of the hydrogel. The amount and size of gas bubbles can be controlled at varying pressures and temperatures to form void spaces and as the hydrogel solidifies, it results in its porous structure (Annabi et al. 2010). Another relevant technique is AM such as 3D printing to control porosity as there is much more precise control in the design aspect of pore size and distribution within the hydrogel's structure (Sultan and Mathew 2018). Typically, porosity can be characterized using microscopy such as SEM or confocal microscopy of a hydrogel's cross-section to quantify pore size, connectivity, and distribution using imaging analysis software (Jamshidi and Falamaki 2021).

5. Biocompatibility

The term "biocompatibility" can be oversimplified when referring to smart hydrogels or polymers that are implanted or injected *in vivo*. There is the potential to misrepresent the high level of complexity and interactions that can occur between biomaterials and biological tissues. As long as the biomaterial is to be implanted, injected, or placed within the body, the biocompatibility characteristics will vary depending on specific biological and clinical factors. Additionally, this statement is further supported by many investigators describing that no single biomaterial can be a "generic biocompatible material" (Williams 2019b). A more holistic approach that does not focus solely on whether a smart hydrogel is biocompatible is to instead support the claim that a biomaterial is "biocompatible" based upon the quantitative evidence collected through its mechanical properties, degradation rate, swelling behavior, rheology behavior, surface properties, and its cell interactions over the short and long term. This will provide a more accurate profile of how the smart hydrogel may behave in a biological context and how it is biocompatable.

Smart hydrogels in oral tissue engineering

The synergistic combination of smart hydrogels and advanced AM techniques has emerged as a promising frontier in dental and craniofacial tissue repair and regeneration. Smart hydrogels offer tailored platforms for controlled drug delivery, guided tissue regeneration, and functional scaffolds for repairing damaged tissues, including teeth, bones, and soft oral structures. By harnessing the responsiveness of smart hydrogels to external stimuli, and the spatial precision offered by AM technologies, researchers aim to develop innovative solutions that mimic the intricate nature of native tissues, enhancing the field's potential to address complex challenges in oral health. In this section, we review the recent advances and practical implementations of smart hydrogels in the realm of oral hard and soft tissue engineering. Specifically, we will focus on their role in the regeneration and revitalization of dental tissues.

Oral hard tissue engineering

Enamel

Tooth enamel is the hardest mineralized tissue in humans and covers the outermost part of the tooth. They are composed of highly organized hydroxyapatite (HA) rods

(95%), water (4%), and low levels of organic components (Li et al. 2014). The meticulously organized and intricate arrangement of HA crystals, enamel rods, and the interrod matrix spans across nano to microscale dimensions, rendering them the toughest bioceramics within the human body (Navin and Prasanna 2014). An imbalance in this homeostasis leads to enamel demineralization and the acellular composition of mature enamel limits its ability to undergo regeneration, and consequently leads to caries formation. Numerous approaches have been employed in the past several decades to repair and regenerate lost enamel. However, the majority of these approaches that have shown promise utilize bioactive materials like fluoride, casein-based products (Reynolds 2008), bioinspired nano(n)HA substitutes (Pepla et al. 2014) and biomimetic components such as amelogenin (Fan et al. 2011).

Smart hydrogels in the context of enamel regeneration have only been investigated in recent years. Hanafy et al. developed a chitosan-inspired biomimetic hydrogel and compared its efficacy with zinc-doped nHA in repairing etched enamel (Hanafy et al. 2019). *In vitro* assessments using SEM revealed the formation of new rod-like HA layers in both the chitosan hydrogels and nHA-treated groups (Hanafy et al. 2019). This study supports the potential of chitosan-based hydrogels as a smart biomimetic system for enamel repair in caries. More recently, Musat et al. reported the use of chitosan-agarose (CS-A) hydrogels to repair acid-etched enamel (Muşat et al. 2021). They evaluated the potential of CS-A hydrogels in enamel remineralization of tooth slices over ten days immersed in artificial saliva. Results show that CS-A hydrogels supported the formation of HA nanorods and provided ECM support to facilitate remineralization (Muşat et al. 2021).

An alternative strategy of using hydrogels for enamel repair includes their role as a vehicle/scaffold for cells or bioactive factors that promote remineralization. Odontogenic ameloblast-associated protein (ODAM) is known to strongly interact with Amelotin (AMTN) during the enamel maturation stage (Lee et al. 2010). Ikeda et al. used a recombinant human ODAM protein-impregnated collagen hydrogel in the presence of simulated body fluid to evaluate mineral deposition. They found calcium phosphate depositions on ODAM hydrogel-treated surfaces compared to the bovine serum albumin-treated collagen hydrogels. The authors concluded that the ODAM-collagen hydrogel matrix has the potential to promote HA nucleation and assist in remineralization of enamel (Ikeda et al. 2018). Similarly, an amelogenesis-inspired hydrogel composite was reported by Liu et al. to promote the remineralization of non-cavitated enamel. The authors developed a novel composite hydrogel using amelogenin-derived peptide QP5 and bioactive glass. The BQ hydrogel-treated enamel surface showed a marked reduction in lesion depth and minimal mineral loss both under *in vitro* and *in vivo* conditions. The authors suggest that QP5-bioactive glass composite hydrogels perform better for non-cavitated enamel repair as opposed to hydrogels containing either QP5 or glass (Liu et al. 2022c). In a recent work by Mohabatpour and colleagues, the authors combined the use of 3D bioprinting with smart hydrogels for enamel engineering *in vitro*. They developed a novel carboxymethyl chitosan (CMC)-alginate (Alg) bioink that can be used as 3D printed scaffolds for enamel regeneration. The CMC-Alg hydrogels maintained high viability of dental epithelial cells and showed functional secretion of alkaline phosphatase after 14 days of culture (Mohabatpour et al. 2022). These printable bioinks can serve as ECM to support cell-driven mineral deposition and enamel-like tissue formation *in vitro* and potentially for clinical use in the future.

Cementum

The cementum forms a thin acellular that covers the root surfaces of the tooth. They play a crucial role in maintaining the periodontal ligament (PDL) health and facilitate attachment of PDL fibers connecting the tooth and alveolar bone (Gonçalves et al. 2005). Loss of cementum due to attachment loss and subsequent root exposure is commonly observed in periodontitis. Currently used strategies to regenerate lost cementum include delivery of cementoblast progenitor cells, co-delivery of secreted cytokines, cell sheet application, and use of micro-nano scale tailored biomaterials (Liu et al. 2019). While research using smart hydrogels for PDL complex is pertinent, there is limited evidence on their use specifically for cementum regeneration. In an earlier work by Park et al., the authors found that the delivery of progenitor cementoblasts using hydrogels, especially fibrin matrix was challenging due to the high levels of protease secretions which caused hydrogel degradation and cementoblast apoptosis (Park et al. 2017). To overcome this challenge, the group introduced aminocaproic acid (ACA) and combined it with chitosan particles to create ACA-releasing chitosan particles (ACP). ACP-fibrin hydrogels prevented matrix degradation and promoted *in vitro* differentiation of cementoblasts. Further, on introducing ACP-fibrin hydrogels into periodontal defects in beagle dogs, a significant amount of new cementum formation was observed when compared to simply using fibrin hydrogels or clinically approved enamel matrix derivatives (EMD). Further, these hydrogels were also shown to promote osteogenesis when used with pre-osteoblasts by increasing bone volume and providing better root coverage *in vivo* (Park et al. 2017). A systematic review of studies utilizing scaffolds, including hydrogels combined with stem cells for cementum regeneration in dogs reported that different sources of stem cells have the potential to promote cementum regeneration (Crossman et al. 2018).

In addition to stem cells promoting cementogensis, the use of cementum-specific proteins such as cementum-derived growth factor (CDGF), cementum attachment protein (CAP), and cementum protein-1 (CEMP1) and growth factors have shown promise towards cementum and bone formation (Arzate et al. 2015). Sowmya et al. fabricated a tri-layered nanocomposite hydrogel scaffold for the simultaneous regeneration of cementum, PDL, and alveolar. To promote cementum, the authors fabricated a chitin-PLGA Poly(lactic-co-glycolic acid) hydrogel scaffold functionalized with CEMP1 and nano-bioactive glass ceramic (nBGC). The addition of growth factors, CEMP-1, FGF-2, and platelet-rich plasma (PRP) promoted closure of periodontal defects in rabbit maxilla with the formation of new cementum, fibrous PDL, and alveolar bone, respectively (Sowmya et al. 2017). In a similar approach, Wang et al. fabricated a calcium phosphate cement (CPC)/propylene glycol alginate (PGA) combined with bone morphogenetic protein-2 (BMP-2)/FGF-2 in a periodontal defect model of non-human primates (NHP) *Macaca fascicularis* (Wang et al. 2019). Three walled periodontal defects were created in the mandibles of NHPs and treated with PGA/FGF + CPC/BMP. Three months post-treatment, there was a significant improvement in cementum and PDL regeneration with epithelial down growth suggesting the potential of PGA/FGF + CPC/BMP-based hydrogels in PDL complex regeneration (Wang et al. 2019). More recently, a similar approach aimed at PDL complex regeneration was proposed by using integrin-binding peptide-functionalized PEG hydrogels (Fraser and Benoit 2022). The dual peptide hydrogels used RGD and GFOGER moieties to regulate periodontal ligament cell (PDLC) activity by ALP regulation and matrix mineralization, respectively. The dual combination of peptides showed the formation of two definitive PDLC phenotypes with distinct gene expression patterns and pyrophosphate concentration. High GFOGER peptides and moderate levels of RGD promoted the expression of cementoblasts and osteoblast-specific markers. The study concluded the use of peptide-functionalized hydrogels as an ideal transplantation scaffold for cementum regeneration in rats with the potential to be used clinically (Fraser and Benoit 2022).

Alveolar bone regeneration

A healthy alveolar bone is crucial for maintaining the homeostasis of the PDL complex. Severe forms of periodontitis lead to loss of soft tissue attachment and thereby bone loss and furcation involvement. The gold standard of treatment in such cases involves the use of patient-derived autologous grafts and more recently, non-human and synthetic bone substitutes (de Grado et al. 2018). In the past decade, the combination of stem cells, hydrogel scaffolds, and biomolecules including growth factors and signaling cues have been largely explored for application in alveolar bone regeneration (Hollý et al. 2021). Pan et al. investigated the use of human PDL stem cells (PDLSCs) encapsulated in gelatin methacrylate (GelMA) hydrogels for alveolar bone regeneration in rat maxillary alveolar bone (Pan et al. 2020). They developed a highly porous and interconnected hydrogel with the potential to promote *in vitro* proliferation, migration, and osteogenic differentiation of PDLSCs. Additionally, micro-CT and histological assessments of critical-sized maxillary defects in rats transplanted with GelMA-encapsulated PDLSCs revealed notable bone regeneration (Pan et al. 2020). To improve the versatility and physicochemical properties of GelMA hydrogels, Chen et al. previously fabricated photocrosslinkable GelMA-nanohydroxyapatite (nHA) microgel arrays for periodontal tissue engineering (Chen et al. 2016). GelMA-nHA displayed superior microstructure, mechanical strength, and surface properties to promote *in vitro* adhesion and proliferation of PDLSCs. Further, subcutaneous transplantation of stem cell encapsulated microgels in nude mice showed an increase in vesiculation and mineralized tissue formation (Chen et al. 2016). Overall, gelatin-based smart hydrogels serve as a versatile matrix for stem cell-induced alveolar bone regeneration.

    In vivo, Bone Morphogenetic Protein (BMP) has demonstrated a strong capacity for inducing bone formation (Cheng et al. 2003). Fostered by their osteoinductive nature, BMPs have been commonly used in conjunction with hydrogels and scaffolds for alveolar bone regeneration. An injectable, thermosensitive chitosan nanoparticle (CSn) and α,β-glycerophosphate (α,β-GP) based hydrogel was encapsulated with BMP-2 plasmid DNA (pDNA-BMP2) to evaluate their potential in alveolar bone regeneration (Li et al. 2017). CS/CSn-GP or CS/CSn-pDNA-BMP2-GP were injected into the rat calvaria to evaluate bone formation via H&E staining. The pDNA-BMP2-GP system promoted the recruitment of osteoblasts with lacunae formation and trabecular bone formation as opposed to connective tissue-mediated repairs in control groups. Further, the group tested the efficacy of CS/CSn-pDNA-BMP2-GP hydrogels in alveolar bone regeneration in a dog periodontitis model which showed a similar pattern of new bone formation (Li et al. 2017). Hamlet et al. used 3D-printed PCL with a hyaluronic acid (HA) hydrogel encapsulated with BMP-7 and osteoblasts to evaluate their osteogenic potential (Hamlet et al. 2017). Functionalized PCL-HA hydrogels served as an ideal platform for osteoblast culture in 3D with the formation of a mineralized collagenous matrix at 6 weeks. Further, subcutaneous transplantation of osteoblast-encapsulated PCL-HA hydrogels with BMP-7 showed a significant amount of mineralized tissue formation compared to BMP-7 free transplants (Hamlet et al. 2017).

Hydrogel encapsulation of bioactive molecules such as exosomes derived from stem cells offers a promising approach toward tissue regeneration (Khayambashi et al. 2021). Jing et al. developed a dynamic bioresponsive DNA hydrogel encapsulated with exosomes derived from apical papilla stem cells (SCAP-Exo) with superior osteogenic and angiogenic properties (Jing et al. 2022). The PEG/DNA hydrogels allowed a controlled release of SCAP-Exo in response to matrix metalloprotease-9 (MMP-9) release in diabetes-associated bone defects. Injectable SCAP-Exo containing PEG/DNA hydrogels largely promoted formation of vascularized bone tissues in mandibular alveolar bone defect models in diabetic rats (Jing et al. 2022). Such bioresponsive smart hydrogels provide remarkable adaptability for application in dynamic disease microenvironments. In such contexts, where specific disease-related biomarker cues directly govern the release of therapeutic molecules, these hydrogels exhibit significant versatility. In addition to promoting bone regeneration via controlled release of stem cells, their derivatives, and other osteoinductive biomolecules, recent research has also focused on the use of smart hydrogels for delivery of antibacterial compounds and drugs (Qi et al. 2022, Xu et al. 2019).

Considerable advancements have been made in the field of smart hydrogel-based oral hard tissue regeneration. Table 2 outlines recent examples of smart hydrogels used in oral hard tissue regeneration. Despite the successful outcomes displayed by individual strategies in terms of *in vivo* hard tissue regeneration, a research gap exists concerning the synergistic amalgamation of these techniques to expedite their clinical application. Furthermore, the establishment of a standardized approach for the fabrication, formulation, and encapsulation of biomolecules within clinical-grade hydrogels is imperative. This standardization would facilitate the assessment of their efficacy in comparison to established benchmarks and other clinically validated biomaterials intended for the regeneration of hard tissues.

Table 2. Smart hydrogels for oral hard tissue regeneration.

Tissue	Hydrogel	Results	Study type	Ref.
Enamel	Chitosan with Ca^{2+} and PO_4^{3-} ions.	Cell-free generation of parallel and elongated enamel-like rods.	*In vitro*	(Hanafy et al. 2019)
	Chitosan-Agarose	7-day remineralization shows HA layer with a 1:64 Calcium/Phosphate ratio and microhardness recovery of 77.4%	*In vitro*	(Muşat et al. 2021)
	ODAM-collagen hydrogel	Needle-like HA crystals formed on surfaces treated with collagen-containing ODAM.	*In vitro*	(Ikeda et al. 2018)
	QP5 (Amelogenin derived peptide)-Bioactive glass composite hydrogel	BQ hydrogels reduce lesion depth, recovers surface microhardness and color, and provide resistance to erosion of the re-mineralized layer.	*In vitro & In vivo*	(Liu et al. 2022c)
	Carboxymethyl chitosan–alginate hydrogels	Alg4% – CMC% bioprinted hydrogels show the highest potential for ameloblast differentiation, calcium and phosphate deposition, and matrix mineralization.	*In vitro*	(Mohabatpour et al. 2022)
Cementum	ACA-chitosan-fibrin hydrogels	ACP-fibrin hydrogels inhibit protease activity and OCCM30 cementoblast apoptosis and promote cementogensis and osteogenesis in beagles.	*In vitro & In vivo*	(Park et al. 2017)
	Chitin/PGLA-CEMP1+ nBGC or Chitin/PGLA-FGF-2 or Chitin/PGLA -PRP	Tri-layered nanocomposite hydrogels promote cementogenic, fibrogenic, and osteogenic differentiation of human dental follicle cells and promotes formation of new cementum, PDL, and bone in rabbit's maxilla.	*In vitro & In vivo*	(Sowmya et al. 2017)
	PGA/FGF-2+ CPC/BMP-2 hydrogels	Hybrid smart hydrogels with growth factors promote significant cementum and PDL regeneration in NHPs.	*In vivo*	(Wang et al. 2019)
	RGD and GFOGER functionalized PEG hydrogels	Peptide functionalized, PDLC encapsulated hydrogels promote cementum and bone formation when transplanted in periodontal defects in rats	*In vitro & In vivo*	(Fraser and Benoit 2022)

Table 2 contd. ...

...Table 2 contd.

Tissue	Hydrogel	Results	Study type	Ref.
Alveolar Bone	GelMA	GelMA encapsulated PDLSCs show *in vitro* proliferation, and osteogenic differentiation and promote new bone formation in alveolar defects of rats.	*In vitro & In vivo*	(Pan et al. 2020)
	GelMA + nHA	GelMA (10%) + nHA (2%) w/v promoted osteogenic differentiation of PDLSCs with elevated expression of ALP, BSP, OCN, and RUNX2 genes and showed mineralized tissue formation and vascularization when transplanted subcutaneously in nude mice.	*In vitro & In vivo*	(Chen et al. 2016)
	CS/CSn-pDNA-BMP2-GP	CS/CSn-pDNA-BMP2-GP hydrogels promoted formation of new mineralized tissue, Sharpey's fibers attachment, and an increase in ALP expression compared to pDNA-BMP2 free CS/CSn gels in the dog periodontitis model.	*In vitro & In vivo*	(Li et al. 2017)
	PCL-HA + BMP7	PCL- HA hydrogels allow a biphasic release of BMP-7 with sustained release for 5 weeks and promoted formation of woven bone-like structures subcutaneously in rats after 4 weeks.	*In vitro & In vivo*	(Hamlet et al. 2017)
	PEG/DNA hybrid hydrogels	SCAP-Exo integrated PEG/DNA hydrogels responds to dynamic release of MMP-9 in diabetic bone defects by promoting osteogenesis and angiogenesis in rat's alveolar bone.	*In vitro & In vivo*	(Jing et al. 2022)

Oral soft tissue engineering

Gingival mucosa

Gingival tissues comprise the segment of the oral epithelium responsible for enveloping the alveolar bone in both the mandibular and maxillary regions. Severe gingival inflammation followed by periodontitis leading to lack of keratinized epithelium around the tooth neck is the most common issue warranting the need for gingival regeneration. Connective tissue grafts and coronally advanced flap surgeries are the treatment standards in severe cases of gingival recession, subsequent to periodontitis and root involvement (Wessel and Tatakis 2008). Hydrogel-inspired biomaterials have been recently explored in gingival tissue engineering and wound healing applications. Hatayama et al. fabricated pH-responsive collagen-based scaffolds (CS) to treat gingival regeneration in beagle dogs (Hatayama et al. 2017). They tested two collagen-based scaffolds, CS-pH 7.4 and CS-pH 3.0 in a surgically created buccal gingival wound in dogs. Results showed a natural and significant recovery of epithelium and submucosal tissues in both collagen scaffolds-covered wounds compared to the controls with no scaffolds. Among the scaffolds, CS-pH 7.4 showed a significant difference in gingival regeneration compared with CS-pH 3.0, further validating the pH-responsive nature that affects the mechanical, physiochemical, and biological activity of collagen hydrogels (Hatayama et al. 2017). Other groups have also utilized collagen hydrogels in combination with chitosan (Rosdiani et al. 2017) or replaced collagen with biomimetic chitosan (Feng et al. 2021) to create mechanically superior and tunable scaffolds for accelerated gingival tissue engineering.

Cell-encapsulated hydrogels are widely assessed for use in oral mucosal tissue engineering. Tabatabaei et al. compared the potential of GelMA and collagen hydrogels as connective tissue scaffolds for fibroblast survival and subsequent oral epithelial cell adhesion (Tabatabaei et al. 2020). The authors showed that fibroblast cultures in collagen hydrogel showed higher cell viability than GelMA. Additionally, oral epithelial keratinocytes showed the formation of stratified keratinized epithelium on the collagen surface and not on GelMA (Tabatabaei et al. 2020). While the authors showcase the superior nature of collagen hydrogels in this work, a more recent work supports the potential of RGD-modified alginate GelMA hydrogels in oral wound healing and soft tissue regeneration. Ansari et al. used gingival mesenchymal stem cells (GMSC) encapsulated alginate-GelMA hydrogels to test MSC viability and collagen deposition (Ansari et al. 2021). *In vivo* transplantation of GMSC encapsulated alginate-GelMA hydrogel in mice showcased angiogenesis enhanced wound healing and suppression of inflammatory biomarkers. The authors concluded that GMSC is a versatile stem cell source and alginate-GelMA hydrogel sheets offer a unique platform for oral soft tissue regeneration including a potential scope of use in reconstructive surgery.

To address the need for tailored tissue substitutes to adapt to various soft oral tissue discrepancies, Yi et al. developed a 3D bioprinted platelet-rich fibrin (PRF) hydrogel to promote oral soft tissue regeneration. Their bioink comprised of alginate and gelatin hydrogels combined with injectable PRF (i-PRF) to facilitate personalized soft tissue augmentation. The iPRF hydrogels exhibited excellent mechanical properties aiding in their bioprintability with superior *in vitro* and *in vivo* biocompatibility (Yi et al. 2022). The group further demonstrated that 3D bioprinted hydrogels encapsulated with human gingival fibroblasts promoted angiogenesis and suppressed inflammation when transplanted subcutaneously in mice models (Yi et al. 2022). Combined, these results show the potential of smart, mechanically tunable 3D bioprinted hydrogels for stem cell encapsulation, growth factor release, and immunomodulation for personalized use in oral soft tissue regeneration. Nevertheless, based on a recent systematic review that investigated the hydrogel-based biomaterials used as scaffolds for gingival regeneration, extensive preclinical data is still needed to ascertain the clinical use of smart hydrogels in oral soft tissue regeneration (Hutomo et al. 2023).

Periodontal ligament

The periodontal ligament (PDL) is a connective tissue composed of different types of highly organized and differentially aligned fibers that connect the cementum to the alveolar bone (Beertsen et al. 1997). The PDL structure consists of cells, fibers, nerves, and blood vessels articulated precisely within a matrix to attenuate occlusal forces and provide tooth retention, nutrition, and sensory perception (Berkovitz 2004). Regeneration strategies in response to the PDL complex disease therefore aim to regulate complex architecture while maintaining their functional characteristics. A range of regenerative treatments for periodontal issues have emerged in the past couple of decades. These therapies encompass techniques like guided tissue regeneration (GTR), enamel matrix derivatives, bone grafts, the controlled release of growth factors, as well as the synergistic use of cells and growth factors in conjunction with scaffold matrices (Kwan et al. 2023). These advancements are aimed at addressing the rejuvenation of compromised tooth-supporting structures, which encompass the PDL, alveolar bone, and cementum (Liang et al. 2020). Smart hydrogels with superior biocompatibility and biodegradation profiles with controlled delivery of stem cells, drugs, and biomolecules to periodontal defects have been investigated recently.

Momose et al. tested the use of fibroblast growth factor-2 (FGF-2) encapsulated collagen microgels for furcation treatment in dogs (Momose et al. 2016). Collagen FGF-2 hydrogels promoted cell and tissue growth with cell recruitment and blood vessel-like structures. Moreover, the periodontal attachment was recovered, comprising of tissue resembling cementum, PDL, and Sharpey's fibers. Scaffold loaded with FGF2 facilitated a self-assembly process, ultimately reinstating the functionality of periodontal structures (Momose et al. 2016). Chien and colleagues formulated a thermo-responsive, chitosan/gelatin/glycerol phosphate hydrogel to incorporate iPSCs and BMP-6 for PDL regeneration in rats (Chien et al. 2018). Hydrogels containing iPSCs and BMP-6 promoted osteogenesis, showing a significant increase in bone volume, with subsequent PDL regeneration in rats when compared to hydrogels with BMP-6 alone. In addition, the synergistic combination also reduced the presence of inflammatory cytokines in the periodontal defects, suggesting their potential in PDL regeneration not only via stem cell recruitment but also by minimizing inflammatory markers (Chien et al. 2018).

With a goal towards PDL complex regeneration, He et al. developed a high stiffness gelatin-transglutaminase (TG) hydrogel for stem cell recruitment and macrophage modulation (He et al. 2019). The TG hydrogels loaded with interleukin-4 (IL-4) and stromal-derived factor (SDF)-1α were transplanted into the periodontal defects created in rats. Micro-CT evaluation at weeks four and eight showed new bone and cementum formation compared to untreated groups. Further, they evaluated the tissues via H&E staining which at week four revealed the formation of better epithelial attachment in groups without IL-4. Combined, the results from this study revealed that the use of both SDF-1α and cytokines provided an ideal microenvironment for PDL complex regeneration via macrophage phenotype modulation, stem cell homing, and osteogenic differentiation (He et al. 2019). More recently, Liu et al. developed a thermosensitive gingipain responsive hydrogel loaded with SDF-1 as a drug delivery platform to inhibit *Porphyromonas gingivalis* and control the associated inflammation for *in situ* PDL regeneration (Liu et al. 2021). *In vitro*, results from their study showed that SDF-loaded hydrogels promoted the proliferation migration and osteogenic differentiation of PDLCs. Additionally, when tested in a rat periodontal defect model, gingipain stimulation of these hydrogels inhibited the growth of *P. gingivalis*, thereby reducing inflammation and promoting stromal cell recruitment for PDL regeneration (Liu et al. 2021). These studies provide valuable evidence for the use of SDF-loaded hydrogels as a prospective approach toward successful PDL complex regeneration.

In conclusion, commonly used biomaterials for PDL complex regeneration include mostly inorganic materials such as tricalcium phosphate, bioactive glasses, and above discussed polymeric materials like collagen and gelatin (Liang et al. 2020). However, given the challenges with interfacial tissue regeneration, that is, the cementum-bone-PDL interface, novel biomimetic scaffolds including hierarchal and heterogeneously graded hydrogels are required to successfully mimic and regenerate the native PDL complex (Zhang et al. 2021).

Dentin-pulp complex

The dentin-pulp complex (DPC) is crucial for maintaining tooth vitality and sensitivity by adapting to different external stimuli through self-protective defenses. A homeostatic dentin-pulp complex responds to different types of injury via modes of dentin regeneration which is their primary biological role (Simon et al. 2011). Great progress has been made towards developing biomaterials for pulp healing and dentin regeneration. Hydrogel-based regeneration of the dentin-pulp complex has generated a lot of interest in recent years (Abbass et al. 2020). Bakhtiar et al. fabricated decellularized human amniotic membrane (HAM)-based hydrogels with varying concentrations of genipin to culture dental pulp stem cells (DPSCs). Their results showed that HAM hydrogels at 30 mg/ml concentration crosslinked with 0.5 mM genipin performed best in terms of modulus of elasticity, cell viability, and the least immunogeneicity when injected into mice (Bakhtiar et al. 2023). The superior biocompatibility offered by HAM in addition to the improved biodegradability of genipin crosslinked hydrogels portray their high potential for DPC regeneration. Wang et al. developed a similar injectable hydrogel composed of methacrylic acid

Table 3. Smart hydrogels for oral soft tissue regeneration.

Tissue	Hydrogel	Results	Study type	Ref.
Gingiva	Collagen	Collagen scaffolds with a pH of 7.4 showed superior mechanical properties and induced better wound healing by regenerating gingival tissues compared to scaffolds with pH3.0	*In vivo*	(Hatayama et al. 2017)
	Alginate-GelMA	GMSC encapsulated alginate-GelMA hydrogels with 14–22 kPA stiffness promoted skin wound healing in mice and collagen deposition by angiogenic activation, removal of inflammatory cytokines TNF-α, and increase in anti-inflammatory marker, IL-10.	*In vitro & In vivo*	(Ansari et al. 2021)
	3D bioprinted Alginate-Gelatin-iPRF	Bioprinted alginate-gelatin-iPRF mixtures allow sustained and prolonged growth factor release, with high HFG viability proportional to iPRF concentration with subcutaneous vascularized soft tissue regeneration.	*In vitro & In vivo*	(Yi et al. 2022)
PDL complex	Collagen-FGF2	FGF-2 loaded scaffolds promoted the regeneration of PDL complex structures with vascular cell recruitment without ankylosis and resorption.	*In vivo*	(Momose et al. 2016)
	Chitosan/Gelatin/Glycerol Phosphate	iPSC and BMP-6 loaded C/G/GP hydrogels promote differentiation of both new PDL tissue and mineralized tissue like cementum and bone in rats.	*In vitro & In vivo*	(Chien et al. 2018)
	Gelatin – TGs with SDF-1α & IL-4	Hydrogels containing SDF-1α & IL-4 lead to attachment recovery, in addition to regeneration of PDL, cementum, and bone via macrophage polarization and stem cell recruitment.	*In vitro & In vivo*	(He et al. 2019)
	PEG-diacrylate-dithiothreitol-functional peptide molecule + SDF-1	PEGPD-SDF-1 hydrogels exerts antibacterial effects towards *P. gingivalis*, reduces inflammation, facilitates CD90+/CD34- stromal cell recruitment and osteogenesis.	*In vitro & In vivo*	(Liu et al. 2021)

Dentin-Pulp complex	HAM-genipin	30 mg/ml HAM crosslinked with 0.5 mM genipin provides optimal conditions for DPSCs culture, and forms pulp-like tissue when implanted into rat calvaria.	*In vitro* & *In vivo*	(Bakhtiar et al. 2023)
	Gel-MA/NGR1	Gel-MA/NGR1 hydrogels promote *in vitro* dentinogenesis, with elevated ALP and OCN levels and show a ~ 175-fold increase in calcification in a modified pulp capping model in rats.	*In vitro* & *In vivo*	(Wang et al. 2021)
	TDM-alginate	Tomographic evaluation showed that TDM hydrogels are superior pulp capping materials compared to biodentine and MTA.	Clinical trial	(Holiel et al. 2021)
	GelMA-DMM	Microgels supplemented with 500 ug/ml and 1000 ug/ml DMM performed better compared to currently used pulp capping materials.	*In vivo*	(Cunha et al. 2023)

functionalized gelatin loaded with notoginsenoside R1 (Gel-MA/NGR1) to promote reparative dentinogenesis (Wang et al. 2021b). Gel-MA/NGR1 hydrogels showed formation of interconnected porous microarchitecture with a modulus of elasticity of 50–60 kPA. Furthermore, NGR1 functionalized hydrogels promoted odontogenic differentiation of mouse dental papilla cells showing elevated levels of alkaline phosphate (ALP) and osteocalcin (OCN). The group successfully tested the potential of Gel-MA/NGR1 in promoting reparative dentin in the alveolar cavity of rats (Wang et al. 2021). The study findings conclude the potential of these hydrogels as effective pulp capping agents for dentin repair.

Treated dentin matrix (TDM) from extracted teeth has served as a promising biomaterial for DPSC attachment, proliferation, and differentiation (Li et al. 2011). Holiel et al. conducted a two-year randomized control trial to test the efficacy of injectable TDM hydrogels as a direct pulp capping material. Their findings show that TDM hydrogels showed better dentin bridge formation compared to clinically successful materials biodentine and mineral trioxide aggregate (MTA) (Holiel et al. 2021). To further optimize the clinical performance of TDM hydrogels, recently, other groups further tested their *in vitro* properties, including synthesis and optimization of composite hydrogels such as gelatin-glycidyl methacrylate/TDM hydrogels (Sedek et al. 2023). To improve spatial control of dentin matrix molecules (DMM) within the hydrogels, 3D-printed GelMA microgels were tested on the maxillary molars of Wistar rats. GelMA-DMM microgels showed formation of organized pulp tissue, minimal necrosis, and greater dentin bridge formation compared to control groups containing MTA or platelet-derived growth factor loaded microgels (Cunha et al. 2023). Current literature provides strong evidence of the potential of both natural and synthetic hydrogels for DPC regeneration (Fukushima et al. 2019). Particularly, stem cell and bioactive factor encapsulated natural hydrogels serve as ideal platforms for clinical pulp repair and dentin regeneration. Table 3 outlines examples of smart hydrogels that have shown great preclinical promise and can be potentially translated for use in oral soft tissue regeneration clinically.

Future perspectives in oral tissue engineering

The past two decades have seen a major climb in the utilization and exploration of intelligent biomaterials for various applications in the biomedical field. Smart hydrogels in the realm of intelligent materials have shown great promise, especially in the fields of tissue engineering and regenerative medicine. The versatility of these bioresponsive hydrogels can be attributed to our improved understanding of their biochemical and physio-chemical properties. In addition, the evolution in 3D/4D printing techniques allows for precise control of smart polymers, allowing the fabrication of novel composite hydrogels for various applications.

In lieu of oral tissue engineering and regeneration, naturally derived smart hydrogels offer a great platform owing to their innate biocompatible and biodegradable nature. They have shown great promise for use in 3D cell culture, cell and biomolecule encapsulation, sustained release of drugs and other growth factors, and limited cytotoxicity to name a few. However, these benefits are accompanied by challenges in regulating the mechanical and physical properties of hydrogels, particularly for use in oral hard or mineralized tissue engineering. Surely, as described in previous sections, combining synthetic, intelligent polymers with natural hydrogels have resulted in novel hydrogel candidates. Many hydrogels have shown the potential to host different types of oral stem cells, by allowing proliferation, migration, and attachment within 3D hydrogels in addition to providing angiogenic and osteogenic cues. However, there is still a need to evaluate and integrate these hydrogels as unique systems specific to each tissue, taking into consideration their dynamic environment, and hierarchical interfaces (such as PDL-bone, and dentin-pulp) to prolong their long-term stability within the oral environment.

Applying AM techniques to fabricate multi-stimulus responsive hydrogels could enhance customization and provide a spatially controlled platform for complex tissue regeneration. Further, heterogeneous hydrogels with the ability to respond to different stimuli could better adapt to dynamic oral environments and facilitate effective tissue regeneration. Moreover, understanding how such smart hydrogel-based constructs perform over extended periods in physiologically relevant preclinical models will provide valuable insights into their durability and longevity. Future work should also focus on a comprehensive approach that not only emphasizes technical innovation but also underscores the importance of ethical considerations, regulatory compliance, and practical clinical implementation. As research in the field progresses, addressing these aspects will be essential to ensure the successful translation of smart hydrogel-based approaches to oral tissue engineering and regeneration.

Conclusions

For the foreseeable future, the structure and function of smart hydrogels will continue to go beyond the limits of our current understanding. The clinical adoption of smart-responsive hydrogels remains a major challenge as our body's internal conditions are always changing and for smart hydrogels to adapt to different external stimuli and mimic our natural tissues is complex. There are many *in vitro* and animal studies on smart hydrogels in the oral cavity to repair and regenerate hard and soft tissues which have shown great progress, but there are even fewer studies in clinical trials. More clinical trials will be required to fully elucidate their clinical efficacy and whether the results are consistent with other independent clinical studies. It is reasonable to predict that smart hydrogels will be persistently optimized until each mechanical, chemical, and physical property is tunable; however, it will be important to quantitatively characterize these properties to help provide future researchers a foundation from which to build upon. Although the study of smart hydrogels spans a wide range of applications in many different fields, based on the literature review, their application within the field of dentistry as a biomaterial is promising and should be further investigated for hard and soft tissue repair and regeneration.

References

Abbass, M.M.S., A.A. El-Rashidy, K.M. Sadek, S.E. Moshy, I.A. Radwan, D. Rady, C.E. Dörfer and K.M. Fawzy El-Sayed. 2020. Hydrogels and dentin–pulp complex regeneration: from the benchtop to clinical translation. Polymers 12(12): 2935.

Ahmad, Z., S. Salman, S.A. Khan et al. 2022. Versatility of hydrogels: from synthetic strategies, classification, and properties to biomedical applications. Gels 8(3).

Ahmed, E.M. 2015. Hydrogel: Preparation, characterization, and applications: A review. J. Adv. Res. 6(2): 105–21.

Ahn, M., W.W. Cho, H. Lee et al. 2023. Engineering of uniform epidermal layers via sacrificial gelatin bioink-assisted 3D extrusion bioprinting of skin. Adv. Healthc Mater: e2301015.

Annabi, N., J.W. Nichol, X. Zhong, C. Ji, S. Koshy, A. Khademhosseini et al. 2010. Controlling the porosity and microarchitecture of hydrogels for tissue engineering. Tissue Eng. Part B Rev. 16(4): 371–83.

Ansari, S., S. Pouraghaei Sevari, C. Chen, P. Sarrion and A. Moshaverinia. 2021. RGD-modified alginate-GelMA hydrogel sheet containing gingival mesenchymal stem cells: a unique platform for wound healing and soft tissue regeneration. ACS Biomater. Sci. Eng. 7(8): 3774–3782.

Arzate, Higinio, Margarita Zeichner-David and Gabriela Mercado-Celis. 2015. Cementum proteins: role in cementogenesis, biomineralization, periodontium formation and regeneration. Periodontology 2000 67(1): 211–233.

Ashammakhi, Nureddin, Amin GhavamiNejad, Rumeysa Tutar et al. 2021. Highlights on advancing frontiers in tissue engineering. Tissue Engineering Part B: Reviews 28(3): 633–664.

Atala, A. 2004. Tissue engineering and regenerative medicine: concepts for clinical application. Rejuvenation Res. 7(1): 15–31.

Bakhtiar, Hengameh, Mohammad Reza Mousavi, Sarah Rajabi et al. 2023. Fabrication and characterization of a novel injectable human amniotic membrane hydrogel for dentin-pulp complex regeneration. Dental Materials.

Baumann, L., E.F. Bernstein, A.S. Weiss, D. Bates, S. Humphrey, M. Silberberg et al. 2021. Clinical relevance of elastin in the structure and function of skin. Aesthet Surg. J. Open Forum 3(3): ojab019.

Beertsen, Wouter, Christopher A.G. McCulloch and Jaroslav Sodek. 1997. The periodontal ligament: a unique, multifunctional connective tissue. Periodontology 2000 13(1): 20–40.

Berkovitz, Barry K.B. 2004. Periodontal ligament: structural and clinical correlates. Dental Update 31(1): 46–54.

Bertsch, P., M. Diba, D.J. Mooney and S.C.G. Leeuwenburgh. 2023. Self-healing injectable hydrogels for tissue regeneration. Chem. Rev. 123(2): 834–873.

Bienkowski, R.S., C.H. Wu and G.Y. Wu. 1986. Limitation on sensitivity of measuring collagen degradation: studies with cultured liver cells. Connect Tissue Res. 14(3): 213–9.

Bisotti, Filippo, Fabio Pizzetti, Giuseppe Storti and Filippo Rossi. 2022. Mathematical modelling of cross-linked polyacrylic-based hydrogels: physical properties and drug delivery. Drug Delivery and Translational Research 12(8): 1928–1942.

Bodugoz-Senturk, Hatice, Celia E. Macias, Jean H. Kung and Orhun K. Muratoglu. 2009. Poly(vinyl alcohol)–acrylamide hydrogels as load-bearing cartilage substitute. Biomaterials 30(4): 589–596.

Borges, Fernando T.P., Georgia Papavasiliou and Fouad Teymour. 2020. Characterizing the molecular architecture of hydrogels and crosslinked polymer networks beyond Flory–Rehner—I. Theory. Biomacromolecules 21(12): 5104–5118.

Brougham, C.M., T.J. Levingstone, N. Shen, G.M. Cooney, S. Jockenhoevel, T.C. Flanagan et al. 2017. Freeze-drying as a novel biofabrication method for achieving a controlled microarchitecture within large, complex natural biomaterial scaffolds. Adv. Healthc. Mater. 6(21).

Brovold, M., J.I. Almeida, I. Pla-Palacín et al. 2018. Naturally-derived biomaterials for tissue engineering applications. Adv. Exp. Med. Biol. 1077: 421–449.

Bustamante-Torres, M., D. Romero-Fierro, B. Arcentales-Vera, K. Palomino, H. Magaña and E. Bucio. 2021. Hydrogels classification according to the physical or chemical interactions and as stimuli-sensitive materials. Gels 7(4).

Caliari, S.R. and J.A. Burdick. 2016. A practical guide to hydrogels for cell culture. Nat. Methods 13(5): 405–14.

Cencer, M., Y. Liu, A. Winter, M. Murley, H. Meng and B.P. Lee. 2014. Effect of pH on the rate of curing and bioadhesive properties of dopamine functionalized poly(ethylene glycol) hydrogels. Biomacromolecules 15(8): 2861–9.

Chavda, H. and C. Patel. 2011. Effect of crosslinker concentration on characteristics of superporous hydrogel. Int. J. Pharm. Investig. 1(1): 17–21.

Chen, Qiang, Hong Chen, Lin Zhu and Jie Zheng. 2015. Fundamentals of double network hydrogels. Journal of Materials Chemistry B 3(18): 3654–3676.

Chen, Xi, Shizhu Bai, Bei Li, Huan Liu, Guofeng Wu, Sha Liu et al. 2016. Fabrication of gelatin methacrylate/nanohydroxyapatite microgel arrays for periodontal tissue regeneration. International Journal of Nanomedicine: 4707–4718.

Chen, Ying, Jie Li, Jiawei Lu, Meng Ding and Yi Chen. 2022. Synthesis and properties of Poly(vinyl alcohol) hydrogels with high strength and toughness. Polymer Testing 108: 107516.

Cheng, Hongwei, Wei Jiang, Frank M. Phillips et al. 2003. Osteogenic activity of the fourteen types of human bone morphogenetic proteins (BMPs). JBJS 85(8): 1544–1552.

Chien, Ke-Hung, Yuh-Lih Chang, Mong-Lien Wang et al. 2018. Promoting induced pluripotent stem cell-driven biomineralization and periodontal regeneration in rats with maxillary-molar defects using injectable BMP-6 hydrogel. Scientific Reports 8(1): 114.

Colter, J., B. Wirostko and B. Coats. 2018. Finite element design optimization of a hyaluronic acid-based hydrogel drug delivery device for improved retention. Ann. Biomed. Eng. 46(2): 211–221.

Crossman, Jacqueline, Maryam Elyasi, Tarek El-Bialy and Carlos Flores Mir. 2018. Cementum regeneration using stem cells in the dog model: A systematic review. Archives of Oral Biology 91: 78–90.

Crupi, A., A. Costa, A. Tarnok, S. Melzer and L. Teodori. 2015. Inflammation in tissue engineering: The Janus between engraftment and rejection. Eur. J. Immunol. 45(12): 3222–36.

Cui, L., Y. Yao and E.K.F. Yim. 2021. The effects of surface topography modification on hydrogel properties. APL Bioeng. 5(3): 031509.

Cunha, Diana, Nayara Souza, Manuela Moreira et al. 2023. 3D-printed microgels supplemented with dentin matrix molecules as a novel biomaterial for direct pulp capping. Clinical Oral Investigations 27(3): 1215–1225.

Dehghani, F. and N. Annabi. 2011. Engineering porous scaffolds using gas-based techniques. Curr. Opin. Biotechnol. 22(5): 661–6.

Derwin, Rosemarie, Declan Patton, Helen Strapp and Zena Moore. 2023. The effect of inflammation management on pH, temperature, and bacterial burden. International Wound Journal 20(4): 1118–1129.

Dobrzyński, P. and E. Pamuła. 2021. Polymeric scaffolds: design, processing, and biomedical application. Int. J. Mol. Sci. 22(9).

Dzobo, K., N.E. Thomford, D.A. Senthebane et al. 2018. Advances in regenerative medicine and tissue engineering: innovation and transformation of medicine. Stem Cells Int. 2018: 2495848.

Fan, Yuwei, James R. Nelson, Jason R. Alvarez, Joseph Hagan, Allison Berrier and Xiaoming Xu. 2011. Amelogenin-assisted *ex vivo* remineralization of human enamel: Effects of supersaturation degree and fluoride concentration. Acta Biomaterialia 7(5): 2293–2302.

Feng, Yanhuizhi, Huai-Ling Gao, Di Wu, Yu-Teng Weng, Ze-Yu Wang, Shu-Hong Yu et al. 2021. Biomimetic lamellar chitosan scaffold for soft gingival tissue regeneration. Advanced Functional Materials 31(43): 2105348.

Fernandez de Grado, Gabriel, Laetitia Keller, Ysia Idoux-Gillet et al. 2018. Bone substitutes: a review of their characteristics, clinical use, and perspectives for large bone defects management. Journal of Tissue Engineering 9: 2041731418776819.

Fisher, M.B. and R.L. Mauck. 2013. Tissue engineering and regenerative medicine: recent innovations and the transition to translation. Tissue Eng. Part B Rev. 19(1): 1–13.

Fraser, David and Danielle Benoit. 2022. Dual peptide-functionalized hydrogels differentially control periodontal cell function and promote tissue regeneration. Biomaterials Advances 141: 213093.

Fukushima, K.A., M.M. Marques, T.K. Tedesco et al. 2019. Screening of hydrogel-based scaffolds for dental pulp regeneration—a systematic review. Archives of Oral Biology 98: 182–194.

Furth, M.E., A. Atala and M.E. Van Dyke. 2007. Smart biomaterials design for tissue engineering and regenerative medicine. Biomaterials 28(34): 5068–73.

Gao, L., R. Beninatto, T. Olah et al. 2023. A photopolymerizable biocompatible hyaluronic acid hydrogel promotes early articular cartilage repair in a minipig model *in vivo*. Adv. Healthc. Mater: e2300931.

Geneva, Ivayla I., Brian Cuzzo, Tasaduq Fazili and Waleed Javaid. 2019. Normal body temperature: a systematic review. Open Forum Infectious Diseases 6(4).

Gonçalves, Patricia Furtado, Enilson Antonio Sallum, Antonio Wilson Sallum, Márcio Zaffalon Casati, Sérgio de Toledo and Francisco Humberto Nociti Junior. 2005. Dental cementum reviewed: development, structure, composition, regeneration and potential functions.

Gopinathan, J. and I. Noh. 2018. Click chemistry-based injectable hydrogels and bioprinting inks for tissue engineering applications. Tissue Eng. Regen. Med. 15(5): 531–546.

Griffith, L.G. and G. Naughton. 2002. Tissue engineering—current challenges and expanding opportunities. Science 295(5557): 1009–14.

Gu, Zhipeng, Keqing Huang, Yan Luo, Laibao Zhang, Tairong Kuang, Zhou Chen and Guochao Liao. 2018. Double network hydrogel for tissue engineering. WIREs Nanomedicine and Nanobiotechnology 10(6): e1520.

Guillen-Carvajal, K., B. Valdez-Salas, E. Beltran-Partida, J. Salomon-Carlos and N. Cheng. 2023. Chitosan, gelatin, and collagen hydrogels for bone regeneration. Polymers (Basel) 15(13).

Hamlet, Stephen M., Cedryck Vaquette, Amit Shah, Dietmar W. Hutmacher and Saso Ivanovski. 2017. 3-Dimensional functionalized polycaprolactone-hyaluronic acid hydrogel constructs for bone tissue engineering. Journal of Clinical Periodontology 44(4): 428–437.

Hanafy, Rania Ahmed, Dawlat Mostafa, Ahmed Abd El-Fattah and Sherif Kandil. 2019. Biomimetic chitosan against bioinspired nanohydroxyapatite for repairing enamel surfaces. Bioinspired, Biomimetic and Nanobiomaterials 9(2): 85–94.

Hatayama, Takahide, Akira Nakada, Hiroki Nakamura, Wakatsuki Mariko, Gentarou Tsujimoto and Tatsuo Nakamura. 2017. Regeneration of gingival tissue using *in situ* tissue engineering with collagen scaffold. Oral Surgery, Oral Medicine, Oral Pathology and Oral Radiology 124(4): 348–354. e1.

He, Xiao-Tao, Xuan Li, Yu Xia, Yuan Yin, Rui-Xin Wu, Hai-Hua Sun et al. 2019. Building capacity for macrophage modulation and stem cell recruitment in high-stiffness hydrogels for complex periodontal regeneration: experimental studies *in vitro* and in rats. Acta Biomaterialia 88: 162–180.

Heinz, A. 2020. Elastases and elastokines: elastin degradation and its significance in health and disease. Crit. Rev. Biochem. Mol. Biol. 55(3): 252–273.

Herrada-Manchón, Helena, Manuel Alejandro Fernández and Enrique Aguilar. 2023. Essential guide to hydrogel rheology in extrusion 3D printing: how to measure it and why it matters? Gels 9(7): 517.

Hilmi, B., Z.A. Abdul Hamid, H. Md Akil and B.H. Yahaya. 2016. The characteristics of the smart polymeras temperature or pH-responsive hydrogel. Procedia Chemistry 19: 406–409.

Hoffman, T., A. Khademhosseini and R. Langer. 2019. Chasing the paradigm: clinical translation of 25 years of tissue engineering. Tissue Eng. Part A 25(9-10): 679–687.

Holiel, Ahmed A., Elsayed M. Mahmoud and Wegdan M. Abdel-Fattah. 2021. Tomographic evaluation of direct pulp capping using a novel injectable treated dentin matrix hydrogel: A 2-year randomized controlled clinical trial. Clinical Oral Investigations 25: 4621–4634.

Hollý, Dušan, Martin Klein, Merita Mazreku, Radoslav Zamborský, Štefan Polák, Ľuboš Danišovič et al. 2021. Stem cells and their derivatives—implications for alveolar bone regeneration: a comprehensive review. International Journal of Molecular Sciences 22(21): 11746.

Holmes, D.F., H.K. Graham and K.E. Kadler. 1998. Collagen fibrils forming in developing tendon show an early and abrupt limitation in diameter at the growing tips. J. Mol. Biol. 283(5): 1049–58.

Hu, Cheng, Li Yang and Yunbing Wang. 2022. Recent advances in smart-responsive hydrogels for tissue repairing. MedComm – Biomaterials and Applications 1(2): e23.

Hu, Weikang, Zijian Wang, Yu Xiao, Shengmin Zhang and Jianglin Wang. 2019. Advances in crosslinking strategies of biomedical hydrogels. Biomaterials Science 7(3): 843–855.

Huanbutta, K., K. Burapapadh, P. Sriamornsak and T. Sangnim. 2023. Practical application of 3D printing for pharmaceuticals in hospitals and pharmacies. Pharmaceutics 15(7).

Hudiță, Ariana and Bianca Gălățeanu. 2023. Polymer materials for drug delivery and tissue engineering. Polymers 15(14): 3103.

Hutomo, Dimas Ilham, Lisa Amir, Dewi Fatma Suniarti, Endang Winiati Bachtiar and Yuniarti Soeroso. 2023. Hydrogel-based biomaterial as a scaffold for gingival regeneration: a systematic review of *in vitro* studies. Polymers 15(12): 2591.

Ikeda, Yuichi, Mehrnoosh Neshatian, James Holcroft and Bernhard Ganss. 2018. The enamel protein ODAM promotes mineralization in a collagen matrix. Connective Tissue Research 59(sup1): 62–66.

Jamshidi, M. and C. Falamaki. 2021. Image analysis method for heterogeneity and porosity characterization of biomimetic hydrogels [version 2; peer review: 2 approved]. F1000Research 9(1461).

Jiang, Pan, Xueru Sheng, Sheng Yu, Haiming Li, Jie Lu, Jinghui Zhou et al. 2018. Preparation and characterization of thermo-sensitive gel with phenolated alkali lignin. Scientific Reports 8(1): 14450.

Jing, Xuan, Si Wang, Han Tang et al. 2022. Dynamically bioresponsive DNA hydrogel incorporated with dual-functional stem cells from apical papilla-derived exosomes promotes diabetic bone regeneration. ACS Applied Materials & Interfaces 14(14): 16082–16099.

Kamali, A. and A. Shamloo. 2020. Fabrication and evaluation of a bilayer hydrogel-electrospinning scaffold prepared by the freeze-gelation method. J. Biomech. 98: 109466.

Kean, T. and M. Thanou. 2010. Biodegradation, biodistribution and toxicity of chitosan. Adv. Drug Deliv. Rev. 62(1): 3–11.

Khademhosseini, Ali and Robert Langer. 2016. A decade of progress in tissue engineering. Nature Protocols 11(10): 1775–1781.

Khaliq, T., M. Sohail, M.U. Minhas et al. 2023. Hyaluronic acid/alginate-based biomimetic hydrogel membranes for accelerated diabetic wound repair. Int. J. Pharm. 643: 123244.

Khayambashi, Parisa, Janaki Iyer, Sangeeth Pillai, Akshaya Upadhyay, Yuli Zhang and Simon D. Tran. 2021. Hydrogel encapsulation of mesenchymal stem cells and their derived exosomes for tissue engineering. International Journal of Molecular Sciences 22(2): 684.

Kwan, J.C., J. Dondani, J. Iyer, H.A. Muaddi, T.T. Nguyen and S.D. Tran. 2023. Biomimicry and 3D-printing of mussel adhesive proteins for regeneration of the periodontium-a review. Biomimetics (Basel) 8(1).

Lee, Hye-Kyung, Dong-Seol Lee, Hyun-Mo Ryoo et al. 2010. The odontogenic ameloblast-associated protein (ODAM) cooperates with RUNX2 and modulates enamel mineralization via regulation of MMP-20. Journal of Cellular Biochemistry 111(3): 755–767.

Lee, Kuen Yong and David J. Mooney. 2001. Hydrogels for tissue engineering. Chemical Reviews 101(7): 1869–1880.

Lei, Lei, Yuhan Hu, Hui Shi, Zhishu Bao, Yiping Wu, Jun Jiang and Xingyi Li. 2022. Biofunctional peptide-click PEG-based hydrogels as 3D cell scaffolds for corneal epithelial regeneration. Journal of Materials Chemistry B 10(31): 5938–5945.

Li, Guo, Hongji Zhang, Daniel Fortin, Hesheng Xia and Yue Zhao. 2015. Poly(vinyl alcohol)–poly(ethylene glycol) double-network hydrogel: a general approach to shape memory and self-healing functionalities. Langmuir 31(42): 11709–11716.

Li, Hui, Qiuxia Ji, Ximin Chen et al. 2017. Accelerated bony defect healing based on chitosan thermosensitive hydrogel scaffolds embedded with chitosan nanoparticles for the delivery of BMP2 plasmid DNA. Journal of Biomedical Materials Research Part A 105(1): 265–273.

Li, Jinhua, Chengtie Wu, Paul K. Chu and Michael Gelinsky. 2020. 3D printing of hydrogels: Rational design strategies and emerging biomedical applications. Materials Science and Engineering: R: Reports 140: 100543.

Li, N., X. Ji, B. Wang, Y. Guo, C. Wang and Y. Chen. 2021. Functional composite hydrogels entrapping polydopamine hollow nanoparticles for highly efficient resistance of skin penetration and photoprotection. Mater Sci. Eng. C Mater Biol. Appl. 128: 112346.

Li, Rui, Weihua Guo, Bo Yang et al. 2011. Human treated dentin matrix as a natural scaffold for complete human dentin tissue regeneration. Biomaterials 32(20): 4525–4538.

Li, Xiaoke, Jinfang Wang, Andrew Joiner and Jiang Chang. 2014. The remineralisation of enamel: a review of the literature. Journal of Dentistry 42: S12–S20.

Liang, Yongxi, Xianghong Luan and Xiaohua Liu. 2020. Recent advances in periodontal regeneration: A biomaterial perspective. Bioactive Materials 5(2): 297–308.

Lin, Chien-Chi and Kristi S. Anseth. 2009. PEG hydrogels for the controlled release of biomolecules in regenerative medicine. Pharmaceutical Research 26(3): 631–643.

Lin, Xuan, Xianwei Zhao, Chongzhi Xu, Lili Wang and Yanzhi Xia. 2022. Progress in the mechanical enhancement of hydrogels: Fabrication strategies and underlying mechanisms. Journal of Polymer Science 60(17): 2525–2542.

Liu, Jin, Jianping Ruan, Michael D. Weir et al. 2019. Periodontal bone-ligament-cementum regeneration via scaffolds and stem cells. Cells 8(6): 537.

Liu, Q., A.M. Rauth and X.Y. Wu. 2007. Immobilization and bioactivity of glucose oxidase in hydrogel microspheres formulated by an emulsification-internal gelation-adsorption-polyelectrolyte coating method. Int. J. Pharm. 339(1-2): 148–56.

Liu, S., Y.N. Wang, B. Ma, J. Shao, H. Liu and S. Ge. 2021. Gingipain-responsive thermosensitive hydrogel loaded with SDF-1 facilitates in situ periodontal tissue regeneration. ACS Appl. Mater. Interfaces 13(31): 36880–36893.

Liu, T., B. Yang, W. Tian, X. Zhang and B. Wu. 2022a. Cryogenic coaxial printing for 3D shell/core tissue engineering scaffold with polymeric shell and drug-loaded core. Polymers (Basel) 14(9).

Liu, Wenxing, Shifen Li, Beiduo Wang, Pai Peng and Changyou Gao. 2022b. Physiologically responsive polyurethanes for tissue repair and regeneration. Advanced NanoBiomed Research 2(10): 2200061.

Liu, Zhenqi, Junzhuo Lu, Xiangshu Chen et al. 2022c. A novel amelogenesis-inspired hydrogel composite for the remineralization of enamel non-cavitated lesions. Journal of Materials Chemistry B 10(48): 10150–10161.

Lv, X., C. Liu, Z. Shao and S. Sun. 2019. Tuning physical crosslinks in hybrid hydrogels for network structure analysis and mechanical reinforcement. Polymers (Basel) 11(2).

Madduma-Bandarage, S.K. Ujith and Sundararajan V. Madihally. 2021. Synthetic hydrogels: Synthesis, novel trends, and applications. Journal of Applied Polymer Science 138(19): 50376.

Mano, J.F., G.A. Silva, H.S. Azevedo et al. 2007. Natural origin biodegradable systems in tissue engineering and regenerative medicine: present status and some moving trends. J. R Soc. Interface 4(17): 999–1030.

Mantha, S., S. Pillai, P. Khayambashi et al. 2019. Smart hydrogels in tissue engineering and regenerative medicine. Materials (Basel) 12(20).

Martin, T.A., S.R. Caliari, P.D. Williford, B.A. Harley and R.C. Bailey. 2011. The generation of biomolecular patterns in highly porous collagen-GAG scaffolds using direct photolithography. Biomaterials 32(16): 3949–57.

Martinelli, A., A. Nitti, R. Po and D. Pasini. 2023. 3D printing of layered structures of metal-ionic polymers: recent progress, challenges and opportunities. Materials (Basel) 16(15).

Martinez-Garcia, F.D., T. Fischer, A. Hayn, C.T. Mierke, J.K. Burgess and M.C. Harmsen. 2022. A Beginner's guide to the characterization of hydrogel microarchitecture for cellular applications. Gels 8(9).

Meier, Yuki A., Kaihuan Zhang, Nicholas D. Spencer and Rok Simic. 2019. Linking friction and surface properties of hydrogels molded against materials of different surface energies. Langmuir 35(48): 15805–15812.

Meyvis, T.K.L., S.C. De Smedt, J. Demeester and W.E. Hennink. 2000. Influence of the degradation mechanism of hydrogels on their elastic and swelling properties during degradation. Macromolecules 33(13): 4717–4725.

Miao, G., L. Liang, W. Li et al. 2023. 3D bioprinting of a bioactive composite scaffold for cell delivery in periodontal tissue regeneration. Biomolecules 13(7).

Miguel Ângelo Cerqueira, Ana C. Pinheiro, Oscar L. Ramos, Hélder Silva, Ana I. Bourbon and Antonio A. Vicente 2017. Chapter Two-advances in food nanotechnology. In Emerging Nanotechnologies in Food Science, edited by Rosa Busquets: Elsevier.

Min, Q., C. Wang, Y. Zhang, D. Tian, Y. Wan and J. Wu. 2022. Strong and elastic hydrogels from dual-crosslinked composites composed of glycol chitosan and amino-functionalized bioactive glass nanoparticles. Nanomaterials (Basel) 12(11).

Mohabatpour, Fatemeh, Xiaoman Duan, Zahra Yazdanpanah et al. 2022. Bioprinting of alginate-carboxymethyl chitosan scaffolds for enamel tissue engineering in vitro. Biofabrication 15(1): 015022.

Momose, Takehito, Hirofumi Miyaji, Akihito Kato et al. 2016. Collagen hydrogel scaffold and fibroblast growth factor-2 accelerate periodontal healing of class II furcation defects in dog. The Open Dentistry Journal 10: 347.

Moon, Seo Hyung, Hye Jin Hwang, Hye Ryeong Jeon, Sol Ji Park, In Sun Bae and Yun Jung Yang. 2023. Photocrosslinkable natural polymers in tissue engineering. Frontiers in Bioengineering and Biotechnology 11.

Muşat, Viorica, Elena Maria Anghel, Agripina Zaharia, Irina Atkinson, Oana Cătălina Mocioiu, Mariana Buşilă et al. 2021. A chitosan–agarose polysaccharide-based hydrogel for biomimetic remineralization of dental enamel. Biomolecules 11(8): 1137.

Navin, H.K. and K.B. Prasanna. 2014. Enamel regeneration-current progress and challenges. Journal of Clinical and Diagnostic Research: JCDR 8(9): ZE06.

Nilasaroya, Anastasia, Alan Matthew Kop and David Anthony Morrison. 2021. Heparin-functionalized hydrogels as growth factor-signaling substrates. Journal of Biomedical Materials Research Part A 109(3): 374–384.

O'Connor, N.A., A. Syed, M. Wong et al. 2020. Polydopamine antioxidant hydrogels for wound healing applications. Gels 6(4).

Olson, J.L., A. Atala and J.J. Yoo. 2011. Tissue engineering: current strategies and future directions. Chonnam Med. J. 47(1): 1–13.

Palomino-Durand, Carla, Emmanuel Pauthe and Adeline Gand. 2021. Fibronectin-enriched biomaterials, biofunctionalization, and proactivity: a review. Applied Sciences 11(24): 12111.

Pan, Jie, Jiajia Deng, Liming Yu et al. 2020. Investigating the repair of alveolar bone defects by gelatin methacrylate hydrogels-encapsulated human periodontal ligament stem cells. Journal of Materials Science: Materials in Medicine 31: 1–12.

Park, Chan Ho, Joung-Hwan Oh, Hong-Moon Jung et al. 2017. Effects of the incorporation of ε-aminocaproic acid/chitosan particles to fibrin on cementoblast differentiation and cementum regeneration. Acta Biomaterialia 61: 134–143.

Pepla, Erlind, Lait Kostantinos Besharat, Gaspare Palaia, Gianluca Tenore and Guido Migliau. 2014. Nano-hydroxyapatite and its applications in preventive, restorative and regenerative dentistry: a review of literature. Annali di stomatologia 5(3): 108.

Petrini, P., S. Farè, A. Piva and M.C. Tanzi. 2003. Design, synthesis and properties of polyurethane hydrogels for tissue engineering. Journal of Materials Science: Materials in Medicine 14(8): 683–686.

Place, Elsie S., Nicholas D. Evans and Molly M. Stevens. 2009. Complexity in biomaterials for tissue engineering. Nature Materials 8(6): 457–470.

Qi, Desheng, Ningning Wang, Yuanqiang Cheng et al. 2022. Application of porous polyetheretherketone scaffold/vancomycin-loaded thermosensitive hydrogel composites for antibacterial therapy in bone repair. Macromolecular Bioscience 22(10): 2200114.

Ravanetti, F., P. Borghetti, M. Zoboli et al. 2023. Biomimetic approach for an articular cartilage patch: combination of decellularized cartilage matrix and silk-elastin-like-protein (SELP) hydrogel. Ann. Anat: 152144.

Reynolds, Eric C. 2008. Calcium phosphate-based remineralization systems: scientific evidence? Australian Dental Journal 53(3): 268–273.

Romischke, J., A. Scherkus, M. Saemann, S. Krueger, R. Bader, U. Kragl et al. 2022. Swelling and mechanical characterization of polyelectrolyte hydrogels as potential synthetic cartilage substitute materials. Gels 8(5).

Rosdiani, Azizah Fresia, Prihartini Widiyanti and Djoni Izak Rudyarjo. 2017. Synthesis and characterization biocomposite collagen-chitosan-glycerol as scaffold for gingival recession therapy. Journal of International Dental and Medical Research 10(1): 118.

Sedek, Eman M., Elbadawy A. Kamoun, Nehal M. El-Deeb, Sally Abdelkader, Amal E. Fahmy, Samir R. Nouh and Nesma Mohamed Khalil. 2023. Photocrosslinkable gelatin-treated dentin matrix hydrogel as a novel pulp capping agent for dentin regeneration: I. synthesis, characterizations and grafting optimization. BMC Oral Health 23(1): 536.

Shafiee, A. and A. Atala. 2016. Printing technologies for medical applications. Trends Mol. Med. 22(3): 254–265.

Sharma, P., P. Kumar, R. Sharma, V.D. Bhatt and P.S. Dhot. 2019. Tissue engineering; current status & futuristic scope. J. Med. Life 12(3): 225–229.

Shen, Jiahao, Heng Zhang, Jingxin Zhu et al. 2022. Simple preparation of a waterborne polyurethane crosslinked hydrogel adhesive with satisfactory mechanical properties and adhesion properties. Frontiers in Chemistry 10.

Shi, X., J. Wu, Z. Wang, F. Song, W. Gao and S. Liu. 2020. Synthesis and properties of a temperature-sensitive hydrogel based on physical crosslinking via stereocomplexation of PLLA-PDLA. RSC Adv. 10(34): 19759–19769.

Shokri, A., K. Ramezani, M.R. Jamalpour et al. 2022. *In vivo* efficacy of 3D-printed elastin-gelatin-hyaluronic acid scaffolds for regeneration of nasal septal cartilage defects. J. Biomed. Mater. Res. B Appl. Biomater. 110(3): 614–624.

Silva, A.C.Q., A.J.D. Silvestre, C. Vilela and C.S.R. Freire. 2021. Natural polymers-based materials: a contribution to a greener future. Molecules 27(1).

Simon, S.R.J., A. Berdal, P.R. Cooper, P.J. Lumley, P.L. Tomson and A.J. Smith. 2011. Dentin-pulp complex regeneration: from lab to clinic. Advances in Dental Research 23(3): 340–345.

Skardal, A., L. Smith, S. Bharadwaj, A. Atala, S. Soker and Y. Zhang. 2012. Tissue specific synthetic ECM hydrogels for 3-D *in vitro* maintenance of hepatocyte function. Biomaterials 33(18): 4565–75.

Song, J.Y., H.S. Lee, D.Y. Kim, H.J. Yun, C.C. Yi and S.M. Park. 2023. Fabrication procedure for a 3D hollow nanofibrous bifurcated-tubular scaffold by conformal electrospinning. ACS Macro Lett. 12(5): 659–666.

Sowmya, S, Ullas Mony, P. Jayachandran et al. 2017. Tri-layered nanocomposite hydrogel scaffold for the concurrent regeneration of cementum, periodontal ligament, and alveolar bone. Advanced Healthcare Materials 6(7): 1601251.

Stojic, M., J. Rodenas-Rochina, M.L. Lopez-Donaire et al. 2021. Elastin-plasma hybrid hydrogels for skin tissue engineering. Polymers (Basel) 13(13).

Stojkov, G., Z. Niyazov, F. Picchioni and R.K. Bose. 2021. Relationship between structure and rheology of hydrogels for various applications. Gels 7(4).

Sultan, Sahar and Aji P. Mathew. 2018. 3D printed scaffolds with gradient porosity based on a cellulose nanocrystal hydrogel. Nanoscale 10(9): 4421–4431.

Sun, K., G. Wu, K. Liang, B. Sun and J. Wang. 2023. Investigation into photolithography process of FPCB with 18 microm line pitch. Micromachines (Basel) 14(5).

Tabatabaei, Fahimeh, Keyvan Moharamzadeh and Lobat Tayebi. 2020. Fibroblast encapsulation in gelatin methacryloyl (GelMA) versus collagen hydrogel as substrates for oral mucosa tissue engineering. Journal of Oral Biology and Craniofacial Research 10(4): 573–577.

Tan, H., A.J. DeFail, J.P. Rubin, C.R. Chu and K.G. Marra. 2010. Novel multiarm PEG-based hydrogels for tissue engineering. J. Biomed. Mater. Res. A 92(3): 979–87.

Tan, Huaping and Kacey G. Marra. 2010. Injectable, biodegradable hydrogels for tissue engineering applications. Materials 3(3): 1746–1767.

Teixeira, M.O., J.C. Antunes and H.P. Felgueiras. 2021. Recent advances in fiber-hydrogel composites for wound healing and drug delivery systems. Antibiotics (Basel) 10(3).

Trinh, K.T.L., N.X.T. Le and N.Y. Lee. 2020. Chitosan-polydopamine hydrogel complex: a novel green adhesion agent for reversibly bonding thermoplastic microdevice and its application for cell-friendly microfluidic 3D cell culture. Lab Chip 20(19): 3524–3534.

Vazquez-Portalatin, N., A. Alfonso-Garcia, J.C. Liu, L. Marcu and A. Panitch. 2020. Physical, biomechanical, and optical characterization of collagen and elastin blend hydrogels. Ann. Biomed. Eng. 48(12): 2924–2935.

Wang, Bing, Simone Mastrogiacomo, Fang Yang et al. 2019. Application of BMP-Bone cement and FGF-Gel on periodontal tissue regeneration in nonhuman primates. Tissue Engineering Part C: Methods 25(12): 748–756.

Wang, Menghan, Jianzhong Bai, Kan Shao et al. 2021a. Poly(vinyl alcohol) hydrogels: the old and new functional materials. International Journal of Polymer Science 2021: 2225426.

Wang, Lei, Hui Fu, Wenwen Wang et al. 2021b. Notoginsenoside R1 functionalized gelatin hydrogels to promote reparative dentinogenesis. Acta Biomaterialia 122: 160–171.

Wessel, Jeffrey R. and Dimitris N. Tatakis. 2008. Patient outcomes following subepithelial connective tissue graft and free gingival graft procedures. Journal of Periodontology 79(3): 425–430.

Williams, D.F. 2019a. Specifications for innovative, enabling biomaterials based on the principles of biocompatibility mechanisms. Front Bioeng. Biotechnol. 7: 255.

Williams, David F. 2019b. Challenges with the development of biomaterials for sustainable tissue engineering. Frontiers in Bioengineering and Biotechnology 7.

Xu, F., C. Dawson, M. Lamb, E. Mueller, E. Stefanek, M. Akbari et al. 2022. Hydrogels for tissue engineering: addressing key design needs toward clinical translation. Front Bioeng. Biotechnol. 10: 849831.

Xu, X., A.K. Jha, D.A. Harrington, M.C. Farach-Carson and X. Jia. 2012. Hyaluronic acid-based hydrogels: from a natural polysaccharide to complex networks. Soft Matter 8(12): 3280–3294.

Xu, Xiaowei, Zhongyi Gu, Xi Chen et al. 2019. An injectable and thermosensitive hydrogel: Promoting periodontal regeneration by controlled-release of aspirin and erythropoietin. Acta Biomaterialia 86: 235–246.

Xue, X., Y. Hu, S. Wang, X. Chen, Y. Jiang and J. Su. 2022. Fabrication of physical and chemical crosslinked hydrogels for bone tissue engineering. Bioact. Mater 12: 327–339.

Yao, Y., D. Guan, C. Zhang et al. 2022. Silkworm spinning inspired 3D printing toward a high strength scaffold for bone regeneration. J. Mater. Chem. B 10(36): 6946–6957.

Yi, Ke, Qing Li, Xiaodong Lian, Yapei Wang and Zhihui Tang. 2022. Utilizing 3D bioprinted platelet-rich fibrin-based materials to promote the regeneration of oral soft tissue. Regenerative Biomaterials 9: rbac021.

Younes, I. and M. Rinaudo. 2015. Chitin and chitosan preparation from marine sources. Structure, Properties and Applications. Mar Drugs 13(3): 1133–74.

Yousefi-Mashouf, Hamid, Lucie Bailly, Laurent Orgéas and Nathalie Henrich Bernardoni. 2023. Mechanics of gelatin-based hydrogels during finite strain tension, compression and shear. Frontiers in Bioengineering and Biotechnology 10.

Yousefzade, O., R. Katsarava and J. Puiggalí. 2020. Biomimetic hybrid systems for tissue engineering. Biomimetics (Basel) 5(4).

Zhan, Yiwei, Wenjiao Fu, Yacheng Xing, Xiaomei Ma and Chunying Chen. 2021. Advances in versatile anti-swelling polymer hydrogels. Materials Science and Engineering: C 127: 112208.

Zhang, Liwen, Lei Fu, Xin Zhang, Linxin Chen, Qing Cai and Xiaoping Yang. 2021. Hierarchical and heterogeneous hydrogel system as a promising strategy for diversified interfacial tissue regeneration. Biomaterials Science 9(5): 1547–1573.

Zhang, Y., Y. Wang, Y. Li et al. 2023. Application of collagen-based hydrogel in skin wound healing. Gels 9(3).

Zhang, Yanyu and Yishun Huang. 2021. Rational design of smart hydrogels for biomedical applications. Frontiers in Chemistry 8.

Zhang, Zhouqiang, Zishuo Ye, Feng Hu, Wenbo Wang, Shoujing Zhang, Li Gao et al. 2022. Double-network polyvinyl alcohol composite hydrogel with self-healing and low friction. Journal of Applied Polymer Science 139(4): 51563.

Zhu, J. 2010. Bioactive modification of poly(ethylene glycol) hydrogels for tissue engineering. Biomaterials 31(17): 4639–56.

Zhu, J. and R.E. Marchant. 2011. Design properties of hydrogel tissue-engineering scaffolds. Expert Rev. Med. Devices 8(5): 607–26.

Zhu, Senbo, Yong Li, Zeju He et al. 2022. Advanced injectable hydrogels for cartilage tissue engineering. Frontiers in Bioengineering and Biotechnology 10.

Index